# AGRICULTURAL DEVELOPMENT IN NORTH-EAST INDIA

## ISSUES AND OPTIONS

# AGRICULTURAL DEVELOPMENT IN NORTH-EAST INDIA

## ISSUES AND OPTIONS

*Editor*

**K.K. BAGCHI**

**ABHIJEET PUBLICATIONS**
DELHI 110094

**ABHIJEET PUBLICATIONS**
2/46 Tukhmeerpur Extension
Delhi 110094
Phones: 011-22960492, 65698474
e-mail: abhijeetpublication@gmail.com
abhijeet_singh1@indiatimes.com

AGRICULTURAL DEVELOPMENT IN NORTH-EAST INDIA

*First Published* 2008



ISBN 978-81-89886-44-8



PRINTED IN INDIA

---

Published by J.K. Singh fo Abhijeet Publications, Delhi 110094, Lasertypeset by Gaurav Graphics, Delhi and Printed as Himanshu Printers, Delhi.

# PREFACE & ACKNOWLEDGEMENTS

This volume was contemplated by the editor considering the crucial importance of agriculture to the life of the people of North-Eastern states. Development experience of countries of the world shows that in the process of structural transformation of an economy, the agricultural sector plays a pivotal role in terms of supply of food to the non-agricultural population, supply of raw materials to agro-based industries, supply of labour and capital for the development of the non-agricultural sector at the initial stages of development. The prosperity of the agricultural sector is, therefore, needed not only for the population engaged in this sector but also for the overall development of an economy. Data and evidences show that the potentialities of agricultural development of the north-eastern states have not been fully exploited. Realisation of the full potential of the agricultural sector is essential for the benefit of the entire North-Eastern states and for that matter for India as a whole. The present volume has discussed the nitty-gritty of the issues and options of agricultural development of the north-eastern states with current data and evidences.

It is hoped that the volume will be proved useful to the planners, policy makers, administrators, the academia and scholars of the region and elsewhere of the country in moving forward the task of agricultural development in these states. It is also expected that the present volume would also stimulate further study on the issues raised in the included papers. It may be noted that while the opinions expressed in the papers are that of the authors of the papers, I am solely responsible for the shortcomings, if any, remaining in the editorial work.

In editing this book I have received co-operation from many individuals. First and foremost, I am extremely greatly to the contributors who responded instantaneously on receipt of the call for paper for the book. I express my sincere gratitude to M/s Abhijeet Publications to bring out this volume in a very short period of time after the submission of the manuscript. Last but not the least, I am also indebted to my family members for ungrudgingly bearing with me while I devoted a good amount of time in editing and correspondence work related to the book.

**K.K. Bagchi**
Department of Economics
University of North Bengal
Darjeeling (W.B.)

# CONTENTS

**Part II**

**State Specific & Crop Specific Issues on Agricultural Development**

# CONTRIBUTORS

1. **Dr. (Mrs). Anuva Saikia** is Professor & Head, Department of Agricultural Economics, Assam Agricultural Economics, Jorhat, 785 013, Assam.
2. **Dr. C. Hazarika** is Associate Professor, Department of Agricultural Economics, Assam Agricultural Economics, Jorhat, 785 013, Assam.
3. **Dr. Arun Pandit** is with Central Potato Research Institute, Shimla.
4. **Dr. N.K. Pandey** is with Central Potato Research Institute, Shimla.
5. **Dr. Rajesh K. Rana** is with Central Potato Research Institute, Shimla.
6. **Dr. K.P. Chandran** is with Central Potato Research Institute, Shimla.
7. **Dr. N.R. Kumar** is with Indian Institute of Vegetable Research, Varanasi.
8. **Dr. A.K. Mishra** is Sr. Scientist (Soil and Water Conservation Engineering), Water Technology Centre, Indian Agricultural Research Institute, New Delhi.
9. **Dr. J.P. Misra** is Professor and Head, Department of Agricultural Economics, N. D. University of Agriculture and Technology, Narendranagar, Faizabad, U.P.
10. **Dr. R.N. Barman,** Associate Professor Department of Agricultural Economics and Fartm Management, B.N. College of Agriculture (AAU), Biswanath Chariali, Assam.
11. **Dr. R. Das,** Associate Professor Department of Agricultural Economics and Farm Management, B.N. College of Agriculture (AAU), Biswanath Chariali, Assam.
12. **Dr. Aftab Uddin Ahmed** is Reader, Department of Economics, Barnagar College, Dist. Barpeta, Assam.

13. **Dr. K.B. Umesh** is Associate Professor of Agricultural Economics, Directorate of Research, GKVK, University of Agricultural Sciences, Bangalore.
14. **Dr. B. M. Shashidhara,** Associate Professor of Agricultural Marketing and Cooperation, Directorate of Research, GKVK, University of Agricultural Sciences, Bangalore.
15. **Dr. B.N. Vènu,** Research Associate (Agricultural Economics), Directorate of Research, GKVK, University of Agricultural Sciences, Bangalore.
16. **Dr. Pravat Kumar Kuri** is Reader, Department of Economics, Burdwan University, Burdwan, West Bengal.
17. **Mr. Bidyut Kumar Ghosh** is Research Scholar, Department of Economics, Burdwan University, Burdwan, West Bengal.
18. **Dr. S.S. Kalamkar** is with Gokhale Institute of Politics and Economics (Deemed to be a University), Pune, Maharashtra.
19. **Dr. Gopakumaran Nair, G.** is Assistant General Manager, Department of Economic Analysis and Research, National Bank for Agriculture and Rural Development, Mumbai.
20. **Dr. R.P. Singh** is with Division of Agricuural Economics, Indian Agricultural Research institute, New Delhi.
21. **Dr. Ranjit Kumar** is with Division of Agricuural Economics, Indian Agricultural Research institute, New Delhi.
22. **Dr. N.P. Singh** is with Division of Agricuural Economics, Indian Agricultural Research institute, New Delhi.
23. **Dr. S.V. Hariharan** is Professor, School of Economics, Madurai Karaj University, Madurai, 625 021, Tamil Nadu.
24. **Mr. M. Anandan** is Research Scholar, Department of Economics, Bhatatidasan University, Trichy, 620 024, Tamil Nadu.
25. **Dr. Ashwani Kumar Sharma,** Sr. Scientist (Ag. Econ.), Research Coordination and management Unit, Indian Institute of sugarcane Research, Lucknow, (U.P.)
26. **Dr. D. V. Yadav,** Principal Scientist (soil Science) and i/c Research Coordination and Management Unit, Indian Institute of Sugarcane Research, Lucknow (U.P.)
27. **Dr. K.C. Talukdar** is Professor, Department of Agricultural Economics, Assam Agricultural Economics, Jorhat - 13, Assam.
28. **Dr. Subhasis Mandal** is Scientist, Division of agricultural Economics & Statistics, ICAR Research Complex for NEH Region, Umroi Road, Umiam, Meghalaya, 793103.

29. **Dr. K.K. Datta** is Principal Scientist, National Centre for Agricultural Economics and Policy Research (NCAP), New Delhi, 110 012.
30. **Dr. M.R. Verma** is Scientist, Division of Agricultural Economics & Statistics, ICAR Research Complex for NEH Region, Umroi Road, Umiam, Meghalaya, 793 103.
31. **Dr. A.K. Tripathi** is Senior Scientist, Division of Agricultural Economics & Statistics, ICAR Research Complex for NEH Region, Umroi Road, Umiam, Meghalaya 793 103.
32. **Dr. Juthika Barman** is Lecturer, Department of Economics, Narengi Anchalik Mahavidyalaya, Guwahati, Assam. Res: C/O Mr. B. C. Barman, Rajgar Road, Opposite Manipur bhawan, Guwahati, 781 003, Assam.
33. **Dr. K.K. Barman** is Professor, Department of Economics, Gauhati University, Guwahati, 781 014, Assam.
34. **Dr. M.P. Bezbaruah** is Professor of Economics, Gauhati University, Guwahati, 781 014, Assam.
35. **Dr. N.C. Roy** is Professor, Department of Economics, Rajiv Gandhi University, Rono Hills, Itanagar, Arunachal Pradesh.
36. **Mr. Subhabrata Chakrabarti** is Lecturer, Department of Economics, Sikkim Government College, Gangtok, Sikkim
37. **Dr. B. Datta Ray** is Secretary, North-East India Council for Social Science Research, B.T. Hostel, Shillong –3, Meghalaya.
38. **Dr. S. Basanta Singh** is Senior Scientist, Division of Agricultural Economics, I.C.A.R. Research Complex for NEH Region, Manipur Centre, Lamphelpat, 795004, Manipur.
39. **Dr. S.V. Ngachan** is with Division of Agricultural Economics, I.C.A.R. Research Complex for NEH Region, Manipur Centre, Lamphelpat, 795 004, Manipur.
40. **Dr. P.M. Paasha** is retired Professor of Economics, Lummawrie, Laitumkhrah, East Khasi Hills, Shillong, 793003, Meghalaya.
41. **Dr. Niranjan Roy** is Reader, Deptt. of Economics, Assam University, Silchar.
42. **Dr. K. C. Bora** is Professor, Deptt. of Economics, Dibrugarh University, Dibrugarh, 786 004, Assam.
43. **Dr. D. K. Chakraborty** is Lecturer, Deptt. of Economics, Dibrugarh University, Dibrugarh, 786 004, Assam.

# INTRODUCTION

*K.K. Bagchi*

Agriculture is the backbone of the economy of north-eastern states of India in terms of employment and its share in the Net State Domestic Product (NSDP). Although the share of agriculture in NSDP is declining gradually, but in terms of employment it is still the dominant sector. This means that per capita income of the population engaged in the agricultural sector is lower than that of the non-agricultural sectors. The state of agricultural development is still not satisfactory in this part of the country, whether it is productivity, diversification and use of technology, use of modern inputs like fertilizer and irrigation etc.

There are regional variations in the state of agricultural development among the north-eastern states. Manipur, Tripura and Assam appears to have made reasonable strides while the remaining states shall have to travel a long way to reach a comfortable state. As far as production of specific crops, vegetables and horticultural crops are concerned performances of individual states vary. Bridging the gaps between the states as well as improvement in the overall performance is highly imperative.

As far as the agricultural development of north-eastern states is concerned, the most remarkable aspect is that, there is a huge gap between the potentiality and actual achievement. Whether it is the yield rates of various crops or the use of more productive inputs or diversification of agricultural activity—much remains to be achieved. It is through the genuine and sincere participation of all concerned which can help to accelerate the pace of change forth the better in all these respects.

The issues, therefore, which are important for agricultural development in the north-eastern states, are:

- Current state of agricultural development.
- Factors affecting agricultural development.
- The gap between potential and actual levels of agricultural development.
- Resource sustainable remunerative alternatives to agricultural development.
- Techno-economic and institutional factors in the development dynamics of agriculture.
- Role of institutional finance for overall development of the north-eastern region.
- Need and scope of diversification of agriculture.
- Role of watershed development in the north-eastern region as a means of economic upliftment.
- Overall strategies for development of agriculture in the north-eastern region.
- Specific strategies for development of agriculture in the north-eastern region.
- Specific strategies for the development of tea coffee, sugar, fruits and vegetables.
- Strategies for overcoming the problems of agricultural development of specific states.

Discussion on the above issues leads to the discussion of options, which could be explored to achieve the desired goal of rapid sustainable agricultural development of the north-eastern states. The papers included in the present volume have explored in considerable detail the issues and options in these regard. Although the discussions on the issues and options contained in this volume are not all comprehensive, but the coverage is quite wide. In what follows we have summarized the essence of each of the paper in order to have a brief idea contained in the papers. It may me mentioned here that Part I of the book contains 13 papers dealing with general issues, while Part II includes another

13 papers concerning specific issues of agricultural development of the north-eastern states.

Saikia in her paper analyses present status of agriculture in the north-eastern (NE) region along with its potentials, problems and strategies for development. According to her development issues of NE region is not only related to economic factors but also to the diverse socio-cultural, ethnic, and environmental factors. She has suggested some pertinent strategies for agricultural development in the NE states. Implementation of these suggestions will definitely accelerate the pace of agricultural development in these states.

Mandal *et al* have found out that agricultural sector in north-eastern states is experiencing a transitional phase, slowly shifting out from traditionally low-income agriculture, but yet to reach the stage of pre-conditions for take-off into self-sustaining growth. Under this stage of economic transition, various sub-sectors of agriculture and allied activities showed high growth prospects, when stimulated through policy initiatives. But the real challenge remains to sustain this growth rates in the long run. The study concludes that the process of economic growth and development for the north-eastern states is primarily dependent on agriculture sector and thereby, development planning should continue to follow agriculture-led-growth.

According to Kalamkar the NE region/states in which rural poverty rates and dependence on agriculture are highest, agriculture shows limited improvement. The NE states are among the most backward in the country and agricultural development is much below its potential in this region. With changes taking place in the domestic and international economies, agricultural strategy in India can have a re-look at the north-east hill economies, learn from their failure to respond to the green revolution and focus on an acceptable path to development of the region. By re-looking into the strategy of agricultural development in NE states of India Kalamkar has made several valuable suggestions for rapid agricultural development of the region.

Mishra and Misra have explored the possibility of resource sustainable and remunerative alternatives to agricultural development in the north-eastern region of India. According to the authors agriculture development in the region till date can be referred to as primitive in nature being practiced for sustenance of its poor, illiterate tribal populace despite ample resource potentials and development opportunities for better, highly favourable economical resources sustainable alternatives. The region falls short of food grains of the order of 1.6 m tonnes/year mainly due to poor productivities of various commodities that in turn depend on scores of limiting factors for the development of modern agriculture. It has been visualized in the paper that even if these shortcomings are overcome, the region cannot compete with other states as far as the productivity of various crops are concerned vis-à-vis natural resources sustainability. Research experiences with resources conservation and watershed management have also confirmed the fact that farmers are on the look for other economical options than agriculture. The paper aims at critically evaluate some of these concerns and suggesting the corrective measures for realising the goals of economic well being to the farmers and artisans of the region.

Bezbaruah's paper deals with an explanation of techno-economic and institutional factors in the development dynamics of agriculture in the NE India. It has also raised several pertinent issues relating to agricultural development in the NE region such as review of research priorities for agriculture in the NE region, techno-economically viable alternatives to precarious hill agriculture systems currently in practice, sustainable exploitation of water resource, problems of agriculture in the flood plains, diversification of rice centric agriculture in the plains, low penetration of technology, institutional barriers in land, and labour markets etc. that call for further investigation and more intensive discussion.

Gopakumaran has stressed on the institutional finance for overall development of NE region. He expresses the view that one of the major reasons for the poor status of agricultural development is the low level of access to institutional credit and its small size in the region—warranting hectic remedial measures. According to him the issue of improving the credit flow needs to be addressed by generating adequate demand by modernizing/diversifying agriculture, strengthening agricultural extension and encouraging contract farming. He has suggested several complementary measures such as promotion of SHGs, Joint Liability Groups, Farm Management Group, strengthening NGO network etc. to improve the access of credit by the rural poor in this region.

Borah and Chakraborty have made an attempt to study the pattern of growth and diversification of agriculture in the NE region based on secondary data collected from various sources. The study finds that the NE region has diversified agro-climatic and soil condition and offers congenial environment for the development of agriculture. But the region is lagging behind in this front from many states of the country. The reasons for low production and productivity of agriculture in the region are many and varied. Among them institutional, technical and natural factors are principal broad factors retarding the growth of agriculture in the region. The overall suggestion of the study is that agricultural policies must take a holistic view in the directions increasing double-cropped area, a diversified cropping pattern, creation of marketing infrastructure, provision of institutional agricultural credit in adequate amounts etc. to enable the economy of the NER to realize its potential.

Pasaah's paper deals with the agricultural growth and diversification of the agrarian economy of the eastern and north-eastern India since Independence and more particularly since 1991 when economic reforms were launched in India. According to him the aim of agricultural development must be to diversify the rural economy of these regions, which are lagging behind the rest of India. The development of activities allied to agriculture would tremendously contribute to the diversification of agriculture in the region. There is also the need to develop rural industries based on agricultural outputs. The diversification would not only lead to higher incomes and employment of the rural population but would also result into even self-sufficiency and exportable surplus of agricultural produce from this region.

Singh et al have highlighted the watershed development in the NER region of India as a means of economic upliftment.

The paper has empirically demonstrated the impact on income and employment generated from experimental watersheds under various farming systems. These watersheds will also arrest the run-offs and soil loss of different farming system. Moreover, the barren and uncultivable land to the tune of 16 per cent in the region will be brought under cultivation by enhanced irrigation potential. This will give enough boost to the economic upliftment and thereby development of the region as a whole. In nutshell, according to the scholars, watershed in the north-eastern region promises a win-win solution by complementing the natural resource conservation and agricultural productivity, thereby alleviating the poverty in a big way.

Umesh et al have assessed the performance and benefits of coffee development programme for the tribal farmers of NE states. According to them the coffee development programmes have created a positive impact on the income, employment and the livelihood among the tribal households. The impact on socio-economic development of tribals is slow, as coffee is taken up as one of the components in different cropping systems. The promotional programmes may take some more time to alter the mindset among the tribal coffee growers with respect to discontinuation of Jhum cultivation. According to the authors, in view of this serious efforts are very much desirable to strengthen further the expansion and consolidation programmes. The prospects of transforming the region into an organic coffee producing zone and connect it with the domestic as well as international market should be explored to benefit the welfare of the tribals in the region.

Sharma and Yadav have explored the current status of sugar sector and prospects of its development in NER of India. They have found that currently the scenario in the sugarcane sector is not encouraging as far as its growth is concerned. They have identified several reasons for this. They have also found that the profit centre of sugar industry has gradually been shifted towards the effective use of by-products because the by-products has more value in cogeneration and alcohol production in areas where sugar recovery is less. Therefore, they suggest that the possibilities for establishing sugar multi-complexes with facilities for cogeneration, ethanol and alcohol production could be explored in Assam. Such developments, according to them, will help generate demand for adequate investment in cane development activities as well as industry oriented research and technology along with a conducive and enabling policy environment. This, in turn, would lead to the development of the sugar sector in the region.

Talukdar's paper deals with the growth and trade prospects of Indian tea industry in the globalised world. By analyzing secondary data the study suggests that Indian tea plantation sector needs massive replanting and replacement planting programme as about half of tea bushes are in the low productive age groups affecting cost and profit of the gardens. Export of value added tea needs priority to match the rising demand of value added tea in the world. According to the study the main challenges of Indian tea in the global competition are high overhead cost, social obligations to labour and high cost of management, pest endemic for which incidence of chemical residues is high, prone to adverse climate with dormancy in production process adding to overhead cost, low productivity and higher cost of production due to old bushes and lack of modernization in the processing industries.

Datta Ray in his paper has set out his reflections on the establishment of Special Economic Zones (SEZs). According to him SEZs are alternatives neither to agricultural development nor to employment creation. It will do more harm than do better. Hence the idea should be buried forever. He opines that till 2/3rds of rural population in NE India being dependent on agriculture, it is imperative that we inject fresh dynamism into agriculture.

The paper by Kuri and Ghosh has attempted to analyze the growth performance of agriculture and the nature of changes in cropping pattern in the State of Assam over a period of 49 years commencing from 1951-52. According to them the overall performance of agriculture in the state does not show any significant trend of prosperity. The increase in yield rates, which is very important under the current circumstances, has been very low and even negative for many crops. In the absence of any appreciable change in relative yield growth rates between crops, the farmers in the State did not make any significant changes in their cropping pattern. Instead, there seems to be a gradual shift of cropping pattern from diversification towards unification favouring food grain crops. They observe that to raise productivity a conscious effort is necessary involving agricultural scientists, extension and development agencies to unveil the untapped potential of agriculture in the State to obtain the targeted growth in yield of crops.

Barman and Barman have made a study on the growth and instability of food grains production in Assam. The paper concludes that in the context of the rapid growth of population and little scope for extension of cultivation to new areas, the future strategy of raising food grains production in Assam should be concentrated on increasing cropping intensity and greater emphasis should be laid on increase per hectare yield of crops. For this purpose farmers of Assam should be provided with all necessary facilities such as quality seeds, adequate fertilizer and water management, credit facilities, plant protection measures, marketing and storage facilities etc. Steps should also be taken by the government to distribute land in such a way that each farmer gets sufficient land for remunerative production. Further, appropriate measures should be taken to control flood, which causes much damage to crops every year in all localities of the state.

Talukdar has made an attempt to examine the pattern of crop diversification and potentialities of its growth, technological assessment in rice crop, the impact of diversification on farm economy and high value alternatives in flood affected areas of Assam. The study suggests that farm strategies may be formulated for commercial cultivation of vegetables, pulses, oilseed crops, spices and condiments for farm income generation and to bridge the demand supply gap of pulses, oilseeds and spices. There is scope to increase productivity of rice with proper use of technology and crop planning. The important constraints to diversification like strengthening of input and output markets, irrigation, roads and communication, mechanization, credit and storage facilities and agro-services should be minimized for increasing cropping intensity and crop diversification in the state. Extension for commercialization of crop cultivation will be highly imperative for generating income and employment in the farm sector.

Hazarika has dealt with a very important aspect of agricultural development of Assam in his study—the problems of post-harvest loss and food security in respect of fruits and vegetables. According to him by reducing the post-harvest losses, one can increase the per capita availability of food with the existing level of production technology. For this, it is necessary to know the quantum of post-harvest losses at various levels and during different operations. In the study he has therefore made an attempt to estimate the post-harvest losses of major perishable horticultural crops in Assam. The paper concludes that in Assam, the post-harvest loss of pineapple, tomato, orange and ginger is very high and proper post harvest management is the need of the hour to reduce these losses. The study has also prescribed policies to be adopted for reducing such losses to the minimum.

Pandit et al have made an empirical study of the status and constraints of potato production in Assam. It has made a case study of Barpeta district that happens to be the highest (18% of the state) potato producing district of Assam. The study reveals that the estimated potato production cost was Rs. 255.36 per quintal and the benefit-cost ratio was 1.57. Further, farmers were not using quality seed of improved varieties and balanced doses of fertilizers. There was lack of irrigation and cold storage facilities in the area. Farmers were also lacking the knowledge of scientific potato production practices. Labour, mulching, plant protection and micronutrients were the important factors affecting potato yield. Further, potato production in the study area was operating under increasing returns to scale. Through Data Envelopment Analysis they have shown that pure technical inefficiency has the greater contribution to the overall inefficiency. The study suggests intensifying the efforts to educate the farmers regarding scientific potato production practices.

Roy's paper deals with farm size and tenurial status of farmer and practice of modern farming in Assam. It finds that the relatively small size of farms has not been found to be a constraint in adoption and use of improved inputs and practices for better farming in the state of Assam. However, tenant farmers have been found to be in disadvantage in extending their area under HYVs and application of fertilisers in adequate quantity. Tenancy has been found to be fairly extensive among sample farmers, nearly a half of sample farmers had leased-in land as part of their operational holdings. The tenancy was generally temporary, usually of one year and also completely informal in the form of share cropping mostly without sharing of cost. The study calls for tenancy reforms in the state. The study finds that despite the provision of the Assam Adhiar Protection and Regulation Act, 1984, sharecropping is still practiced largely informally in the state of Assam. In this context he suggests the adoption of a 'crash Programme' similar to Operation Barga (that implemented effectively the provisions of the West Bengal Land Reforms Act, 1955) in the state of Assam also so that the provisions of the legislations aimed at protecting the interest of the tenant farmers are enforced in letter and spirit.

Barman and Das have focused on the problems of technology adoption and constraints of development of 'char' areas of north bank plain zone in Assam. The 'Char' and 'Chapari' lands are such riverine land forms that are periodically formed and eroded due to meandering and course changing of the rivers. These are flat lands surrounded by river waters on one or more sides and are subjected to frequent flash flooding for varying length of time as well as to varying depth. The authors find that most of the Char areas have been suffering from laggardness as far as adoption and spread of new technology are concerned. Since not many studies have been done on the problem, the scholars expect that the findings of their study would help the policy makers to formulate suitable policies for the economic upliftment of the Char farmers.

Ahmed in his paper has explored into the factors and constraints that affect the adoption of new agricultural technology in Assam. In this paper based on empirical research he has made an attempt to identify the factors which influence the adoption of new technology in agriculture in the study blocks and to find out the constraints to adoption of new agricultural technologies by different categories of farmers i.e., large, medium and small framers. The study concludes that some constraints like lack of capital, low price of agricultural produce, problem of insufficient cold storage, inadequate institutional credit, problem of soil and water testing facility, inadequate irrigation facility, high cost of fertilizers, high rental charges of implements and machines, lack of information and problem of regulated market are the most important ones and mainly due to these constraints that the process of adoption of new agricultural technology in Barpeta district has been slow and interrupted.

Singh and Ngachan in their paper have made an attempt to study problems and prospects of watershed development projects in Manipur. The study finds that the watershed project altered the land use system favorably to horticultural crops mostly fruits with little attention to the development of field crops and livestocks including fisheries. The project could increase the income and employment opportunities of the households in the watershed. However, the most crucial thing is that these increased incomes were not distributed uniformly to all sections of people in the area satisfactorily. It necessitates proper attention to the landless, marginal and small farmers while planning for watershed development projects. Some of the policy options for successful implementation of watershed schemes suggested by the study are, empowerment of local people and village institutions, site specific research and a multi-disciplinary perception of issues by research and development agencies, institutional support such as extension and credit, enactment of suitable land reforms, joint operational partnership, flexibility in watershed guidance and attention on post watershed period.

Hariharan and Anandan's paper analyses mainly the growth of agricultural sector in Meghalaya and compares it with the growth of other sectors based on the available data. Further, the position of Meghalaya in the Indian economy in terms of NSDP and the growth of NSDP (in current and constant prices) of Meghalaya are also analysed. In the last part, the performance of the agricultural sector of Meghalaya state is compared with other North-Eastern States through a few indicators. The authors conclude that the agricultural sector of Meghalaya State has a good potential for development. However, the available statistics reveal that the agricultural development of the state is not impressive. They suggest that the Central and State governments need to concentrate more on the development programmes to stimulate the process of development of the agricultural sector in Meghalaya.

Roy's paper highlights the factors determining the agricultural growth in the state economy of Arunachal Pradesh. First he discusses the relative position of agriculture in the state's economy. Next he describes the techno-institutional changes underlying the agricultural growth and constraints and conceptual framework, which is tested in the third section. The study finds that the growth of labour productivity is the outcome of use of new inputs. Growth rate has turned negative in recent years. Regaining of momentum of growth requires strengthening of new institutions by conducting the much-wanted cadastral survey and ensuring a larger flow of bank credit to agriculture in the state. Diversification of the jhum cultivator's income through orchardization can reduce the incidence of atrophied jhuming.

Sangwan's paper deals with the problem of processing of horticultural crops in Arunachal Pradesh. In the paper he has made an account of the present status of horticulture in the state, an assessment of the potential of horticulture and suggested practicable strategy for agro-processing to sustain the present tempo of horticulture growth. One of his main suggestions for the development of agro-processing sector in the state is that the state government has to step in the shoes of an investor in the short run to realise the true potential of horticulture and other resources. Specifically he has suggested a 'Build and Lease out Policy' to be adopted by the state government.

Chakrabarti's paper deals with the status and role of agriculture in the economic development of Sikkim. It has used time series data for this purpose. It shows that the share of agriculture has declined over the years in Sikkim. But the employment pattern in agriculture has not maintained a balanced share of this sector in SDP. The study finds that the fall in the proportion of the agricultural workers is highly insignificance compared to the fall of the share of the agricultural sector in SDP. According to the scholar, this fact has become a source of economic concentration or regional disparity in the state.

# PART I

# GENERAL ISSUES ON AGRICULTURAL DEVELOPMENT

# 1

# PRESENT STATUS, POTENTIALS, PROBLEMS AND STRATEGIES FOR AGRICULTURAL DEVELOPMENT IN NORTH-EASTERN STATES

*Anuva Saikia*

## Introduction

The seven states of N.E. region cover 285.09 lakh ha of geographical area, which is 7.76% of total land area of the country with a population of 384.95 lakh. Hill areas (upto 5000 metres), riverine plains, swamps, tilla land, char areas are the agro-ecological situations. Forest area covers 116.9 lakh ha (45.88%). Arunachal Pradesh has highest area under forest (62.09%) followed by Mizoram (61.81%) while barren and uncultivable land is highest in Manipur (63.54%). Net sown area which is the basic resource for agricultural production increased from 36.17 lakh ha to 38.39 lakh ha in the last three decades while gross cropped area increased from 45.35 lakh ha to 53.49 lakh ha. Due to difficult terrains, hills and barren land, only 2.21% of total geographical area in Arunachal Pradesh, 5.17% in Mizoram is under cultivation while in Manipur, Meghalaya, Nagaland it is 6.26%, 9.18% and 12.72% respectively. In Assam it is 35.07%. The region receives abundant rainfall of 2600 mm annually but due to concentration in about six months, there is problem of flood in rainy season and water stress in the dry period. Mowsynram in Meghalaya receives highest rainfall in the world at 12000 mm while rain shadow belt exists in Lanka and Hojai areas of Assam receiving around 1200 mm rainfall. The climate is per humid to humid. The soil is mostly acidic. The paper analyses present status of agriculture in the N. E. region, potentials for development, problems and strategies. Development issues of N.E. region relate not only to economic factors but the diverse socio/cultural, ethnic, traditional beliefs and thoughts, environmental factors affect the economic issues. Many theories have been advanced to explain underdevelopment of N.E. region (Mruthyunjaya 2001).

## Present Status of Agriculture

### *Changes in Cropping Pattern*

Rice continues to remain as the major crop in all states covering 34.32 lakh ha (64.17% of gross cropped area). During the last three decades the most prominent changes were the expansion of rice area in Mizoram (1262.0%), Meghalaya (364.41%),

Nagaland (141.66%) and Arunachal Pradesh (100.0%). In Assam the expansion was less at14.0%, remained same in Manipur and declined in Tripura (-2.59%). Assam is the major rice growing state in the region and the major changes were tremendous expansion (83.8%) of area under Summer rice (Boro) due to creation of irrigation facilities through STWs and higher yield (2.0 tons/ha) under irrigated condition, decline of area (-26.5%) under autumn rice (Ahu) due to stagnation in yield (0.9 tons/ha). The changes in area under the dominant rice crop Winter rice (Sali) were marginal over the years but the growth of productivity is around 1.6% per annum, i.e. from 1.1 ton/ha during 1970s to 1.5 tons/ha during 2003-04. The major factors contributing to yield improvement are expansion of area under HYV rice from about 15.5% during 1975-76 to 54.0% during 2003-04 and increased use of fertilizer from less than 2.0 kg/ha to 47.56 kg/ha during the period.

Wheat is grown mainly in Assam but the area is declining since 1984-85 from 1.0 lakh ha to 0.70 lakh ha, due to lack of irrigation facilities, problems of processing, pre-harvest sprouting in the event of pre-monsoon showers and lower yield around 1.0 tons/ha. Wheat is grown in small pockets in Arunachal Pradesh (0.03 lakh ha) and Tripura (0.02 lakh ha), Maize is important crop in the hill districts of Arunachal Pradesh (0.34 lakh ha), Nagaland (0.30 lakh ha) while in Assam the area is stagnant around 0.19 lakh ha.

Oilseeds are next important crops covering 6 – 8% of gross cropped area except in Manipur (1.70%), Tripura (2.41%) and Meghalaya (3.68%), Rape and mustard were the only oilseed crops during 1970s but since 1980s new oilseed crops like linseed, castor, sunflower, groundnut entered into the cropping pattern.

Total pulse area is increasing from 0.95 lakh ha during 1970-71 to 1.63 lakh ha due to new entries in the cropping patterns of Arunachal Pradesh, Mizoram, Nagaland and Tripura since 1980s. Local pulses are grown in hill states. In Assam pulse area is stagnant around 1.1 lakh ha due to problems of soil moisture stress during sowing season, poor management, problems of disease and pest and lower yield.

Jute and sugarcane are two important cash crops of the region. Assam is the major jute producing state while jute area also exists in Meghalaya and Tripura. Jute area is declining in all these states due to problems of retting, processing and marketing. Other competing crops mainly rice is substituting jute area. Cotton is grown particularly in Meghalaya. In other states its area is declining.

Potato is gaining priority in the cropping patterns of all N.E. States except Mizoram.

Major Horticultural crops particularly banana, arecanut, coconut, papaya, tapioca, sweet potato, orange, lemon, mango, guava, litchi, jack fruit and a number of vegetables are grown in all states while apple is grown in Arunachal Pradesh and Nagaland. Spices and condiments of various types are grown throughout the region. Total area under major fruits and vegetable crops are 101.8 lakh ha with the production of 1483.0 lakh tons. (Table 1)

### *Shifting Cultivation*

In the North East more than 70% of total geographical area is covered by hills where shifting cultivation (Jhum cultivation) is the common practice of agricultural production. Annual area under shifting cultivation is 3869 sq. km. Minimum area under shifting cultivation in one time is 14,660 sq. km and 4.43 lakh families are practicing it. (Table2 ). The jhum cycle varies from 3 to10 years but due to pressure of population the cycle comes down to2-3 years. Shifting cultivation has its own danger of soil erosion, soil degradation, loss of valuable forest products and flooding in the river valley down below. Mixed cropping of rice maize, pulses and vegetables are grown with very low yield (Borthakur 1992). However, shifting cultivation provides the type of food of choice with less risk and uncertainty (Das Gupta 2001).

### *Growth of Productivity of Crops*

The N.E. states produce 58.75 lakh tons (2000-01) of food grains and 2.36 lakh tons of oilseeds which is about 3.0% and 1.28% respectively of total national production. Rice production is 54.80 lakh tons which is 6.45% of total national production. The growth of production of food grains from 30.55 lakh tons in 1972-73 to 58.75 lakh tons (40.80%) in 2000-01, is more due to improvement is productivity (52.59%) rather than expansion of area under crops (26.04%). As compared to the triennium ending 1974-75, the growth of productivity of foodgrains in the N.E. states was 44.24% in the triennium ending 2000-01, and it was 46.12% for rice. Mizoram (121.52%), Nagaland (117.54%), Tripura (113.71%) showed higher growth but these states cover only 17.15% of total rice area in the region. Yield of rice is the highest in Manipur (2.2 tons/ha). Assam has the highest share of rice area (76.49%) and the growth of productivity is 39.28% due to several factors such as diverse land types and soil fertility status for rice cultivation, lower yield of upland Autumn rice (Ahu) around 0.9 tons/ha (the crop covers 5.5 lakh ha and the area is declining), problems of flood affecting upto 4.5 lakh ha of crop land, slow adoption of improved crop production technologies, rainfed situations etc. resulting in situation wise yield variations from 1.02 tons/ha upto 2.25 tons/ha.

The annual average growth of productivity indicates that growth of yield is negative for wheat, and in case of maize it is higher in Meghalaya (165.0%), Nagaland (104.76%), Manipur (75.04%). In case of pulse crops growth is higher in Mizoram (125.44%), Nagaland (68.50%), Tripura (57.80%)

and lower in other states. Nagaland shows highest growth of productivity (127.80%) of oilseed crops and yield is highest at 1003 kg/ha followed by Tripura (83.65%) and Mizoram (62.90%). Growth is lower in Assam (7.33%), Meghalaya (24.80%) and negative in Manipur (-19.32%). Arunachal Pradesh (-0.55%). About 85% of oilseed crops are cultivated under rainfed conditions with low nutrient, which affect the yield.

Growth of yield of jute is high (84.93%) in Tripura, 44.72% in Meghalaya 20.97% in Assam and negative in Nagaland (-6.6%), but the area is declining in all states and rice is substituting jute (Table 3). The major problems of jute are absence of organized market in the region, dominance of middlemen and low price received by farmers, unscientific rattling affecting quality and lower grade of end product, lack of training and knowledge on utilization of jute for value added products etc. Sugarcane is not exhibiting satisfactory growth of productivity. Closure of the sugar mills and lack of incentives to farmers are the major reasons for decline of the crop.

The annual average growth of productivity indicate higher growth rates during 1980-81 to 1990-91 as compared to the previous decade except wheat, oilseeds and non food crops such as 1.67% for rice, 0.23% for wheat, 2.15% for maize 5.64% for pulses 1.86% for oilseeds, 1.46% for potato, 1.33% for jute and 1.78% for sugarcane. In the next decade of 1990-2001 the growth rates declined to 1.51% for rice, 0.21% for wheat, 1.98% for maize, (-) 0.75% for pulses, 1.39% for oilseeds, (-) 0.16% for sugarcane while for jute (6.67%) and potato (1.91%) it was higher. In Manipur and Meghalaya the growth of yield was higher for rice at 3.05% and 3.95% respectively as compared to the previous decade. Low seed replacement rate for paddy in Assam (3.5%), decline of wheat as cereal crop, stagnation/decline in yield of pulse crops along with technology gaps are the major factors.

The whole N.E. region is rich in horticultural crops like fruits and vegetables spices and condiments, floriculture particularly orchids, medicinal and aromatic plants. Commercial cultivation of vegetables and potato are coming up extensively with high growth rates, due to adoption of improved seed-fertilizer technology. Yield of potato doubled during the last thirty years from 4.3 tons/ha to 8.9 tons/ha and it is highest in Tripura (17.75 tons/ha). Although most of the fruit crops are grown without any improved management practice, due to suitable soil and climatic condition the yield is high for some fruit crops such as citrus crops in Nagaland (9.6 tons/ha). Assam (9.2 tons/ha) against all India average of 8.8 tons/ha, pineapple in Nagaland (28.58 tons/ha). Tripura (23.48 tons/ha). Assam (15.43tons/ha) against all India average of 15.6 tons/ha. Banana is grown in all states covering, 1.8 lakh ha (37.5% of total banana area of the country) but due to poor management, yield is less (11.26 tons/ha) which is less than half of the all India average at 33.48 tons/ha.

## Status of Adoption of Improved Technology

Crop production practices are mainly traditional under rainfed conditions in all the N.E. states. Under shifting cultivation local crop varieties are used and no plant nutrient is applied. In the plains districts use of HYV and plant nutrients is applied mainly for rice and vegetable cultivation. Development of irrigation facilities are scanty and cover around 16.3% of gross cropped area. Manipur has the highest proportionate area under irrigation (40.76%) followed by Nagaland (33.3%) while in other states it is less than 15.0% of GCA. Area under HYV rice is highest in Tripura (89.0%) followed by 54.0% in Assam, Manipur (44.3%). Mizoram (32.35%), Arunachal Pradesh (29.66%), Meghalaya (27.85%), Nagaland (24.13%). Fertilizer consumption is the highest in Manipur (105.0 kg/ha) followed by Assam (47.56 kg/ha), Tripura (32.3 kg/ha), while it is lower in Meghalaya (11.7 kg/ha), Mizoram (7.6 kg/ha), Arunachal Pradesh (3.68 kg/ha) and Nagaland (2.3 kg/ha). (Table 4)

## Livestock Status

Agricultural production systems in the North Eastern states are highly dependant on the draft energy supplied by livestock mainly bullocks and buffaloes as the use of machineries for agricultural production is still at a low level. In the rural households economy of N.E. states, bovine and ovine population are considered inseparable component of the farming systems and livestock are kept for milk, meat, manure, draught energy and transport purposes. 'Draft animal contributes 3607.84 mega joule of energy per ha of gross cropped area which is about 30% of total energy utilized in crop production' (Halim et. al 1999). The stocking density of livestock per 100 ha of gross cropped area is 200.95 for cattle and 18.28 for buffalo in Assam, 186.41 and 4.52 respectively in Arunachal Pradesh, 231.0 and 43.18 in Manipur, 156.0 and 14.63 in Nagaland, 284.0 and 6.92 in Meghalaya, 30.27 and 4.58 in Mizoram and 267.53 and 3.92 in Tripura. 'The eastern region sustained the highest number of draught animal per 100 ha of cropped land followed by southern, western and northern region' (Pandey, 1995). However majority of bovine and ovine population are of non-descript low productive and the animal husbandry practices are mostly traditional. 'Indigenous cattle outnumbered cross bred cattle across regions and at all India level during various censuses. Indigenous cattle population was the highest in the eastern region in all the censuses while cross bred cattle were concentrated in the eastern and southern region' (Pandey, 1995).

Total livestock in the N.E. states as per 2002 census is 194.92 lakhs, out of which cattle population is 113.43 lakhs, buffalo 9.11 lakhs, goat 39.89 lakhs, pig 30.67 lakhs. Compositional structure of livestock indicates dominance of cattle (58.19%), followed by goat (20.46%), pigs (15.73%) and buffaloes (4.67%). Total poultry population is 320.30 lakhs (Table 5).

The N.E. states now produce 1075 thousand tons of milk, 8948 lakh numbers of eggs and 112.25 thousand tons of meat. The overall growth of livestock population and production of milk, egg and meat is positive over the years in the N.E. states. Negative growth rate was observed in milk in Manipur and Mizoram, egg in Manipur, Assam during 1992-97. The growth of meat production is positive in all N.E. states (Table 6).

## Potential of Agricultural Development

Potentials exist for increasing production and productivity of the major crops particularly rice, wheat, pulses and oilseeds and horticultural crops. Both area and productivity can be raised for these crops through expansion of area under double/multiple cropping and adoption of improved crop management practices. Creation of irrigation facility is the prime requirement. Widening the network of input delivery system, credit facilities, and marketing infrastructure are the other basic requirements for achieving the potentials of agricultural development.

In Mizoram, Meghalaya, Nagaland, Arunachal Pradesh area expansion of rice is taking place in the last three decades. In these states programmes for productivity improvement of rice can be taken up. In case of Assam the major strategy to improve rice economy is to be directed towards area expansion and yield improvement for Summer Rice (Boro) in typical low lying areas. In non-traditional areas, Summer Rice cultivation can be taken up under irrigated conditions, as farmers would be interested to grow the crop due to higher yield prospects. Yield improvement of Winter Rice (Sali) particularly in nine low productivity districts of Assam is necessary. Yield improvement of Bao (Deep water rice) is possible through adoption of improved disease resistant varieties and control of diseases particularly Ufra which is the major pest of deep water rice. Potentials exist for growing scented rice, glutinous rice, and semi-glutinous rice for value addition. Glutinous rice has demand for industrial use and this possibility needs to be explored further. The yield gap analysis shows that for rice more that 60% of the potential yield is yet to be achieved. The yield gap ratio is also high for pulses (60%) oilseeds (59%) and horticultural crops (30 to 40% for coconut, orange and 50 to 80% for other crops). There is scope to increase yield through better crop management practices. Development of floriculture particularly orchids, fruits crops of commercial potential, spices and condiments, medicinal and aromatic plants should receive priority.

The important fruit crops of commercial potential are

banana, pineapple mandarin orange, lemon, other citrus crops, apple of Arunachal Pradesh, jack fruit, etc Except vegetables most of the fruit crops are grown in poor management conditions. These crops need technology intervention to improve crop productivity, post harvest management, agro-processing services for value addition, storage and packaging and marketing facilities.

**Problems of Agricultural Development**

- Slow growth of crop productivity due to dominance of traditional production technologies under rainfed conditions, such as use of local varieties of seeds, low rate of application of plant nutrients, absence of cultural management of crops, dependence on monsoon, lack of water management practices, disease and pest problems due to high rainfall and humidity.
- Shifting cultivation in hill areas with very low productivity of crops, danger of soil erosion of 60-70% (Borthakur, 1992), soil degradation, loss of valuable forest products and flooding in river valley down below.
- Farm mechanization is at very low stage. Availability of farm power to the farmers is barely 0.56 KW/ha against 1.02 KW/ha for all India average. In the hills districts farmers use traditional tools and implements for land preparation and use of animal power is less.
- Dominance of indigenous cattle with low levels of animal power of 0.5 H.P., low productivity of local cows at 120-150 litres of milk/lactation.
- Poor maintenance of health and hygiene of animals affecting productivity.
- Fisheries are not developed on scientific management basis inspite of existence of large water bodies.
- Problems of flood in Assam affecting 4.75 lakh ha chronically causing damage of standing crops, stored crops in granneries, livestock and also human lives, silting crop land with sand, problem of soil acidity and iron toxicity.

- Uneconomic and fragmented holdings in all N.E. states except Nagaland, which keep the farms at subsistence, level hindering investment for agricultural development. About 59.0 % of the holdings are below 1.0 ha and 21.12 % are 1.0-2.0 ha. (Table 7).
- Low capital formation in agriculture both in public and private sectors.
- Non-development of irrigation and drainage systems causing water stress conditions in the rabi season and flooding during monsoon.
- Non availability of suitable and economically feasible crop production technologies for rainfed upland situations, short duration semi-dwarf HYV rice for flood affected areas, HYV rice for hill areas, deep water situations with water level upto 50 cm, rice varieties tolerant to low temperature during reproductive stage. (Barua et al, 1989), cropping systems for double/ multiple cropping and riverine 'Char' areas. Early maturing varieties of wheat to escape pre monsoon showers, standardization of agro-technique for fruit crops of high commercial value, development of improved farm tools and implements, improved ploughs, techniques for post harvest management of fruits and vegetables, spices and condiments are the technology gaps.
- Uncertainty in the availability of critical inputs like HYV seeds, fertilizers and plant protection chemicals, lack of extension support, non-development of market infrastructure and weak support of agricultural credit facilities.
- Lack of strong extension support for transfer of available technologies to the farmer's fields, supervision of pest and disease attack and timely control measures.
- Market infrastructure is not developed in the whole region and marketing of agricultural produce is dominated by middlemen. Regulated markets are without infrastructure and not functioning in true spirit except collecting chess.

- Agricultural credit is not playing effective role for promoting agricultural development in the region. The proportion of agricultural sector advances to total advances in the N.E. region is 8.4% for commercial banks, 16.2% for Regional Rural Banks and 21.9% for cooperative banks while the CD ratio is 27.10% in the region and it is declining from 39.3% in 1994. The farmers are also reluctant to go to commercial banks for availing credit facilities due to official formalities, time involved in getting loans and prefers to depend on private money lenders who still provide about 30-40% of the total borrowing.

**Strategies for Agricultural Development**

- Supply of HYV seeds of the major crops of rice, wheat, maize, pulses and oilseeds, seeds/planting materials of vegetables and fruit crops and motivation of the farmers to use these seeds/planting materials through front line demonstrations, on-farm trials, technology missions are the major strategies. Production of certified seeds through registered seed growers/private entrepreneurship to be strengthened in all N.E. states. In Assam there are 88 seed farms and progeny orchard along with 383 seed villages. Such facilities to be developed in other N.E. states. Private entrepreneurship in production of certified seeds and planting materials need to be encouraged through training and financial support. The State Agricultural Universities and Department of Agriculture of the N.E. states should strengthen programmes for production of breeder seeds and foundation seeds of the major crops respectively.
- Development of single window input delivery system in interior rural areas, through input agencies, cooperatives, Self Help Groups, Field Management committee is necessary. Timely supply of inputs to be ensured.
- Development of irrigation facilities particularly STWs and lift irrigation along with well laid out field cannel system is necessary. Rainwater harvesting and use in the dry period should be started.

## Table 1: Changes in Cropping Pattern in N.E. States

| *Area under crops (lakh ha)* | *Period* | *Arunachal Pradesh* | *Assam* | *Manipur* | *Nagaland* | *Tripura* | *Mizoram* | *Meghalaya* |
|---|---|---|---|---|---|---|---|---|
| *1* | 2 | 3 | 4 | 5 | 6 | 7 | 8 | 9 |
| Rice | 1970-71 | 0.60 | 21.37 | 1.50 | 0.60 | 2.65 | 0.05 | 0.34 |
| | | (46.15) | (73.97) | (80.21) | (58.82) | (74.44) | (56.41) | (17.08) |
| | 1980-81 | 0.73 | 22.16 | 1.68 | 0.91 | 2.14 | 0.49 | 1.01 |
| | | (47.09) | (64.04) | (70.0) | (55.48) | (50.70) | (72.06) | (43.91) |
| | 1997-98 | 1.20 | 25.26 | 1L579 | 1.45 | 2.578 | 0.681 | 1.579 |
| Wheat | 1970-71 | - | 0.13 | - | - | - | - | - |
| | 1980-81 | 0.02 | 0.84 | - | - | 0.104 | - | 0.03 |
| | | (1.54) | (2.43) | | | (1.12) | | (1.30) |
| | 1997-98 | 0.038 | 0.85 | - | 0.018 | 0.02 | - | 0.043 |
| Maize | 1970-71 | - | 0.22 | 0.06 | - | - | - | - |
| | | | (0.76) | (3.21) | | | | |
| | 1980-81 | 0.19 | 0.23 | 0.09 | 0.15 | - | 0.02 | 0.17 |
| | | (12.26) | (0.66) | (3.75) | (9.14) | | (1.22) | (7.11) |
| | 1997-98 | 0.345 | 0.19 | 0.036 | 0.30 | 0.02 | 0.082 | 0.172 |
| Total Pulses | 1970-71 | - | 0.90 | - | - | 0.03 | - | 0.02 |
| | | | (3.11) | | | (0.84) | | (1.0) |
| | 1980-81 | - | 1.06 | 0.06 | 0.05 | 0.05 | 0.01 | 0.02 |
| | | | (3.06) | (2.5) | (3.08) | (1.18) | (1.47) | (0.86) |
| | 1997-98 | 0.065 | 1.18 | - | 0.164 | 0.092 | 0.041 | 0.092 |

| | | | | | | | | |
|---|---|---|---|---|---|---|---|---|
| Total Oilseeds | 1970-71 | - | 1.73 | - | 0.02 | 0.05 | 0.02 | 0.05 |
| | | | (5.3) | | (1.96) | (1.40) | (2.56) | (2.51) |
| | 1980-81 | 0.08 | 2.33 | 0.05 | 0.03 | 0.07 | 0.02 | 0.08 |
| | | (5.3) | (6.73) | (2.08) | (1.83) | (1.66) | (2.94) | (3.48) |
| | 1997-98 | 0.241 | 3.07 | 0.031 | 0.285 | 0.103 | 0.091 | 0.091 |
| Jute | 1971-72 | - | 1.34 | - | Neg. | 0.08 | - | -0.7 |
| | | | (4.64) | | | (2.55) | | (3.52) |
| | 1980-81 | - | 1.12 | - | Neg. | 0.04 | - | 0.06 |
| | | | (3.23) | | | (0.94) | | (2.51) |
| | 1997-98 | 0.945 | 0.95 | - | 0.040 | - | - | 0.042 |
| Sugarcane | 1970-71 | - | 0.34 | - | 0.01 | 0.02 | - | 0.002 |
| | | | (1.17) | | (0.98) | (0.56) | | (0.10) |
| | 1980-81 | - | 0.48 | 0.02 | 0.04 | 0.02 | 0.01 | 0.002 |
| | | | (1.38) | (0.83) | (2.44) | (0.47) | (1.47) | (0.08) |
| | 1997-98 | 0.31 | 0.31 | 0.013 | 0.01 | 0.011 | 0.013 | 0.001 |
| Potato | 1970-71 | - | 0.27 | - | - | 0.03 | 0.01 | 0.16 |
| | | | (0.93) | | | (0.84) | (1.28) | (8.04) |
| | 1980-81 | 0.01 | 0.38 | 0.02 | 0.04 | 0.02 | 0.006 | 0.17 |
| | | (0.64) | (1.09) | (0.83) | (2.44) | (0.47) | (0.88) | (7.11) |
| | 1997-98 | 0.045 | 0.75 | 0.032 | 0.038 | 0.049 | 0.007 | 0.203 |
| Cotton | 1970-71 | - | 0.04 | - | - | 0.02 | 0.01 | - |
| | | | (0.14) | | | (0.56) | (1.28) | |
| | 1980-81 | - | 0.04 | 0.007 | - | 0.01 | 0.02 | 0.07 |
| | | | (0.22) | (0.29) | | (0.24) | (2.94) | (3.04) |
| | 1997-98 | NA | 0.02 | 0.001 | 0.002 | 0.011 | 0.01 | 0.076 |

- Watershed management approach to utilize the water resources of rivers and streams can be initiated.
- Larger coverage of farmers under Kishan Credit Card facilities through DRDA is necessary. Commercial banks and cooperative banks should popularize loan schemes to farmers and provide agricultural credit as priority sector.
- Shifting cultivation cannot be eliminated from hill agriculture, as it provides them type of food of choice with less risks and uncertainty (Das Gupta 2001). Improved methods to be popularized among Jhumia families to increase farm returns and protect soil erosion and environmental degradation. Popularization of improved model among Jhumia families include growing of field crops (rice + sesamum + maize) along the slope, inclusion of perennial horticultural crops (Pineapple and mandarin orange) and use of perennial grasses for conservation of soil and soil fertility. The model is suitable for hills with gentle slope (Pathak 2001).
- Settled cultivation as done in Karbi Anglong district of Assam, Meghalaya, pilot project of land reclamation and improvement in Mizoram, pilot project of terrace cultivation in Nagaland, improved method of Jhum cultivation developed by tribals in Manipur (Singh 2001, Singh 2001) should be popularized.
- Changes in attitude of farmers from traditional practices to modern technologies necessary.
- Research and development efforts should be directed towards for evolving technologies for specific situations of flooded plains, hill areas, rainfed upland, deep water situations, cropping systems suitable for 'farm diversification, improved farm tools and implements, post harvest management.
- Popularization of bio-fertilizers and integrated pest management would result in increased productivity of crops. Indigenous Technical Knowledge (ITK) of farmers in control of disease and pest to be collected and evaluated.

- Development of allied agricultural activities like bee keeping, mushroom production, sericulture, floriculture, value addition through processing of agricultural produce to be taken up in rural areas for providing additional income and employment.
- The strategy for development of horticulture should include supply of seeds and planting materials, support of agro-processing services, market link-up from producers to consumers, rationalizing access of traders and middlemen in marketing, to discourage pre-harvest contract, strengthen off-season vegetable production, develop agricultural produce marketing systems.

**Table 2: Shifting Cultivation in N.E. Region**

| *States* | *Annual Area under shifting cultivation (sq Kms)* | *Fallow period (in years)* | *Minimum Area under shifting cultivation one time or other (Sq Kms)* | *No. of families practicing shifting cultivation* |
|---|---|---|---|---|
| Arunachal Pradesh | 700 | 3.10 | 2,100 | 54,000 |
| Assam | 696 | 2.10 | 1,392 | 58,000 |
| Manipur | 900 | 4.7 | 3,600 | 70,000 |
| Meghalaya | 530 | 5.7 | 2,650 | 52,290 |
| Mizoram | 630 | 3.4 | 1.890 | 50,000 |
| Nagaland | 190 | 5.8 | 1,913 | 1,16,046 |
| Tripura | 223 | 5.9 | 1,115 | 43,000 |
| | 3,869 (1.5 per cent) | | 14,660 (5.7 per cent) | |

Source: Basic Statistics for N.E. India – 2000, NEC.

## Table 3: Growth of Productivity of Crops in N.E. Region (Yield kg/ha)

| State | Triennium ending 1974-75 | | | | | | | Triennium ending 2000-01 | | | | | | |
|---|---|---|---|---|---|---|---|---|---|---|---|---|---|---|
| | Rice | Wheat | Maize | Total pulse | Total Food-grains | Total oilseeds | Potato | Rice | Wheat | Maize | Total pulse | Total Food-grains | Total oil-seeds | Potato |
| Arunachal Pradesh | 874.6 | 933.33 (ending 78-79) | 1143 (ending 76-77) | 886.6 (ending 93-94) | 878.6 | 984 (ending 86-87) | 6402.6 (ending 86-87) | 1085.0<br>16.25 | 1389.66<br>21.58 | 1429.46<br>25.06 | 1000.0<br>12.80 | 1137.6<br>29.47 | 978.66<br>-0.55 | 7039<br>9.97 |
| Assam | 1003 | 1287 | 546.6 | 456 | 985.3 | 450 | 4103 | 1397.0<br>39.28 | 1287.0<br>. 0.0 | 732.3<br>33.97 | 557.3<br>22.21 | 1350.3<br>37.04 | 483.0<br>7.33 | 8393<br>104.55 |
| Manipur | 1333 | 972.33 (discon-tinued | 2088.6 | 404.6 (ending 75-76) | 1351.3 | 483 | 4097.3 (ending 75-76) | 2237.1<br>67.82 | - | 3348.3<br>60.31 | - | 2276.3<br>68.45 | 389.66<br>-19.32 | 5076.23<br>20.89 |
| Meghalaya | 1102.6 | 944.33 | 612 | 630.6 | 1027.3 | 543 | 4277 | 1461.56<br>32.55 | 1653.33<br>75.07 | 1455.2<br>138.0 | 937<br>48.59 | 148.5<br>41.0 | 677.66<br>24.80 | 9677<br>126.25 |
| Mizoram | 804 | - | 1745.6 | 583.3 (ending 76-77) | 842.6 | 558 | 3433.3 | 1781.0<br>121.52 | - | 2052.2<br>17.56 | 1315<br>125.44 | 1786.26<br>112.0 | 909<br>62.90 | 5142<br>49.77 |
| Nagaland | 581.3 | 2427.6 (ending 93-94) | 610 | 497.6 | 590.3 | 440.3 | 4000 (ending 75-76) | 1264.6<br>117.54 | 1861.0<br>-23.33 | 1112.3<br>82.34 | 838.5<br>68.50 | 1177.26<br>99.43 | 1003<br>127.80 | 8524<br>113.1 |
| Tripura | 984.3 | 1578.13 (ending 93-94) | 520 | 369 | 980.3 | 414 | 8550 | 2103.53<br>113.71 | 1966.6<br>24.60 | 950.0<br>82.69 | 582.3<br>57.80 | 2038.66<br>107.96 | 760.33<br>83.65 | 17755<br>107.66 |
| Total N.E. | 1009 | 1101.3 | 961.6 | 455.13 | 992.3 | 454 | 4370 | 1474.33<br>46.12 | 1339.10<br>21.58 | 1342.6<br>39.62 | 579.6<br>27.34 | 1431.33<br>44.24 | 576.66<br>27.0 | 8912<br>103.93 |
| All India | 1088.6 | 1939.3 | 1002.3 | 452 | 821.3 | 527.6 | 9455 | 1896.76<br>74.23 | 2630.6<br>35.64 | 1760.5<br>75.64 | 579.6<br>28.23 | 1600.2<br>94.83 | 858.33<br>62.68 | 17446<br>84.51 |

No wheat and pulse in Manipur during nineties. Lower figures indicate P.C. changes.

### Table 4: Area under HYV rice and rate of fertilizer consumption of N.E. Region, 1997-98

| *States* | *Total rice area (lakh ha)* | *HYV rice area (lakh ha)* | *P.C. to total rice area* | *Fertilizer consumption per ha of Gross Cropped Area* |
|---|---|---|---|---|
| Arunachal Pradesh (2000-01) | 1.18 | 0.35 | 29.66 | 3.68 |
| Assam (2003-04) | 25.29 | 13. 66 | 54.01 | 47.56 |
| Manipur (1997-98) | 1.58 | 0.70 | 44.30 | 105.38 |
| Meghalaya | 1.58 | 0.44 | 27.85 | 11.7 |
| Mizoram | 0.68 | 0.22 | 32.35 | 7.6 |
| Nagaland | 1.45 | 0.35 | 24.13 | 2.3 |
| Tripura (2000-01) | 2.41 | 2.14 | 89.0 | 32.3 |

### Table 6: Production of Milk, Egg, Meat in N.E. States, 2003-04

| *States* | *Milk (000 tons)* | *Egg (Lakh Nos)* | *Meat* (000 tons)* |
|---|---|---|---|
| Arunachal Pradesh | 46 | 86 | 17.4 |
| Assam | 727 | 5138 | 22.35 |
| Meghalaya | 69 | 936 | 28.8 |
| Manipur | 71 | 794 | 18.05 |
| Mizoram | 15 | 307 | 7.45 |
| Nagaland | 63 | 682 | 18.2 |
| Tripura | 84 | 1005 | NA |
| Total | 1075 | 8948 | 112.25 |

* In case of meat, for 1997-98 in all States and for 2003-04 for Assam.

**Table 5: Livestock and Poultry Population of N.E. States: ('000 Nos)**

| *States* | *Cattle* | *Buffaloes* | *Sheep* | *Goats* | *Pigs* | *Horses and Ponies* | *Yaks* | *Total Livestock* | *Total poultry* |
|---|---|---|---|---|---|---|---|---|---|
| Arunachal Pradesh | 453 | 11 | 28 | 183 | 275 | 6 | 9 | 965 | 1267 |
| Assam | 8419 | 677 | 154 | 2986 | 1533 | 11 | - | 13780 | 21763 |
| Manipur | 508 | 95 | 8 | 33 | 388 | 2 | - | 1034 | 3055 |
| Meghalaya | 738 | 18 | 17 | 280 | 351 | 2 | - | 1406 | 2152 |
| Mizoram | 33 | 5 | 1 | 16 | 163 | 2 | - | 220 | 1307 |
| Nagaland | 383 | 36 | 2 | 161 | 571 | 1 | - | 1154 | 2444 |
| Tripura | 1228 | 18 | 6 | 639 | 237 | 2 | - | 2130 | 3595 |
| Total N.E. | 11343 | 911 | 146 | 3989 | 3067 | 27 | 9 | 19492 | 32030 |
| | | | | | | | | (4.07) | (9.24) |
| All India | 1997708 | 88832 | 56766 | 120774 | 13581 | 827 | 54 | 478542 | 346693 |

*Source:* Basic Statistics, NEC, 2002.

**Table 7: Number of Operational Holdings 1995-96 (000 Nos.)**

| *States* | *Average size of operational holding* | *Marginal >1.0 ha* | *Small 1.0 -2.0 ha* | *Semi medium 2.0-4.0 ha* | *Medium 4.0-10.0 ha* | *Large <10.0 ha* | *Total* | *% of marginal +small holdings* |
|---|---|---|---|---|---|---|---|---|
| Arunachal Pradesh | 3.31 | 21 | 20 | 29 | 28 | 6 | 104 | 39.42 |
| Assam | 1.17 | 1669 | 561 | 351 | 96 | 5 | 2683 | 83.11 |
| Manipur | 1.22 | 69 | 49 | 21 | 3 | Neg | 143 | 82.51 |
| Meghalaya | 1.33 | 72 | 44 | 39 | 6 | Neg | 160 | 72.50 |
| Mizoram | 1.29 | 28 | 26 | 12 | Neg | Neg | 66 | 81.82 |
| Nagaland | 4.82 | 9 | 21 | 40 | 64 | 14 | 149 | 20.13 |
| Tripura | 0.60 | 247 | 40 | 12 | 1 | Neg | 301 | 95.35 |
| Total | 1.96 | 2115 | 761 | 504 | 198 | 25 | 3603 | 79.82 |
| All India | 1.41 | 71179 | 21643 | 14261 | 7092 | 1404 | 115580 | 80.30 |

## References


Barua, A.R., Bhattacharya, H.: Constraints of Crop Production in Assam. *Agriculture in Assam*, Ed. P.C. Goswami, Assam Institute of Development Studies, Guwahati, 1989.

Borthakur, D.N.: *Agriculture of the North Eastern Region*, Bee Cee Prakashan, Guwahati, 1992.

Das Gupta, Malabika : Status of Jhum Cultivation in the North East, Proceedings of Fourth Conference of North East Economic Association, held on 7-8 September, 2001, in Shillong.

Halim, R.A., Saikia H.C., Bhowmick B.C.: Energy Use In Crop Production, *Agricultural Economics Research Review*, Vol. 12, Nov. 1, 1999.

Mruthyunjaya: Prioritization of Strategies for Agricultural Development in the N.E. India – A Synoptic Overview, Proceedings of National Workshop On Prioritization of Strategies for Agricultural Development in North Eastern India, held in February 27-28, 2001 at AAU, Jorhat, Ed. B.C. Barah published by NCAP, New Delhi.

Pathak, A.K.: Technology Development and Its Impact on Farmers Fields, Proceedings of Workshop on Prioritization of Strategies For Agricultural Development in North Eastern India held at AAU, Jorhat on 27-28, 2001, published by NCAP, New Delhi. Ed. B.C. Barah.

Pandey, U.K.: The Livestock Economy of India – A Profile, Indian *Journal of Agricultural Economics*, Vol. 50, No. 3, July–Sept. 1995.

Singh, N. Mohendro: Jhuming in Manipur – A Few Critical Issues, Proceedings of Fourth Conference of North East Economic Association held on 7-8 September 2001, in Shillong.

Singh, E. Iboyaima: Jhum Cultivation in Manipur – An Aspect, Proceedings of Fourth Conference of North East Economic Association, held on 7-8 September 2001, in Shillong.

# 2

# ASSESSMENT OF AGRICULTURAL DEVELOPMENT IN THE NORTH EAST

*Subhasis Mandal, K.K. Datta, M.R. Verma and A.K. Tripathi*

## Introduction

Indian economy has recorded spectacular growth in recent past and in fact, is one among the fastest growing economies of the world. However like other countries around the world, this economic growth is not always ensuring the balanced economic development in terms of reducing regional imbalances in some parts of our country. The north eastern region is the most prominent in this list, which comprises of a bunch of less developed states with poor economic performance. The NEH Region of India comprising of the states of Arunachal Pradesh, Assam, Manipur, Meghalaya, Mizoram, Nagaland, Sikkim and Tripura, representing a distinct agro-climatic zone of the country. The region occupies nearly 8 per cent of total geographical area and supports nearly 3.8 per cent of the country's population as par the 2001 census. More than 80 per cent of the people in these states are dependent on agriculture and allied activities, but with poor resources base and low economic status with limited investment capacity (Mandal *et al*, 2005). Since independence, continuous efforts have been put forth and in particular during preceding decade several aggressive policy initiatives (e.g. spending 10 per cent of the budget for every ministries) has been taken up for all-round development of theses states. Though there are some success stories in few fronts (e.g., high literacy rate, nutritional status among children is higher than national average etc.), but overall the region still has to bear the stigma of 'outlays without outcome'. Several reasons can be attributed to this economic laggardness, such as, difficult geographical terrain, improper perspective planning failing to have a holistic road map covering the various sectors, problem of leakages in most of the developmental projects due to lack of non-imposition of strict accountability and finally, failure to ensure the synergies between developmental goals and efforts with security needs, which must go hand in hand to achieve the developmental targets. Despite these difficulties, the region has enormous potentials due to its rich endowment with natural wealth and locational advantages. Perceived weakness of this region can be converted into strength through proper planning and implementation of the developmental projects. Income-level of rural families can be substantially enhanced particularly on a mass scale by harnessing the vast readily available potential in the field of agriculture, horticulture, medicinal and aromatic herbs and bamboo, water, wind power and minerals, to name just a few sectors (Singh, 2005).

Agriculture in NEH region is predominated by subsistence farming, where shifting cultivation still remains important. Subsistence agriculture is an uncertain activity and therefore, risky, particularly when survival is at stake, and is the dominant factor not to change the traditional farm practices even in the face of opportunities. The poor farmers are naturally risk averter and prefer to be safe than sorry; they tend to prefer an inferior outcome that is relatively certain to the prospect of a higher average return with a greater degree of risk attached (Thirwall, 1999). The sectoral transformation is occurring at a slow pace, where the agriculture sector is playing passive and supportive role in the process of economic development. However, to cater the economic development, the agriculture sector must be strengthened to play much more active and indispensable role in any overall strategy of economic progress (Todaro & Smith, 2004). Since agriculture sector is dominant in this region, development policy for these states needs to assess with reference to its performance, contribution and role.

## Agricultural Development Status

### *Land Utilisation and Distribution of Farm Lands*

Agricultural performance of the region remains sluggish on account of a number of factors. For one, the ratio of irrigated area to total area under all crops in the NER has generally remained much lower than the all-India figure of 39 per cent (in 1998-99). Moreover, the total area in the NER covered under HYV seeds formed only around 0.25 per cent of the all-India total in 1996-97. The northeastern states are severely constrained by difficult terrain with 70 per cent area under hills, and forest covering ranges between 42 and 76 per cent of the area of each states (Mohan, 2003). Out of total geographical area of over 25 million ha, only 24 per cent area is suitable for agricultural operation i.e., arable land as compared to 59 per cent at national level (Table 1). Assam is having nearly 43 per cent of total geographical area suitable for agricultural operation, followed by Nagaland (38 per cent), Tripura (30 per cent), Mizoram (21 per cent), Sikkim (15 per cent) and Manipur (7 per cent). Out of the total arable land (64 million ha) a little over 60 per cent (4 million ha) area has been brought under actual cultivation (Net Sown Area/ NSA) at the North East Region level, while at national level nearly three-fourth (73 per cent) area has been brought under actual cultivation (NSA). Among the north eastern states, Tripura ranked first in terms of percentage of NSA to total available arable land (89 per cent), followed by Sikkim (86 per cent), Manipur (85 per cent), Assam (82 per cent), Arunachal Pradesh (63 per cent), Nagaland (34 per cent), Mizoram (25 per cent) and Meghalaya (19 per cent). The Gross Cropped Area (GCA) of the region is over 5 million ha, out of which Assam alone is accounting for more than 70 per cent of the area. The cropping intensity of the region is marginally high (138 per cent) as compared to the nation as a whole (132 per cent). However, the cropping intensities varying widely, from maximum of 154 per cent in Tripura to as low as just 100 per cent in Mizoram. Cropping intensity in Assam was calculated to be 142 per cent followed by Arunachal Pradesh (132 per cent), Manipur (130 per cent), Sikkim (127 per cent), Meghalaya (119 per cent) and Nagaland (108 per cent). Overall, the development of irrigation facility is very poor and nearly 84 per cent of the total cropped area is being cultivated as rainfed condition. The percentage of Gross Irrigated Area (GIA) to Gross Cropped Area (GCA) is highest in Manipur, but with a low magnitude i.e., 41 per cent only, followed by Nagaland (32 per cent), Meghalaya (18 per cent), Arunachal Pradesh (15 per cent), Assam (15 per cent), Tripura (14 per cent), Sikkim (13 per cent) and Mizoram (8 per cent), which indicates the need of creation of massive water harvesting and storing structures suitable to the physiographic situation of the region. It is important to note that the north eastern farmers are not much depending on the external input use, fertilizer, is an added advantage for the region to explore the potential of organic agriculture which has been given policy thrust recently. The fertilizer consumption in NEH Region showed that the consumption per ha is hovering only around 35 kg (2001-02). Manipur is the exception among north eastern states in terms of using fertilizer at higher level (105 kg/ha). Next to Manipur, Assam is using maximum amount of fertilizer per ha (39 kg/ha) followed by Tripura (30 kg/ha), Mizoram (14 kg/ha), Meghalaya (17 kg/ha), Sikkim (10 kg/ha), Arunachal Pradesh (3 kg/ha), and lowest in Nagaland (2.2 kg/ha) observed during 2000-01. Also, the trend in use of fertilizer during recent period (2001-02 to 2003-04) does not show any significant difference within the north eastern states. The average size of farm holdings among north eastern states was observed to be highest in Nagaland (4.82 ha), followed by Arunachal Pradesh (3.31 ha), Sikkim (1.65 ha), Meghalaya (1.33 ha), Mizoram (1.29 ha), Manipur (1.22 ha), Assam (1.17 ha) and lowest in Tripura (0.6 ha). At Regional level, three out of eight states from this region, namely, Arunachal Pradesh, Nagaland and Sikkim has the higher farm holding size as compared to the national average (1.41 ha). However, in spite of possessing larger farm holdings, the hill farmers can not enjoy the economies of scale due to limited substitutability among the factors of production (e.g., replacing human labour with machine labour), hence, the scope for reduction in cost of production is also restricted.

The land utilization pattern for agriculture in NEH Region is clearly indicating the existing wide gap between the amount of land available for agriculture (arable land) and the land actually brought under cultivation (NSA). Mostly the lands are used for a single season and only 38 per cent of NSA is being utilized for cultivation of more than once. Despite of receiving highest rainfall, the region is predominated by rainfed agriculture; the agricultural operation is severely restricted by either heavy rainfall or lack of soil moisture. The pace of agricultural development in NEH Region, in terms of land intensification, is quite slow, perhaps the major factor that the region is still carrying rich biodiversity. The development of agriculture plan should be given top priority to protect the natural wealth, therefore, needs to be addressed from the purview of multifunctional (e.g, contribution to ecology, maintaining rich biodiversity etc) nature of agriculture.

### Trends in Foodgrains Productivity

The magnitude of change in trends in foodgrains productivity was higher during 1980-81 to 1990-91 period as compared to 1990-91 to 2002-03 period in various north eastern states, except for Meghalaya (Table 2). The improvement in foodgrains productivity was highest in Nagaland (36 per cent), during 1980 to 1990 followed by Tripura (32 per cent), Mizoram (24 per cent), Manipur (19 per cent) and Arunachal Pradesh (11 per cent). During later decade (1990-2003), Mizoram recorded highest increment in productivity (44 per cent) followed by Meghalaya (42 per cent), Nagaland (35 per cent), Manipur (31 per cent), Tripura (23 per cent), Assam (12 per cent) and Arunachal Pradesh (5 per cent). It is notable from the Table 2 that the trend in foodgrains productivity was almost similar to the trend of change in rice productivity for all north eastern states. The reason being, rice alone accounted for more than 70 per cent of the cropped area in these states and also has almost equal share to the total foodgrains production. Therefore, productivity of rice observed to be almost determining the productivity of foodgrains as a whole. Given the positive improvement of foodgrains production, the major challenge is to continue this improvement, particularly in view of the complex, diverse and risk prone (CDR) production environment existing in these hilly states. Rice in these states are grown traditionally and characterized by low yield, particularly at upland situation. Moreover, some of the northeastern states possess huge rice diversity where advancement of modern varieties is almost insignificant. The rice in this region is cultivated under varied topographical situation and production environment; for example, Assam has been dominated by rainfed production system, which covered 79 per cent of total cropped areas. The rainfed areas were further classified into upland and shallow land, which occupied 60 per cent of total rice area, deep water 14 per cent and areas with intermittent water (30-100 cm) was 26 per cent. Rice in Assam has a dubious distinction, having grown to diverse land types in three seasons in a year. This diversity of production ecosystems made the understanding of rice production system more complex (Bhowmick, 2005). The NEH Region, the area under modern rice variety is less and a large number of traditional varieties are being grown by the farmers. The traditional farming systems are being followed and most of the farming practices are operating in subsistence to semi-commercial stage. Additionally, the farming practices are not responsive to the market demand, as most of the landraçes are traded through small-scale informal marketing channels. The primary objective of growing rice is not for profit maximization but to meet the household requirements to maintain households' food security. The opportunity of *in-situ* on farm conservation may be quite high and feasible in this region because the farmers are continuing to grow a large number landraces since ages. So, conservation of the traditional varieties can be viewed as a unique opportunity for this region, which has great importance for our society (public value) to secure the future varietals development programme. The efforts to conserve plant genetic resource have long been emphasized and northeastern states particularly, Arunachal Pradesh, Nagaland, Mizoram and Meghalaya have the right potential for exploring these opportunities.

## Employment and Output from Agriculture

Agriculture and allied activities are the major employment provider for the workers in this region, which is accounted for nearly 50-70 per cent of the total workforce across the various states (Table 3). But the contribution of agricultural output is quite low, ranging in between 27-37 per cent only, which is almost half of the share of labour force, engaged in this sector, implying the low level of labour productivity. The efficiency of labour input must be increased through accelerated agricultural development, for which identification and prioritization of the potential sub-sectors is extremely important. Also the linkages between agricultural sector to other sector needs to be analysed to employ the surplus labour gainfully. The agricultural output can be increased through promotion of possible horizontal and vertical integration of the agriculture and allied activities. It is important to note from the Table 3, that the female workers are mostly engaged in agricultural activities and in fact, nearly 70-80 per cent of the female workers are earning their wages from this sector, whereas, nearly 50-60 per cent of the male workers are engaged in agricultural activities. Women provide an important source of labour for cash crop production, growing food for household consumption, raising livestock and generating additional income through diversified activities. However, many of this work remain statistically 'invisible'. Nevertheless, the women group needs to be given special attention to increase their efficiency, which in turn will lead to improvement in the overall labour productivity and will increase the share of agricultural output to Net State Domestic Product (NSDP). When credit and resources are unavailable, reducing the variability of household earnings generally entails choosing less efficient method of production and thus realizing low income. This trade-off occurs most often in female headed household where resource constraints are greatest, thus as a consequence of their restricted range of choices, women tend to retain traditional modes of economic activities (Todaro and Smith, 2004).

Value of agricultural output per workers (cultivators plus agricultural labourers) indicated some encouraging results and in fact, values for some of the states exceed the respective per capita income when compared at current prices of 2000-01. For example, per capita income (state level) at current prices of 2000-01, was below of their respective value of agricultural output per workers for the states, Arunachal Pradesh (Rs. 15640), Assam (Rs. 10467), Manipur (Rs. 10658), Meghalaya (Rs. 14654) and Tripura (Rs. 15253). This implied that even with the low level of labour productivity, the agricultural activities continued to provide relatively better employment opportunity as compared to overall activities available in these states. However, this may have another implication also, that is the opportunities of gainful employment in non-agricultural activities are restricted or risky. Therefore, the results indicated that agricultural activities is continuing to provide employment opportunities to majority of the workers and might be given priority for overall economic development of these states.

### Food Security and Per Capita Income Growth

Attainment of food security is most important to progress and sustain the economic development of any region or state of the country. To achieve the food security, the growth rate for foodgrains production should exceed the population growth rate at individual state level, which is not happening for the states, Arunachal Pradesh, Manipur, Meghalaya and Sikkim (Table 4). However, rest of the states in this region, Assam, Mizoram, Nagaland and Tripura has been able to maintain higher foodgrains production growth as compared to population growth rate, which is a welcome development. The performance of foodgrains production growth was experiencing to be alarming for Sikkim (-1.58 per cent), which needs to be looked into seriously. Similarly, also the per capita output from agriculture (excluding livestock and fisheries) was observed to decline for Sikkim (from Rs. 3549 at 1991 to Rs. 2372 at 2001) and Arunachal Pradesh (from Rs. 5062 at 1991 to Rs. 2333 at 2001). Due to non-availability of desired data set the detailed analysis was not included in the present study. As far as per capita food production is concerned, Arunachal Pradesh, Sikkim and Tripura failed to maintain the food production level during 2001 as compared to 1991. High population growth rate is the major factor for this decline, however, the magnitude of decline was very high for Arunachal Pradesh (from 679 g to 539 g per capita production) and Sikkim (from 718 g to 539 g) needs to be looked into seriously.

The achievement of higher growth rate in foodgrains production over population growth rate is an essential condition for food security, but, simultaneously sustaining the per capita income growth is also a must to ensure the food availability at household level. The growth of per capita income reflects the purchasing power of the individuals which is indispensable to ensure the distribution of available foodgrains. Therefore, foodgrains production growth ensures the availability of foods in one hand and per capita income growth guarantees the accessibility of food produced on other hand. The results in Table 4 indicated that growth rate in per capita income (during 1993-94 to 2002-03 at 1993-94 prices constant prices) was highest in Tripura (8 per cent) followed by Meghalaya (5 per cent), Manipur (4 per cent), Sikkim (3.5 per cent), Nagaland (3 per cent), Assam (1 per cent) and Arunachal Pradesh (0.9 per cent). Results further indicated that half of the north eastern states (Assam, Mizoram, Nagaland and Tripura) showed foodgrains production growth over population growth and also the per capita income growth for all the states (except Arunachal Pradesh) indicated positive improvement. Therefore, the problem of food security for these states can be attributed partly to the deficiency in food production and also due to the distribution failure. Besides improving the foodgrains productivity, efficient public distribution system is equally essential to achieve the desired food security for this region.

## Value of Agricultural Output and Growth

Annual compound growth rate of value of agricultural output observed to be quite encouraging for the period of 1990-90 to 2002-03 (Table 5). The growth performance was even better than national level (2 per cent) in all northeastern states except, Arunachal Pradesh (-4.3 per cent), Sikkim (0.31 per cent). The growth rate observed to be highest for Nagaland (12.57 per cent) followed by Tripura (5.22 per cent), Mizoram (4.72 per cent) and positive for all other states. The growth rate of Nagaland can be attributed to the high growth rate achieved for production of spices and condiments (24 per cent) and fruits and vegetables (20 per cent). Growth rate for value of paddy output continues to be positive for all the states at varying magnitude, but output growth for value of pulse production shows some dismal figures, except for Manipur, Nagaland, Arunachal Pradesh and Meghalaya, where the growth rates were positive. Pulse production can be increased in the state with adequate technological back-up and extending complete package of practices in a mass scale. Similar to output growth for paddy, total cereal production also indicated positive growth for the northeastern states, except Sikkim. The growth rate for value of output from fruits and vegetables as well as spices and condiments were observed to be highly encouraging (except for Arunachal Pradesh and Sikkim). This calls for highly focused policy initiation to realize the full potentialities. Major task would be to increase the marketable surplus through horizontal integration and then, promoting vertical integration for value addition of these products. To facilitate this integration, there is need for credible database at micro level on these products giving detailed of area wise volume of production and their movement, cost of production and possible value addition. Once this integration is successful, thereafter, the possibility of air lifting for these value-added products can be thought of to reduce the time for transportation.

## Value of Output from Livestock and Growth

Similar to agriculture, the output growth for livestock product also observed to be encouraging for the northeastern states (Table 6). The growth rate was varying from highest in Nagaland (6.63 per cent), followed by Tripura (6.37 per cent), Meghalaya (5.93 per cent), Mizoram (5.61 per cent), Sikkim (1.83 per cent), Arunachal Pradesh (1.54 per cent) and Assam (0.70 per cent). Four out of eight northeastern states has even recorded much higher output growth as compared to all India level (3.79 per cent). The major sources of growth were realised from meat and milk production during this period (1990-91 to 2002-03). Value of output growth from egg showed both positive as well as negative sign. The northeastern states are highly suitable for rearing diverse livestock population which includes, cattle, buffaloes, goat, sheep, poultry, pig and in addition mithun and yak. The livestock sectors needs to be made more vibrant through enhancing the productivity, particularly, in pockets where the livestock densities are very high with respect to existing population base, available workers or net sown areas. It is essential to increase the volume of marketable surplus for trading of livestock products within the countries as well as to neighbouring countries, where the demand for livestock products are always exceeding supply. However, to achieve the surplus production suitable technological imperatives as well as horizontal integration among the primary producers is a must.

## Organic Farming: The Future Harvest?

The North Eastern Region is probably the most suitable area of India for organic agricultural farming because of its number of advantages. This region is said to be, by and large, organic by default or the farmers of this region are *de facto* organic producers. The farmers of this region are using very insignificant amount of inorganic inputs, fertilizer and pesticides. Almost in all states of this region, farmers are practicing traditional agriculture since thousands of years. Recently, in few pockets of the region, the use of inorganic inputs has been started but the level of use is almost negligible (except Manipur). This low input use status (inorganic input) is the biggest strength for the possibility of organic agriculture in this hilly region. The conversion period of organic from inorganic agriculture is to be much lower and easier in this region as compared to other parts of the country. Moreover, the conversion process will be highly supported by the readily available huge amount of forest biomass and the residues of diverse livestock population. Experiments show that during the transition period, the farmers faces the problem of yield losses which will also be minimum in this region as the farmers are using very low level of inorganic inputs. Usually the organic products usually fetch higher prices, nearly 20-25 per cent premium as compared to the conventional products, which, may compensate the problem of yield losses during early period of conversion and expected to increase the level of income. The NEH Region is having typical hilly geographical terrain, and the shifting of the produce takes long time to reach even in the domestic markets. As the international market (particularly USA and European Union) is the major buyer of the organic produce, the supply of perishable product from this region will be one of the difficult tasks. So, the government should give much focus on the domestic market to create the demand for organic produce to facilitate and secure the selling of the surplus production of NEH Region. Behavioral study of consumer awareness and preference towards the organic product is very much essential.

Considering these advantages recently the region has been identified as the most potential area for organic agriculture in the country and accordingly, policy thrust has been laid for necessary internalization of the region. Being the 'hot spot' (one of the 18$^{th}$ mega biodiversity centers of the world) in terms of biodiversity, and now identified as the potential organic agricultural production zone, it is necessary to analyse the possible relationship between these two issues, i.e. management of biodiversity and possibility of organic agriculture. So far, most of our initiatives for this region have been focused separately either on biodiversity or on organic farming, rather missing the inherent link between these two. Perhaps the north east region of India has the unique opportunity to conserve vulnerable plant genetic resources and simultaneously have the strong potential to become the leading center for organic food production. Therefore, to achieve the goal of agro-biodiversity management which is the most important component of natural resources in north eastern region, through organic agriculture (as an option), several multidisciplinary research are to be carried out to formulate the proper policy framework. In depth analysis are required for ethno-botanical and socio-economic studies are required to understand the farmers' knowledge. Selection criteria of germplasm and their innovative management are important to know. Such indigenous knowledge also warrants due share and benefit when it is applicable to advance technology. Dynamics of the local landraces (traditional varieties) and farmers' varieties needs to be understand in detail and, simultaneously, appropriate scientific know-how should be infused with the traditional agricultural practices for optimum exploitation of natural resources and ensuring the synergistic effect of biodiversity-organic agriculture in NEH Region.

## Conclusion and Policy Implications

Agricultural sector in northeastern states is experiencing the transition phase, slowly shifting out from traditionally low-income agriculture, but yet to reach the stage of pre-conditions for take-off into self-sustaining growth. Under this stage of economic transition, various sub-sectors particularly under agriculture and allied activities showed high growth prospects, when stimulated through policy initiatives. But the real challenge remains to sustain this growth rates in long run. Under this traditional economy, it is difficult but extremely important to prioritise the sub-sectors which have short-run as well as long-run growth prospects. Sometimes, some of the sub-sectors may even show misleading growth rate pretending to have high or low potential but actually it is not, particularly due to the non-availability of reliable data-base (e.g., growth rate for foodgrains production in Mizoram can be observed to be as high as greater than 10 per cent when calculated for the period of 1984-85 to 1997-98), emphasizing the need for credible database for various sectors. Capacity building and empowering with technical knowledge might be given top priority for the human resource development in this region. The traditional family farmers resist change towards modern technologies and it is necessary to analyze whether their behaviour is stubborn and irrational (e.g., low adoption rate for high yielding variety seeds for rice cultivation) or they are acting rationally within the context of their particular economic environment, and the agro-technologies should be appropriated accordingly to increase the adoption level of these innovations. Efficient utilization of plan outlays after prioritizing the sub-sectors and minimising the leakages should be emphasized for self-propagation of growth. The focus should be to increase the efficiency of production along with the volume of production;

therefore, returns to investment should be the criteria for flow of fund to various projects. Enhancing the productivity of women labour engaged in agricultural activities needs to be looked into more positively (e.g., organizing training for women groups at their own villages, giving preferential treatment for women group in obtaining institutional credit etc). Strict monitoring and evaluation should be followed for implementation of agricultural developmental project. The performance of output growth for various sub-sector (crop, fruits, vegetables, spice, fishery, livestock, etc) under agricultural sector indicated that combination of these activities will facilitate to augment the farm income rather relying on any single activities, which is true even at farm household level, thereby suggesting promotion of integrated farming system approach at massive scale in this region. Expansion of market through horizontal integration among the primary producers' to increase the volume of marketable surplus and then, vertical integration for value addition of these products is extremely important to realise the potential of this region. The promising areas like expansion of meat production, increasing fish output, growing off-season vegetables, fruits, spices and condiments and promoting organic products should be given priority. Market should not be viewed as 'master' rather; it should be used as our 'slave' and demand creation (e.g., organic rice) for the products from northeastern states should be done with innovations and appropriate strategic planning (e.g., expansion of trade with neighbouring countries through establishing formal trade point at border areas). The volume of marketed surplus from agricultural production must be increased and value addition should be done at least at primary level to make the product price competitive at domestic as well as international market. The process of economic growth and development for the northeastern states is primarily dependent on agriculture sector and thereby, development planning should continue to follow agriculture-led-growth.

## References

Bhowmick, B C., Barah, B C., Pandey, Shushil and Barthakur, N. (2005), "Changing Pattern of Rice Production Systems and Technology in Assam - A Spatio-temporal Analysis of Performance and Prospects", Policy Paper 22, National Centre for Agricultural Economics and Policy Research (NCAP), New Delhi.

Census of India (1991), Ministry of Home Affairs, Government of India, New Delhi.

Census of India (2001), Ministry of Home Affairs, Government of India, New Delhi. (www.censusindia.net)

Centre for Monitoring Indian Economy Private Limited/CMIE (1997 & 1998), *Agriculture*, Mumbai.

Central Statistical Organization/CSO, *Net State Domestic Product (1993-94 series)*, National Accounts Division, Ministry of Statistics and Programme Implementation, New Delhi, (www.mospi.gov.in).

Central Statistical Organization (2003), *Statistical Abstract-2003*, Ministry of Statistics and Programme Implementation, New Delhi, (www.mospi.nic.in).

Government of India (2003), *17th Indian Livestock Census 2003 – Districtwise*, Directorate of Animal Husbandry and Dairying and Fisheries, Ministry of Agriculture, New Delhi. (http://dms.nic.in/ami/home.htm).

Government of India (2004), *Agricultural Statistics at a Glance: 2004*, Directorate of Economics and Statistics, Ministry of Agriculture, New Delhi.

Government of India, *Economic Survey*, Various Issues, Ministry of Finance, New Delhi.

Mandal, Subhasis, Datta, K. K., Mohanty, Suchitra and Mahapatra, B. K. (2005), "Risk of Financing Agriculture in the North East Hill Region of India with Special Reference to Meghalaya", *Agricultural Economic Research Review*, Vol. 18 (Conf. no), pp. 87-102.

Mohan, Rakesh (2003): 'Economic Development of the North East Region: Some Reflections', *Reserve Bank of India Bulletin*, December.

Planning Commission, *Ninth Five Year Plan (1997-2002)-Outlays and Expenditures*, Government of India, (www.planningcommission.nic.in.

Reserve Bank of India (2005), *Handbook of Statistics on Indian Economy*, (www.rbi.org.in).

Reserve Bank of India (2005), *National Income, Saving and Employment*, (www.rbi.org.in).

Singh, Manmohan (2005), "North East 2020", *Yojana*, Decempber, pp.2-6.

Thirwall, A. P. (1999), *Growth and Development (6th Edition)*, Macmillan Press Ltd, London.

Todaro, Michael P. and Smith, Stephen, C. (2004), *Economic Development (8th Edition)*, Pearson Education (Singapore) Pte. Ltd.

**Table 1. Land Utilisation Pattern in North Eastern Hilly States (1995-96)**

| *State* | *Total Geogr-aphical Area (000'ha)* | *Arable Land (000'ha)* | *Net Sown Area (000'ha)* | *Gross Cropped Area (000'ha)* | *Cropping Intensity (per cent)* | *Gross Irrigated Area (000'ha)* | *Fertilizer Use (Kg/ha, 2001-02)* | *Average Holding Size (ha)* |
|---|---|---|---|---|---|---|---|---|
| (1) | (2) | (3) | (4) | (5) | (6) | (7) | (8) | (9) |
| Arunachal Pradesh | 8374.3 | 293 (3.5) | 185 (63.14) | 244 | 132 | 36 (14.75) | 2.88 | 3.31 |
| Assam | 7843.8 | 3387 (43.18) | 2780 (82.08) | 3938 | 142 | 572 (14.53) | 38.81 | 1.17 |
| Manipur | 2232.7 | 164 (7.35) | 140 (85.37) | 182 | 130 | 75 (41.21) | 104.94 | 1.22 |
| Meghalaya | 2242.9 | 1074 (47.88) | 206 (19.18) | 247 | 120 | 45 (18.22) | 17.16 | 1.33 |
| Mizoram | 2108.1 | 445 (21.11) | 109 (24.49) | 109 | 100 | 9 (8.26) | 13.72 | 1.29 |
| Nagaland | 1657.9 | 626 (37.76) | 211 (33.71) | 228 | 108 | 72 (31.58) | 2.13 | 4.82 |
| Tripura | 1048.6 | 310 (29.56) | 277 (89.35) | 426 | 154 | 60 (14.08) | 30.45 | 0.60 |
| Sikkim | 710 | 110 (15.49) | 95 (86.36) | 121 | 127 | 16 (13.22) | 9.72 | 1.65 |
| NEHR | 26218.3 | 6409 (24.45) | 4003 (62.46) | 5495 | 137 | 885 (16.11) | 35.04 | - |
| All India | 328726.3 | 194680 (59.22) | 142215 (73.05) | 186561 | 131 | 71510 (38.33) | 90.12 | 1.41 |

*Source:* CMIE Report on Agriculture 1997 & 1998 and *Economic Survey* - Various Issues.

*Note:* Figures in parentheses in column (3), (4) and (7) indicate the percentage of arable land to total geographical area, percentage of net sown area to total arable area and percentage of irrigated area to gross cropped area, respectively for the north eastern states.

## Table 2. Trends and Progress in Foodgrains Productivity in NEH Region of India (1980-81 to 2002-03)

(q/ha)

| States/Year | 1980-81 | 1990-91 | 2002-03 | Change (%) 1980-81/ 1990-91 | Change (%) 1990-91/ 2002-03 |
|---|---|---|---|---|---|
| Arunachal Pradesh | | | | | |
| Rice | 10.6 | 11.7 | 12.2 | 10.38 | 4.27 |
| Foodgrains | 10.5 | 11.7 | 12.3 | 11.43 | 5.13 |
| Assam | | | | | |
| Rice | 11.0 | 13.1 | 14.7 | 19.09 | 12.21 |
| Foodgrains | 10.7 | 12.7 | 14.2 | 18.69 | 11.81 |
| Manipur | | | | | |
| Rice | 14:5 | 17.4 | 23.8 | 20.00 | 36.78 |
| Foodgrains | 14.5 | 17.7 | 23.1 | 22.07 | 30.51 |
| Meghalaya | | | | | |
| Rice | 13.4 | 11.6 | 17.2 | -13.43 | 48.28 |
| Foodgrains | 14.3 | 11.5 | 16.3 | -19.58 | 41.74 |
| Mizoram | | | | | |
| Rice | 9.8 | 12.4 | 19.1 | 26.53 | 54.03 |
| Foodgrains | 1045 | 13.0 | 18.7 | 23.81 | 43.85 |
| Nagaland | | | | | |
| Rice | 9.0 | 12.3 | 14.9 | 36.67 | 21.14 |
| Foodgrains | 851 | 11.6 | 15.7 | 36.47 | 35.34 |
| Tripura | | | | | |
| Rice | 13.6 | 18.3 | 22.7 | 34.56 | 24.04 |
| Foodgrains | 13.5 | 17.8 | 21.9 | 31.85 | 23.03 |
| Sikkim | | | | | |
| Rice | | | 14.4 | - | - |
| Foodgrains | | | 13.2 | - | - |

*Source: Compiled from Basic Statistics of North Eastern Region – 2002 and Various Issues,* CSO Database - *Statistical Abstract-2003.*

## Table 3. Employment and Output from Agriculture Sector in NEH Region (2001)

| *Particulars* | *Employment in Agriculture (% to total workers)* | | | *Share of Agricultural Output to NSDP (%)* | *Value of Agricultural Output per worker* (Rs./year)* | |
|---|---|---|---|---|---|---|
| | *Male* | *Female* | *Total* | | *Current prices (2001)* | *Constant prices (1993-94)* |
| Arunachal Pradesh | 50 | 81 | 62 | 37 | 21395 | 12356 |
| Assam | 51 | 57 | 53 | 38 | 24669 | 13213 |
| Manipur | 56 | 60 | 57 | 35 | 14347 | 11667 |
| Meghalaya | 61 | 73 | 66 | 27 | 17347 | 8478 |
| Mizoram | 54 | 68 | 60 | 29 | 16847 | ** |
| Nagaland | 59 | 80 | 68 | 31 | 18381 | 12995 |
| Tripura | 46 | 63 | 51 | 28 | 23924 | 8655 |
| Sikkim | 47 | 71 | 56 | 28 | 14211 | 12702 |

*Source:* Calculated from Census of India 2001, RBI (2005) – Handbook of Statistics on Indian Economy, CSO Database – Net State Domestic Product (1993-94 Series)

*Notes:* *Includes cultivators and agricultural labourers, ** Data was not available.

## Table 4. Growth Rate of Foodgrains Production, Population, Per Capita Income vis-à-vis Change in per capita Foodgrains Availability and per capita Agricultural Output

| *Particulars* | *Population growth (%)* | *Foodgrains production growth (%)* | *Per capita availability of foodgrains* | | *Per capita Income Growth (at 1993-94 constant prices)* | *Per capita Output (Rs.) from Agriculture (at 1993-94 constant prices)* | |
|---|---|---|---|---|---|---|---|
| | *1991-2001* | *1984-85 to 2002-03* | *1991* | *2001* | *(1993-2003)* | *1991* | *2001* |
| Arunachal Pradesh | 2.33 | 0.98 | 679 | 501 | 0.94 | 5062 | 2333 |
| Assam | 1.73 | 2.36 | 421 | 429 | 1.09 | 2087 | 2260 |
| Manipur | 2.63 | 1.80 | 425 | 434 | 4.38 | 1457 | 1641 |
| Meghalaya | 2.62 | 2.41 | 236 | 241 | 4.84 | 1455 | 1575 |
| Mizoram | 2.56 | 6.86 | 304 | 381 | - | 1398 | 2151 |
| Nagaland | 4.97 | 6.55 | 447 | 382 | 2.80 | 1669 | 3210 |

| | | | | | | | |
|---|---|---|---|---|---|---|---|
| Tripura | 1.46 | 2.01 | 511 | 449 | 8.02 | 1949 | 2616 |
| Sikkim | 2.72 | -1.58* | 718 | 539 | 3.51 | 3549 | 2372 |

Source: Estimated based on data from *Basic Statistics for North Eastern region – 2002, CSO Database on Statewise Estimate of Value of Output from Agriculture and Livestock,*

Notes: * Data pertains to the period of 1989-90 to 2002-03.

### Table 5: Annual Compound Growth Rate of Value of Agriculural Output in NEH States of India (1990-91 to 2002-03)

*(per cent)*

| *State/Commodity* | *Paddy* | *Pulses (Total)* | *Cereals (Total)* | *Fruits & Vegetables* | *Spices & Condiments* | *Agriculture (Total)* |
|---|---|---|---|---|---|---|
| Arunachal Pradesh | 0.39 | 3.25 | 0.42 | -11.13 | 13.51 | -4.32 |
| Assam | 1.88 | -0.62 | 1.82 | 3.46 | 0.87 | 2.63 |
| Manipur | 1.62 | 6.63 | 1.61 | 4.16 | 5.45 | 2.46 |
| Meghalaya | 5.56 | 2.04 | 5.12 | 6.67 | 3.32 | 2.22 |
| Mizoram | 3.40 | -3.77 | 3.30 | 15.26 | 14.55 | 4.72 |
| Nagaland | 2.99 | 6.68 | 4.67 | 20.33 | 23.58 | 12.57 |
| Tripura | 1.86 | -1.49 | 1.79 | 21.02 | 8.53 | 5.22 |
| Sikkim | 0.69 | -4.66 | -0.12 | 5.55 | -3.97 | 0.31 |
| India | 1.04 | -0.35 | 1.26 | 14.71 | 4.36 | 2.02 |

*Source:* Estimated from *CSO Database on Statewise Estimate of Value of Output from Agriculture and Livestock* (at 1993-94 constant prices).

### Table 6: Annual Compound Growth Rate of Value of Output from Livestock and Livestock-production in NEH States of India (1990-91 to 2002-03)

*(per cent)*

| *State/Commodity* | *Milk* | *Meat* | *Egg* | *Dung* | *Livestock (Total)* |
|---|---|---|---|---|---|
| Arunachal Pradesh | 1.17 | 0.50 | -6.31 | 5.72 | 1.54 |
| Assam | 0.81 | 1.72 | 0.65 | -0.47 | 0.70 |
| Manipur | -1.84 | 4.59 | -1.19 | -6.27 | 2.09 |
| Meghalaya | 2.85 | 7.69 | 2.02 | 1.85 | 5.93 |
| Mizoram | 6.84 | 5.34 | 5.31 | 3.39 | 5.61 |
| Nagaland | 2.69 | 8.97 | 3.78 | 3.08 | 5.63 |
| Tripura | 8.77 | 9.64 | 11.53 | 2.47 | 6.37 |
| Sikkim | 3.09 | -3.84 | -3.00 | -1.72 | 1.83 |
| India | 4.14 | 3.44 | 13.32 | 0.62 | 3.79 |

*Source:* Estimated from *CSO Database on Statewise Estimate cf Value of Output from Agriculture and Livestock* (at 1993-94 constant p

# 3

# AGRICULTURAL DEVELOPMENT IN NORTH-EASTERN REGION OF INDIA

*S.S. Kalamkar*

## Introduction

Indian agriculture has witnessed tremendous changes during the last four decades following adoption of green revolution technology during late 1960s. Initially impact of green revolution was confined to wheat and rice only and the regions with good irrigation facilities. As a consequence, much of the growth, which took place after adoption of HYV seeds-fertiliser-irrigation technology, was confined to a limited number of states like Punjab, Haryana, Utter Pradesh and coastal Andhra Pradesh (Dantwala, 1986; Bhalla and Singh, 2001). As the spread of green revolution was highly skewed in favour of certain states and regions, this led to high growth in agricultural output in selected regions while the other regions suffered from stagnancy or poor growth in agricultural output (Chand and Chauhan, 1999), which brought in its wake uneven development across regions and crops and also resulted in widening the regional as well as interpersonal disparities (Bhalla and Alagh, 1979; GOI, 2005). The North Eastern (NE) region/states in which rural poverty rates and dependence on agriculture are highest, agriculture shows limited improvement. The NE states are among the most backward in the country and agricultural development is much below its potential in this region. With changes taking place in the domestic and international economies, agricultural strategy in India can have a re-look at the north-east hill economies, learn from their failure to respond to the green revolution and focus on an acceptable path to development of the region (Ghosh, 2003). As the Working Group on Agricultural Development in Eastern and North Eastern India for the Formulation of the Tenth Five Year Plan has categorically emphasised, *"Looking at the predominance of agriculture as the source of livelihood in the region, a breakthrough in the human development graph of the region can be achieved only by addressing the key issues of the agriculture and allied sectors, "* (GOI, 2001). The number of studies has looked into the extent of development among the 17 major states; however, not many studies have attempted to focus on the agricultural development of North Eastern region/ states of India. Against this background, an attempt has been made in this paper to study the agricultural development in NE region/states of India.

## Data and Methodology

The study is based on secondary data collected from the different sources. The states in the North Eastern (NE) region suffer from a paucity of time series data. Therefore, the data for the study has been compiled from different sources such as government publications, reports and websites. The entire data related to area and production of principal crops have been compiled from *Fertliser Statistics (various years)* published by the Fertiliser Association of India, New Delhi. Data on land holdings, net and gross cropped and irrigated area have been compiled from *Agriculture* published by the Centre for Monitoring Indian Economy (CMIE, various years), Mumbai. Data on inputs use such as fertiliser consumption, cropwise irrigated area, area under HYV have been collected from various reports of the *Fertiliser Statistics* (FAI) and *Indian Agriculture in Brief* published by Agricultural Statistics Division, Directorate of Economics and Statistics, Department of Agriculture and Cooperation, Ministry of Agriculture, Government of India. Besides tabular analysis, annual compound growth rates were calculated to indicate an increase or decrease in various parameters.

### Present Scenario

The NE region of India comprises of the eight states of Arunachal Pradesh, Assam, Manipur, Meghalaya, Mizoram, Nagaland, Sikkim and Tripura. The NE region has a total geographical area of 2.62 lakh sq.km and a population of 39.04 million representing 8.28 percent total area and 3.81 percent of the total population of the country. Assam is the most populous state and accounts for about 70 per cent of the region population. The region has a high concentration of tribal population. The states of Arunachal Pradesh, Meghalaya, Mizoram and Nagaland are mostly inhabited by a number of native tribes. Various tribal groups constitute all most 30 percent of the total population of the region (GOI, 2006). Majority of the population, predominantly tribal, are dependent on agriculture and land-based activities. The agricultural production system in the region is predominantly rainfed, mono-cropped at subsistence level.

The NE region has been declared as an economically backward area and is considered as a special category by the Planning Commission. The people in rural NE live in a region virtually cut-off from the rest of India. Pressure on rural land, hilly terrain, tribal population, economic backwardness of the area, low technological levels, low productivity coupled with low per capita income are the major features this area (NCAER, 2004). Because of their remoteness and relative inaccessibility the socio-economic systems have retained the traditional character to a large degree. Even compared to other hilly regions including Jammu and Kashmir which has its own and different problems and Himachal Pradesh, which is greatly advanced in a conventional scale, the states stand out in need of policy attention. The NE region has remarkable advantages of fertile and organically rich soils, ample rainfall and water resources, river valley, swamps and streams and great climatic diversity supporting diverse cropping possibilities. On the other hand the slopes and heavy rain makes soil matters unstable and acidic and the slope and the conditions favorable to rapid vegetative proliferation make agriculture and land management tedious and highly labour intensive process (Ghosh, 2003).

Agriculture is the main economic activity in the region and despite major impact of green revolution in the irrigated areas of the country; modernization of agriculture has escaped this region as evidenced by poor adoption of modern technologies, low consumption of fertilizers and other indicators of growth. One of the most important reasons for this slow growth is the lack of location-specific and system-based technologies (NAAS, 2001). The region offers scope for cultivation of a wide variety of agricultural crops because of its diversities in topography, altitude and congenial agro-climatic environments (Birthal et al., 2006). The pattern of agricultural growth has however remained uneven across regions and crops. The NE region continues to be a net importer of foodgrains even for its own consumption. In spite of covering 8.28 percent of the country's total geographical area, NE region accounts only 2.8 percent of the country's total foodgrain production in 2003-04. Although cereals dominate the rainfed hill ecosystem, dependence on livestock and horticulture as an alternative source of income is significant (GOI, 2001). The region receive heavy rainfall with annual precipitation being more than 1000 mm. Productivity of foodgrains is however far below the national average. While the traditional method of cultivation is prevalent in the plains of the region, *Jhum* cultivation is prevalent in the hilly areas and mono cropping is largely prevalent in the entire area (GOI, 2005). However, the cultivable area is limited and land-man ratio has become increasing adverse. As a result the *jhum* cycles have shrunk, making the practice uneconomic and environmentally unsustainable. The transport and communication system is poorly developed and as a result majority of the areas in the region remain still inaccessible. The infrastructure facilities like supply of input, marketing, institutional credit and extension services are inadequate.

The socio-economic indicators of NE states are presented in Table 1. The socio-economic indicators show huge variations among the NE states. The population density of NE region was very low (149/sq.km) as compared to all India average

**Table 1: Socio-Economic Indicators of North Eastern States of India**

| States | Population density (2001) | Per capita NSDP at current prices (1999-00) (in Rs.) | Literacy rate (2001) % | Growth rate of SDP (9$^{th}$ plan) % | HDI (1991) rank | SEII (1999) | Poverty (%) (1999-00) | Per capita foodgrains prod. (kg) (2002-03) | IDI (%) (2000) | Road density (1996-97)/ sq .km. of area |
|---|---|---|---|---|---|---|---|---|---|---|
| Arunachal Pra. | 13 | 13352 | 54.34 | 4.4 | 29 | 69.71 | 33.47 | 219.5 | 71.89 | 168 |
| Assam | 340 | 9720 | 63.25 | 2.1 | 26 | 77.72 | 36.09 | 146.1 | 104.39 | 872 |
| Manipur | 107 | 12721 | 70.53 | 6.4 | 9 | 75.39 | 28.54 | 164.1 | 83.50 | 490 |
| Meghalaya | 103 | 12083 | 62.56 | 6.2 | 24 | 75.49 | 33.87 | 95.3 | 77.60 | 379 |
| Mizoram | 42 | 14909 | 88.80 | n.a. | 7 | 82.13 | 19.47 | 145.2 | 84.49 | 229 |
| Nagaland | 120 | 12594 | 66.59 | 2.6 | 11 | 76.14 | 32.67 | 195.6 | 89.89 | 1107 |
| Sikkim | 76 | 14751 | 68.81 | 8.3 | 18 | 108.99 | 36.55 | 188.6 | 83.01 | 258 |
| Tripura | 304 | 13195 | 73.19 | 7.4 | 22 | 74.87 | 34.44 | 174.4 | 92.85 | 1405 |
| All India | 324 | 14292 | 64.84 | 5.3 | - | - | 26.10 | 169.3 | 100.00 | 749 |

*Notes:* NSDP Net State Domestic Product and SDP- State Domestic Product; n.a.- not available,: HDI- Human Development Index; SEII- Social and Economic Infrastructure Index; IDI- Infrastructure Development Index.

*Sources:* CMIE (2000, 2006), EPWRF (2003), GOI (2002), GOI (2006), GOM (2005), www.censusindia.com

(324/sq.km.). As NE region has 3.81 percent of total population of country, Assam has highest (2.60 percent) and Sikkim has lowest (0.55 percent) share in country's total population in 2001. The per capita income of the NE region on an average was Rs. 12918 (as per NSDP), as compared with the national average of Rs. 17823/- at current prices at 2001-02. Though per capita income of NE states except Assam are close to all India average, the rural poverty ratio is still higher level at 40.04 percent as compared to national average of 27.09 percent (GOI, 2002). However, in-spite of the fact the literacy rate was reasonably higher NE region as compared to national average. Though the growth in SDP during ninth plan period was higher in Manipur, Meghalaya, Sikkim and Tripura, the percent share of agriculture SDP in gross state domestic product was lower in these states as compared to other NE states in 2000-01. The low infrastructure development index shows the reason for backwardness of this region. The road network is poor in most of the states of this region. The road density ranges between 168-490 km/ thousand sq. km., except Assam, Nagaland and Tripura which have comparatively a much better road network (GOI, 2006). However, as per human development index (HDI), Manipur ranks 7$^{th}$ position and Arunachal Pradesh on 29$^{th}$ position, shows huge variations among the states.

## Plan Expenditure on Agriculture and Allied Sector

The special efforts have been made on the economic development of the NE region during various plan periods. The sectoral composition of actual plan expenditure of NE states is presented in Table 2. Investment in agriculture is essential for increasing as well as sustaining the overall growth of agriculture. It is observed that though actual plan expenditure on above mention two heads was higher in NE states as compared at national average, but the proportion has declined steeply in these states as compared to all India average during 1980-82 to 1996-98. The expenditure on infrastructure development in NE states has also been declined. The plan expenditure on rural and social sector has increased significantly. Therefore, in order to increase the income level

**Table 2: Sectoral Composition of Actual Plan Expenditure of NE States (Percentage)**

| *States* | *Agriculture & Irrigation* | | | *Rural and Social Sector* | | | *Social Sector* | | | *Infrastructure* | | |
|---|---|---|---|---|---|---|---|---|---|---|---|---|
| | 81-82 | 91-92 | 97-98 | 81-82 | 91-92 | 97-98 | 81-82 | 91-92 | 97-98 | 81-82 | 91-92 | 97-98 |
| Arunachal P | 26.18 | 19.22 | 14.40 | 28.16 | 33.87 | 39.23 | 24.59 | 25.90 | 29.65 | 45.66 | 47.07 | 46.37 |
| Assam | 23.20 | 27.50 | 20.36 | 24.25 | 42.49 | 57.53 | 17.97 | 34.31 | 45.89 | 52.56 | 29.97 | 22.12 |
| Manipur | 34.13 | 31.46 | 21.07 | 35.94 | 31.39 | 40.35 | 29.07 | 24.58 | 32.44 | 29.93 | 37.31 | 38.57 |
| Meghalaya | 15.06 | 20.39 | 15.30 | 39.85 | 40.62 | 51.12 | 28.97 | 29.32 | 37.83 | 45.09 | 38.99 | 33.58 |
| Mizoram | 21.56 | 15.74 | 9.02 | 32.51 | 48.64 | 49.88 | 26.54 | 25.68 | 30.35 | 45.93 | 35.38 | 41.10 |
| Nagaland | 19.47 | 13.93 | 8.14 | 41.20 | 55.48 | 71.99 | 26.65 | 24.38 | 36.73 | 39.33 | 30.42 | 19.87 |
| Sikkim | 24.80 | 19.70 | 12.73 | 33.78 | 36.23 | 54.38 | 19.83 | 28.09 | 45.38 | 41.42 | 47.07 | 32.90 |
| Tripura | 32.89 | 26.87 | 13.70 | 39.58 | 46.86 | 63.97 | 28.69 | 30.16 | 43.18 | 27.53 | 26.11 | 22.34 |
| All India | 7.20 | 5.30 | 3.60 | 15.40 | 21.80 | 23.40 | 8.80 | 12.50 | 14.80 | 77.40 | 72.90 | 73.00 |

Notes: Actual plan expenditure by major heads of development has been clubbed as per the following heads:- Agriculture and Irrigation: Agriculture and Allied activities and Irrigation and Flood Control; Rural and Social Sector: Rural Development, Special Areas Programmes, General and Economic Services and Social Sector; Social Sector: Education, Health, Water Supply and Sanitation, Urban development, Information , Welfare and Labour; Infrastructure: Energy, Industry and Minerals, Transport, Communication, Science, Technology and Environment. Data for 1981-82 are an average of 1980-82, 191-92 an average of 1990-63 and 1997-98 an average of 1996-98.

Sources: NHDR, 2002 (various plan documents, Planning commission, Government of India).

Table 3: Land Use Statistics of NE States of India, 2001-02

| States | Percentage to reporting area (%) | | | | | | | | | Area sown more than once (lakh ha) | Total Cropped Area (lakh ha) | Cropping intensity (%) |
|---|---|---|---|---|---|---|---|---|---|---|---|---|
| | Forests | Area put to non-agricultural uses | Barren & unculturable land | Permanent Pastures & other grazing lands | Land under misc. tree crops & groves | Culturable Waste Land | Fallow land | Current fallow | Net Sown Area | | | |
| Arunachal P | 93.7 | 0.1 | 0.4 | 0.1 | 0.7 | 0.7 | 0.9 | 0.5 | 3.0 | 1.0 | 2.7 | 163 |
| Assam | 24.6 | 13.6 | 18.6 | 2.1 | 3.0 | 1.0 | 0.8 | 1.4 | 34.8 | 12.2 | 39.6 | 145 |
| Manipur | 27.2 | 1.2 | 64.2 | * | 1.1 | * | - | - | 6.3 | 0.8 | 2.2 | 154 |
| Meghalaya | 42.7 | 3.9 | 6.1 | - | 7.0 | 19.8 | 7.3 | 2.9 | 10.3 | 0.5 | 2.8 | 120 |
| Mizoram | 77.1 | 5.8 | 0.4 | 0.5 | 0.9 | 0.2 | 7.7 | 1.7 | 5.6 | - | 1.2 | 100 |
| Nagaland | 53.7 | 4.2 | 0.0 | - | 7.7 | 3.8 | 4.6 | 5.4 | 20.7 | 0.5 | 3.8 | 114 |
| Sikkim | 36.2 | 13.7 | 24.4 | 9.7 | 0.7 | 0.1 | 1.3 | 0.6 | 13.4 | 0.4 | 1.3 | 139 |
| Tripura | 57.8 | 12.5 | 0.3 | * | 2.6 | 0.1 | 0.1 | 0.1 | 26.7 | 1.4 | 4.2 | 151 |
| **NE Region** | **51.5** | **6.9** | **13.8** | **1.1** | **2.7** | **2.7** | **2.2** | **1.4** | **17.6** | **16.8** | **57.7** | **141** |
| All India | 22.6 | 7.9 | 6.3 | 3.5 | 1.1 | 4.4 | 3.3 | 4.8 | 46.2 | 489.3 | 1902.8 | 135 |

Note: * - included under the head 'Land under miscellaneous tree crops and groves'.

Source: FAI (2005).

of farmers of NE region, substantial investment has to be made in agriculture and allied sector as well as in infrastructure development. The 10$^{th}$ plan sectoral outlays for the NE states including NE council shows that out of total plan expenditure, 25 percent outlay was allocated for above mention two heads, which shows that the government commitment to accelerating the pace of agricultural development of the region (GOI, 2003).

### Land Resource Use Pattern

The pattern of land-use of a country at any particular time is determined by the physical, economic & institutional framework taken together. Contrary to common belief, it is seen that so far as agricultural land for field crops is concerned, the region does not enjoy any special advantage as compared to the rest of India. The NE region has total 56.19 lakh hectares of gross cropped area which constitutes 3.19 percent of the country's gross cropped area in 2002-03. As the region is predominately agricultural and rural, its man-resource relationship has to be judged mainly in terms of availability of agricultural land. The overall geographical land-to-man ratio for the NE region (0.81 ha/person) is much higher than the national average (0.39 ha/person). Among the NE States, the geographical land-to-man ratio is highest in Arunachal Pradesh (9.69 ha/person) and lowest in Assam (0.35 ha/person).

The land use statistics of NE states is presented in Table 3. The NE region has a total geographical area of 23.2 million ha of which about 52 percent is under forest cover though there are large variation from minimum of 24.6 percent in Assam to about 94 percent in Arunachal Pradesh in 2001-02. About 21 percent land is classified as not for cultivation and 7.4 percent as other uncultivated land. The state like Arunachal Pradesh and Mizoram represent merely 3.0 and 5.6 percent net sown area of the reporting area respectively. On the whole, the net sown area of NE region was hardly 18 percent of the total reported area, lowest in Arunachal Pradesh (3 percent) and highest in Assam (35 percent), whereas the relevant figure for India was 46.25 percent. The area sown more than once was highest in absolute term in

Assam followed by Tripura. The cropping intensity of NE region was higher than all India average indicates the scope for agricultural development of the region. However, among the states, the cropping intensity in Mizoram, Nagaland and Meghalaya was low as compared to NE region.

### Rural Workforce Distribution

The rural workforce distribution of NE region is presented in Table 4. It can be seen from the table that about 85 percent population of NE region were living in rural areas, which are directly or indirectly dependent on agriculture and allied activities. The highest rural population were in Sikkim (88.90 percent) and lowest in Mizoram (50.5 percent). The percentage of total workers to rural population was around 39 percent in NE region and 42 percent at all India level. It can be seen that percentage of cultivators and agricultural laboures to main workers were less in NE region as compared to national average. Altogether about 54 percent of rural population is directly dependent on agriculture as compared to 75 percent at national level. The other workers population was almost one third of total main workers, shows that the non-farm activities are relatively good and it provides employment to almost 35 percent of main workers of the region.

### Land Holding Distribution

The land holding pattern of NE states in presented in Table 5. Land holding pattern of NE states is somewhat similar from the all-India level. The share of marginal and small holdings in the total number of holdings is almost same in NE region as compared to all-India level. About 81 percent of the farmers in the NE region belongs to the small and marginal category. Moreover, with increase in population, the average size of holdings is gradually reducing over the years. These marginal and small farmers hold only 35 percent of land. The average size of operational holdings is found to be lower in NE region as compared to the same at the all-India level. Among the NE states, the average size of land holding was highest in Nagaland (4.83 ha) and lowest in

**Table 4: Rural Workforce Distribution in NE states, 2001 (%)**

| *State* | *Rural Population (% to total)* | *Percentage to Rural Population* | | | *Percentage to total main workers* | | | |
|---|---|---|---|---|---|---|---|---|
| | | *Total Workers* | *Main Workers* | *Marginal Workers* | *Cultiva-tors* | *Agricult-ural Labourers* | *House-hold Industry* | *Other Workers* |
| Arunachal P. | 79.66 | 46.47 | 39.28 | 7.19 | 46.47 | 4.27 | 0.72 | 26.75 |
| Assam | 87.28 | 36.45 | 26.15 | 10.30 | 36.45 | 15.12 | 3.54 | 37.41 |
| Manipur | 75.88 | 46.72 | 32.69 | 14.03 | 46.72 | 11.76 | 8.17 | 26.58 |
| Meghalaya | 80.37 | 44.58 | 33.96 | 10.63 | 44.58 | 20.16 | 1.95 | 23.42 |
| Mizoram | 50.50 | 57.22 | 44.94 | 12.27 | 57.22 | 3.92 | 0.84 | 17.45 |
| Nagaland | 82.26 | 45.08 | 37.08 | 8.00 | 45.08 | 4.45 | 1.99 | 20.61 |
| Sikkim | 88.90 | 49.75 | 39.57 | 10.17 | 49.75 | 7.08 | 1.28 | 36.67 |
| Tripura | 82.98 | 37.11 | 28.08 | 9.03 | 37.11 | 27.97 | 3.10 | 37.52 |
| NE Region | 84.53 | 38.69 | 28.44 | 10.25 | 38.69 | 14.87 | 3.44 | 34.07 |
| All India | 72.20 | 41.97 | 31.03 | 10.94 | 41.97 | 33.20 | 3.77 | 22.90 |

*Source:* Census of India, 2001(www.censusindia.com).

Tripura (0.60 ha). About 32 percent holding are of size more than 4 ha in NE region as compared to 40 percent at national level. This is primarily because hilly terrain constitutes nearly two-thirds of the region's geographical area, and large sized holdings are not feasible. The average plot size is too small for mechanisation of agriculture and adoption of modern farming practices. Subsistence farming, therefore, is predominant in the NE region and there is hardly any commercial surplus.

## Gross Cropped Area (GCA) and Shift in Cropping Pattern

The trend in gross and net cropped area and cropping intensity of NE states over the period of time is presented in Table 6. It is observed that net sown area (NSA) of NE region continued to grow till date touching the figure 18.2 percent of reporting area in 2002-03 from 13.5 percent in 1970-71. Thus, 4.7 percent more area has been brought under cultivation during last three decades. The area expansion was a source of agricultural growth in NE states, in contrast to the all India pattern where the net sown area has ceased to grow after 1970 (Bhalla and Singh, 2001). The higher growth in net sown area was observed in the states of Nagaland, Assam, Mizoram and Sikkim during corresponding years. The cropping intensity of NE states is almost same to national average. Among the states, Arunachal Pradesh recorded highest cropping intensity, followed by Assam and Manipur. Infact cropping intensity of Mizoram, Meghalaya and Tripura was declined in during last decade. This may be due to physical constraints like weak and underdeveloped irrigation network, extremely cold weather in the higher elevations limit the crop season to summer months only (NAAS, 2001).

The cropping pattern in the region, with the exception of Sikkim, is characterised by predominance of rice as the lead crop, however, maize is the dominant crop in Sikkim. Food crops account for more than 66 percent of the gross cropped area in 2003-04, which is suggestive of prevalence of subsistence agriculture and lack of crop diversification (see, Table 7). About 64 percent of the gross cropped area is accounted for by cereals alone. Within the region, however,

## Table 5: Distribution of Land Holdings, 1995-96

| State | Marginal | | Small | | Semi-Medium | | Medium | | Large | | Total | | Av. holdings |
|---|---|---|---|---|---|---|---|---|---|---|---|---|---|
| | No | Area | No | Area | No | Area | No | Area | No | Area | No | Area | (ha) |
| Arunachal P | 6300 | 2904 | 2262 | 3229 | 1395 | 3736 | 563 | 3231 | 83 | 1273 | 10603 | 14374 | 3.31 |
| Assam | 1669 | 621 | 561 | 770 | 351 | 924 | 96 | 495 | 5 | 328 | 2683 | 3138 | 1.17 |
| Manipur | 69 | 39 | 49 | 67 | 21 | 54 | 3 | 14 | 0 | 0 | 143 | 174 | 1.22 |
| Meghalaya | 72 | 35 | 44 | 55 | 39 | 94 | 6 | 27 | 0 | 2 | 160 | 213 | 1.33 |
| Mizoram | 28 | 17 | 26 | 36 | 12 | 28 | 0 | 4 | 0 | 0 | 66 | 85 | 1.29 |
| Nagaland | 9 | 5 | 21 | 23 | 40 | 104 | 64 | 381 | 14 | 206 | 149 | 720 | 4.83 |
| Sikkim | 24 | 10 | 10 | 14 | 6 | 18 | 3 | 18 | 1 | 13 | 44 | 73 | 1.66 |
| Tripura | 247 | 81 | 411 | 56 | 12 | 30 | 1 | 6 | 0 | 9 | 301 | 181 | 0.60 |
| NE Region | 8418 | 3712 | 3013 | 4250 | 1876 | 4988 | 736 | 4176 | 103 | 1831 | 14146 | 18957 | 1.34 |
| All India | 71179 | 28121 | 21643 | 30722 | 14261 | 38953 | 7092 | 41398 | 1404 | 24163 | 115580 | 163357 | 1.41 |

Notes: Number of holdings in '000 and Area operated in '000 hectares.

Source: CMIE, 2006.

# Table 6: Growth in Gross and Net Cropped Area in NE states, 1970-71 to 2002-03

| States | Net Sown Area (NCA) ' 000 ha | | | | Gross Cropped Area (GCA) ' 000 Ha | | | | NSA % to reporting area | | | | Cropping Intensity (%) | | | |
|---|---|---|---|---|---|---|---|---|---|---|---|---|---|---|---|---|
| | A | B | C | D | A | B | C | D | A | B | C | D | A | B | C | D |
| Arunachal P | 115 | 112 | 150 | 164 | 130 | 141 | 255 | 255 | 2.0 | 2.0 | 2.7 | 3.0 | 113 | 126 | 170 | 155 |
| Assam | 2236 | 2696 | 2706 | 2793 | 2793 | 3373 | 3837 | 3964 | 28.5 | 34.3 | 34.5 | 35.6 | 125 | 125 | 142 | 142 |
| Manipur | 178 | 140 | 140 | 155 | 187 | 219 | 187 | 219 | 8.0 | 6.3 | 6.3 | 8.2 | 105 | 156 | 134 | 141 |
| Meghalaya | 163 | 193 | 201 | 230 | 203 | 223 | 239 | 248 | 7.2 | 8.6 | 9.0 | 10.3 | 125 | 116 | 119 | 108 |
| Mizoram | 39 | 77 | 65 | 90 | 40 | 106 | 102 | 90 | 1.9 | 3.7 | 3.1 | 4.4 | 103 | 138 | 157 | 100 |
| Nagaland | 101 | 148 | 204 | 324 | 101 | 158 | 228 | 370 | 6.1 | 13.7 | 13.2 | 20.3 | 100 | 107 | 112 | 114 |
| Tripura | 240 | 251 | 263 | 280 | 345 | 405 | 443 | 346 | 22.9 | 24.0 | 25.1 | 26.7 | 144 | 161 | 168 | 124 |
| Sikkim | - | 86 | 95 | 116 | - | 91 | 125 | 127 | - | 12.0 | 13.4 | 17.3 | - | 106 | 132 | 109 |
| NE Region | 3072 | 3703 | 3824 | 4152 | 3799 | 4716 | 5416 | 5619 | 13.5 | 16.2 | 16.5 | 18.2 | 124 | 127 | 142 | 135 |
| All India | 140267 | 140002 | 142509 | 132798 | 165791 | 172630 | 185487 | 175988 | 45.2 | 46.0 | 45.7 | 41.6 | 118 | 121 | 131 | 133 |

Notes: A- 1970-71, B- 1980-81, C- 1992-93 and D- 2002-03.

Source: CMIE (1998, 2006).

## Table 7: Changes in cropping pattern in NE region (% to GCA)

| States | 1990-91 | | | | | | 2003-04 | | | | | |
|---|---|---|---|---|---|---|---|---|---|---|---|---|
| | Rice | Maize | Cereals | Food-grains | Oilseeds | Fruits & Veg. | Rice | Maize | Cereals | Food-grains | Oilseeds | Fruits & Veg.* |
| Arunachal P | 49.3 | 15.2 | 74.0 | 74.0 | 8.6 | 13.4 | 44.1 | 14.3 | 68.2 | 70.9 | 9.8 | 15.2 |
| Assam | 65.6 | 0.5 | 68.6 | 71.6 | 8.4 | 5.5 | 64.0 | 0.5 | 66.4 | 69.3 | 7.6 | 7.4 |
| Manipur | 87.4 | 2.6 | 90.0 | 90.0 | 1.4 | 8.6 | 71.4 | 2.0 | 73.4 | 76.0 | 0.9 | 12.2 |
| Meghalaya | 42.7 | 7.6 | 53.4 | 54.7 | 3.7 | 23.5 | 39.5 | 6.1 | 46.9 | 48.6 | 3.5 | 25.9 |
| Mizoram | 69.3 | 8.9 | 78.2 | 79.9 | 7.4 | 12.2 | 50.3 | 8.9 | 59.3 | 63.4 | 6.4 | 17.0 |
| Nagaland | 60.7 | 11.4 | 76.5 | 81.0 | 7.9 | 4.3 | 39.8 | 9.5 | 54.8 | 61.5 | 12.5 | 8.7 |
| Sikkim | 12.7 | 26.3 | 53.4 | 64.0 | 8.9 | 4.6 | 11.0 | 27.5 | 48.4 | 53.7 | 7.4 | 7.5 |
| Tripura | 61.6 | 0.0 | 62.4 | 64.9 | 3.2 | 18.9 | 56.2 | 0.4 | 56.9 | 58.9 | 1.3 | 24.3 |
| NE Region | 62.5 | 2.8 | 68.4 | 71.3 | 7.5 | 7.9 | 58.5 | 2.8 | 63.8 | 66.9 | 7.1 | 10.3 |
| All India | 23.0 | 3.2 | 55.6 | 68.8 | 12.9 | 3.6 | 22.3 | 3.8 | 52.5 | 64.8 | 12.5 | 4.3 |

Note: * refers to the year 2002-03.

Sources: FAI (2005), http://agricoop.nic.in.

there are exceptions. Nagaland records 12.5 percent and Manipur 0.9 percent of the gross cropped area under oilseeds. These state level variations are a manifestation of the differences in the status of agricultural development prevailing in the constituent units of the region. Over the period of time, the cropping pattern has changed marginally and area under rice as well as foodgrains has declined and replace with fruits and vegetables. A remarkable feature of shifting cultivation is that a wide variety of crops is grown in the *jhum* fields. *Jhum* paddy is the dominant crop and is mixed with maize, millets, beans, tapioca, sweet potato, ginger, cotton, tobacco, chillies, sesamum and leafy vegetables. The production units are more or less self contained each family growing crop according to its needs. Major food crops of the region are cereals and pulses which are cultivated in about 67 percent of the cropped area. The cropping pattern however suggests that the known tendency of concentrate on foodgrains, rice as a food crop, mainly driven by urge for subsistence. The crops like wheat and sugarcane are grown where conditions are suitable but the region stands out in its emphasis on fruits and vegetables.

Horticultural crops are grown in all the states of the NE region. Potato is an important crop in Meghalaya, Tripura, Arunachal Pradesh and Manipur. Spices like turmeric and chillies are also grown in this region. Tea is grown in Tripura. Area under fruits and vegetable crops in Meghalaya, Tripura and Mizoram are around 26, 24 and 17 percent of the gross cropped area in the respective states. Pineapple and citrus fruits like orange, lemon are dominant fruit crops of the region. Apple was introduced in Manipur and Arunachal Pradesh during the last decade. Mulbery occupies a small area in Arunachal Pradesh, Nagaland and Meghalaya (Deshpande, et. al., 2004). Plantation of tea and rubber are two major cash crops of the area. Assam itself has around 800 tea plantation estates and about 60 percent foreign exchange earning of the country from tea is from Assam products (Banerjee and Kar, 1999). Realising the importance of horticulture in NE region, Government is also giving due attention by implementing various schemes of horticulture development of this region.

## Table 8: Growth in Net Irrigated and Gross Irrigated Area in NE states

| States | Net Irrigated Area (000 ha) | | | | Gross Irrigated Area (000 ha) | | | | NIA to NSA (%) | | | | GIA to GCA (%) | | | |
|---|---|---|---|---|---|---|---|---|---|---|---|---|---|---|---|---|
| | 1970-71 | 1980-81 | 1992-93 | 2002-03 | 1970-71 | 1980-81 | 1992-93 | 2002-03 | 1970-71 | 1980-81 | 1992-93 | 2002-03 | 1970-71 | 1980-81 | 1992-93 | 2002-03 |
| Arunachal P | 23 | 24 | 36 | 42 | 24 | 24 | 36 | 43 | 20.0 | 21.4 | 24.0 | 25.6 | 18.5 | 17.0 | 14.1 | 16.9 |
| Assam | 572 | 572 | 572 | 174 | 572 | 572 | 572 | 216 | 25.6 | 21.2 | 21.1 | 6.2 | 20.5 | 17.0 | 14.9 | 5.4 |
| Manipur | 65 | 66 | 65 | 40 | 75 | 75 | 75 | 40 | 36.5 | 47.1 | 46.4 | 25.8 | 40.1 | 34.2 | 40.1 | 18.3 |
| Meghalaya | 37 | 49 | 45 | 59 | 37 | 50 | 45 | 76 | 22.7 | 25.4 | 22.4 | 25.7 | 18.2 | 22.4 | 18.8 | 30.6 |
| Mizoram | 2 | 8 | 8 | 16 | 2 | 8 | 8 | 17 | 5.1 | 10.4 | 12.3 | 17.8 | 5.0 | 7.5 | 7.8 | 18.9 |
| Nagaland | 12 | 59 | 60 | 65 | 12 | 61 | 62 | 81 | 11.9 | 39.9 | 29.4 | 20.1 | 11.9 | 38.6 | 27.2 | 21.9 |
| Tripura | 17 | 36 | 50 | 40 | 22 | 36 | 50 | 53 | 7.1 | 14.3 | 19.0 | 14.3 | 6.4 | 8.9 | 11.3 | 15.3 |
| Sikkim | - | 10 | 16 | 9 | - | 10 | 16 | 15 | - | 11.6 | 16.8 | 7.8 | - | 11.0 | 12.8 | 11.8 |
| NE Region | 728 | 824 | 852 | 445 | 744 | 836 | 864 | 541 | 23.7 | 22.3 | 22.3 | 10.7 | 19.6 | 17.7 | 16.0 | 9.6 |
| All India | 31103 | 38720 | 36101 | 53131 | 38195 | 49775 | 66144 | 72575 | 22.2 | 27.7 | 35.2 | 40.0 | 25.0 | 28.8 | 35.7 | 41.2 |

Source: CMIE (1998, 2005).

## Irrigation Development

The normal way of irrigation in the hills is by diverting small streams. In other areas, the irrigation is provided from natural sources only during rainy months. The hydro-geomorphology of the hilly region is such that ground water resource cannot be easily utilised for irrigation purpose. The growth in net and gross irrigated area in NE region is presented in Table 8. The net irrigated area (NIA) in the region does not constitute a significant proportion of net area sown. In all the states, percentage of NIA falls far below the all India average. It can be seen from the table that against our expectation, the percentage of gross irrigated area to gross cropped area in NE region is declined over the period of time, from 19.6 percent in 1970-71 to 9.6 percent in 2002-03. During last decade, irrigated area has declined by almost 7 percent. The steep decline in NIA was observed in the states of Assam, Arunachal Pradesh, Nagaland and Manipur. Irrigation intensity was low as the mountain terrain has somehow made it difficult to exploit the growing water potential an in most states sources other than canals and well supply most of all the irrigation water. In such areas techniques for harvesting rainwater appears to be essential.

The cropwise coverage under irrigation is presented in Table 9. It is observed that the irrigation coverage was very low in NE region as compared to all India average. In case of rice, except Sikkim, all other states have less irrigation coverage than national average (53.9 percent). The picture was almost same in all the crops.

**Table 9**

**Coverage of Irrigated Area under Principal Crops (% of irrigated area to total area under crop) 2000-01**

| *States* | *Rice* | *Maize* | *Total Cereals* | *Total Foodgrains* | *Total Oilseeds* | *All Crops* |
|---|---|---|---|---|---|---|
| Arunachal P [b] | 36.3 | - | 23.7 | 22.9 | - | 16.4 |
| Assam [a] | 8.2 | 5.0 | 7.9 | 7.6 | 0.6 | 5.6 |
| Manipur [c] | 47.8 | - | 39.1 | 44.6 | - | 36.0 |
| Meghalaya | 44.0 | - | 39.1 | 38.1 | 68.9 | 22.6 |
| Mizoram | 16.3 | 3.4 | 14.8 | 15.4 | 47.2 | 14.0 |

| | | | | | | |
|---|---|---|---|---|---|---|
| Nagaland [d] | 43.4 | - | 34.4 | 30.5 | 13.2 | 24.9 |
| Sikkim [a] | 100.0 | 7.0 | 25.7 | 23.6 | - | 14.3 |
| Tripura [c] | 18.7 | - | 18.6 | 18.3 | 6.3 | 13.9 |
| All India | 53.6 | 22.4 | 49.8 | 43.4 | 23.0 | 40.2 |

*Notes:* a – 1990-91, b- 1995-96, c- 1992-93, d- 1998-99 and e –ad hoc estimates; (-) Zero or negligible

*Source:* FAI (2005).

## Growth in Input Use

An analysis of NSSO survey data on cultivation practices shows that the north eastern hill states have retained traditional practices and show an indication towards organic agriculture (Ghosh, 2003). It is observed from Table 10 that fertilser consumption in the region worked out to be as low as 39 kg/ha as against the national average of about 92 kg /ha in 2002-03. The low level of fertiliser consumption is attributed to heavy loss through run off of rain water during the kharif season and to the slow and limited spread of fertilser based technology in the region. The spread of HYVs has been poor as evidenced from the fact that only 54 percent of rice area has been brought under hybrid seeds in NE region against about 77 percent in the country in 1996-97. Of the total quantity of improved seeds required by the farmers, roughly 40 percent is met from within the region leaving a huge gap of 60 percent in its present requirement (NAAS, 2001). In case of electricity consumption also, region lags behind. Only 2.36 percent of total electricity was used for agricultural purposes against 25.33 percent at national level in 2001-02. Tripura state recorded highest consumption of electricity for agricultural purposes. The availability of credit facility in NE region seems to be relatively poor as compared to national average. Among the states, the agricultural credit disbursed per ha was relatively higher in Meghalaya, Mizoram, Nagaland and Sikkim as compared to national average.

**Table 10: Use of Fertiliser and other Inputs in NE Region**

| *States* | *Fertiliser Consumption (kg/ha)* | | | *HYV Coverage Paddy (%) 1996-97* | *% Electricity Consumption for Agril. Purpose 2001-02* | *Agricultural Credit /ha of NSA (Rs.)* |
|---|---|---|---|---|---|---|
| | *1992-93* | *2000-01* | *2002-03* | | | |
| Arunachal | P1.9 | 2.3 | 2.9 | 30.2 | 0.00 | 132 |
| Assam | 6.8 | 34.6 | 44.1 | 57.1 | 0.52 | 276 |
| Manipur | 47.3 | 104.5 | 122.8 | 43.3 | 0.43 | 115 |
| Meghalaya | 14.0 | 13.7 | 19.9 | 40.1 | 0.07 | 328 |
| Mizoram | 7.8 | 13.5 | 20.7 | 7.7 | 0.00 | 338 |
| Nagaland | 3.1 | 1.4 | 1.5 | 1.4 | 0.08 | 316 |
| Sikkim | 12.6 | 8.6 | 9.8 | 43.8 | 0.00 | 481 |
| Tripura | 23.7 | 21.5 | 26.8 | 86.1 | 18.74 | 253 |
| NE Region | 9.6 | 31.0 | 39.2 | 53.8 | 2.36 | 282 |
| All India | 65.5 | 88.9 | 91.5 | 76.9 | 25.33 | 284 |

*Sources:* CMIE (2000), NABARD (2001), GOI (2002a), http://agricoop.nic.in

## Growth in Productivity of Major Crops of NE Region

As the region is predominantly rural, modernisation of agriculture and consequent increase in agricultural productivity would act as the initial growth booster of its economy. Unfortunately, green revolution seems to have bypassed the region. This is evident from low productivity in respect of rice, which is main crop of the region. Table 11 shows that yield rate of rice in all the states of NE region except Manipur was substantially below the all India average in 2003-04. As Assam is the major producer of rice, which accounts about 8 percent of total country area as well as 75 percent area of NE region, productivity was only 1534 kg/ha. Infact, all crop and crops group productivity of NE region was lower than that of national level. During the period of 1990 to 2004, the productivity of all the crops of NE region has increased but at low rate as compared to the rate of growth at national level. Though there was increase in productivity level at regional level, inter-state picture gives mixed trend. Most of the states have experienced declined in productivity of pulses and oilseeds. The wide range of disparities among the states is one of the reasons behind the slow agricultural

**Table 11: Productivity of Major Crops of NE region (kg/ha)**

| States | Paddy | | Maize | | Total Cereals | | Total Pulses | | Foodgrains | | Oilseeds | |
|---|---|---|---|---|---|---|---|---|---|---|---|---|
| | 90-91 | 03-04 | 90-91 | 03-04 | 90-91 | 03-04 | 90-91 | 03-04 | 90-91 | 03-04 | 90-91 | 03-04 |
| Arunachal P | 1170 | 1297 | 1215 | 1412 | 1173 | 1284 | - | 1068 | 1173 | 1276 | 934 | 1030 |
| Assam | 1313 | 1534 | 626 | 719 | 1303 | 1512 | 428 | 552 | 1266 | 1472 | 530 | 524 |
| Manipur | 1742 | 2346 | 2478 | 2295 | 1763 | 2345 | - | 517 | 1763 | 2282 | 500 | 474 |
| Meghalaya | 1155 | 1830 | 1196 | 1533 | 1157 | 1769 | 750 | 745 | 1147 | 1733 | 607 | 653 |
| Mizoram | 1244 | 1914 | 1682 | 1933 | 1294 | 1917 | 1417 | 878 | 1296 | 1849 | 1182 | 720 |
| Nagaland | 1227 | 1377 | 933 | 1489 | 1161 | 1403 | 1160 | 830 | 1161 | 1341 | 891 | 1068 |
| Sikkim | 1218 | 1442 | 1500 | 1556 | 1409 | 1443 | 1019 | 958 | 1344 | 1395 | 1067 | 747 |
| Tripura | 1830 | 2173 | - | 895 | 1830 | 2174 | 578 | 614 | 1783 | 2122 | 715 | 717 |
| NE Region | 1361 | 1618 | 1222 | 1442 | 1347 | 1590 | 559 | 560 | 1315 | 1546 | 601 | 634 |
| All India | 1751 | 2077 | 1524 | 2039 | 1573 | 1987 | 576 | 637 | 1382 | 1731 | 769 | 1067 |

Source: FAI (various years).

development of NE region. Therefore, efforts need to made for location specific research and putting more emphasis on agricultural extension services.

## Diversification of Agriculture

As is well known, agricultural development and crop diversification go hand in hand. The level of diversification of crop enterprise reflects the extent of economic development in the rural sector. If carried out appropriately, diversification can be used as a tool to augment farm income, generate employment, alleviate poverty and conserve precious soil and water resources (Joshi, et al., 2004). It can be seen from the Table 12 that area under foodgrains has declined in three states; however, increased heavily in remaining four states, indicating low level of diversification in these states. Infact, percentage area under foodgrains in NE region has declined marginally by 3 percent. Because of predominant cultivation of foodgrains crops such as rice, which is also, stable food and as such the region is not self sufficient in foodgrains, the area under foodgrains may not decline in near future.

**Table 12**

**Growth in area under foodgrains and non-foodgrains**
***(percentage share in gross cropped area)***

| *States* | *Foodgrains* | | | | *Non- Foodgrains* | | | | *Growth in area under foodgrains in 2002-03 over 1970-71* |
|---|---|---|---|---|---|---|---|---|---|
| | *1970-71* | *1980-81* | *1992-93* | *2002-03* | *1970-71* | *1980-81* | *1992-93* | *2002-03* | |
| Arun. P. | 43.08 | 88.65 | 71.76 | 77.65 | 56.92 | 11.35 | 28.24 | 22.35 | 34.57 |
| Assam | 74.90 | 74.74 | 71.25 | 69.37 | 25.10 | 25.26 | 28.75 | 30.63 | -5.53 |
| Manipur | 77.01 | 91.78 | 81.72 | 79.45 | 22.99 | 8.22 | 18.28 | 20.55 | 2.44 |
| Meghalaya | 59.11 | 55.61 | 55.23 | 54.03 | 40.89 | 44.39 | 44.77 | 45.97 | -5.08 |
| Mizoram | 40.00 | 31.13 | 70.59 | 76.67 | 60.00 | 68.87 | 29.41 | 23.33 | 36.67 |
| Nagaland | 60.40 | 78.48 | 82.46 | 67.03 | 39.60 | 21.52 | 17.54 | 32.97 | 6.63 |
| Tripura | 78.55 | 73.58 | 58.92 | 77.17 | 21.45 | 26.42 | 41.08 | 22.83 | -1.38 |
| Sikkim | - | 0.00 | 62.40 | 56.69 | - | 100.00 | 37.60 | 43.31 | 56.69 |
| NE Region | 72.65 | 72.65 | 70.18 | 69.62 | 27.35 | 27.35 | 29.82 | 30.38 | -3.03 |
| All India | 74.98 | 73.37 | 66.39 | 64.70 | 25.02 | 26.63 | 33.61 | 35.30 | -10.28 |

*Source:* CMIE (1998, 2005).

## Changes in Gross Value of Agriculture Output

The growth in gross value of output from agriculture of the NE region is presented in Table 13. It can be seen from the table that the gross value of output from agriculture of NE region has been decelerated from Rs. 20684 crore in TE 1992-93 to Rs. 9154 crore in TE 2002-03, by 55.74 percent over base period, at the rate of 7.83 percent per annum. The output from agriculture of Arunachal Pradesh, Assam and Sikkim states has declined, however other states have recorded positive growth over base period. NE region contributes around 4 percent in all India output of agriculture. It is clearly indicates that the share of NE states in national agriculture output was very low and not significant. Among the NE states, Nagaland state has recorded highest growth in output from agriculture during 1990-2003.

**Table 13: Gross Value of Output from Agriculture (GVOA) 1990-91 to 2002-03**

| *States* | *TE 1992-93* | | *TE 2002-03* | | *+/- over base* | *Annual Compound* |
|---|---|---|---|---|---|---|
| | *Rs in Crore* | *% Share* | *Rs in Crore* | *% Share* | *period 91-93* | *Growth Rate (%)* |
| Arun. P. | 415 | 0.22 | 267 | 0.11 | -35.77 | -4.31 |
| Assam | 18747 | 9.71 | 6187 | 2.64 | -67.00 | -10.49 |
| Manipur | 274 | 0.14 | 374 | 0.16 | 36.58 | 3.16 |
| Meghalaya | 255 | 0.13 | 397 | 0.17 | 55.28 | 4.53 |
| Mizoram | 122 | 0.06 | 195 | 0.08 | 60.09 | 4.80 |
| Nagaland | 221 | 0.11 | 742 | 0.32 | 236.10 | 12.88 |
| Sikkim | 129 | 0.07 | 95 | 0.04 | -26.71 | -3.01 |
| Tripura | 521 | 0.27 | 898 | 0.38 | 72.53 | 5.60 |
| NE Region | 20684 | 10.71 | 9154 | 3.91 | -55.74 | -7.83 |
| All India | 193049 | 100.00 | 234294 | 100.00 | 21.36 | 1.96 |

*Source:* CSO (2005).

## Incidence and Extent of Indebtedness

The income, expenditure and indebtedness of farmers' households in NE region is presented in Table 14. The average annual income of farmers households (which includes cultivation, wages, farming of animal and non-farm business income) of all NE states except Tripura as well as NE region as such was higher than all India average. The average annual income of farmer household from farm business (cultivation income-cultivation expenses) was also significantly higher in all NE states. If you looked at annual income and expenditure pattern of farmers households in NE region, it is observed that Manipur, Nagaland and Tripura farmers' households' income was not adequate to meet the consumption expenditure of the households. This clearly suggests that farmers are in severe distress. The data shows that about 20 percent of the households in NE region are in debt, as compared to about 49 percent at all India level. The incidence of indebtedness (proportion of household reporting debt) as well as extent of indebtedness (debt per farmer household in Rs.) is found to be higher where consumption expenditure exceeds the income of farmers' households. It is seen that Tripura has highest number of households in debt (50 percent), followed by Nagaland (36.5 percent), Sikkim (32.8 percent), Manipur (24.8 percent) and Mizoram (23.6 percent). The average debt/household varies from Rs. 72/- in Meghalaya to Rs. 2977/- in Tripura. Though the incidence of indebtedness was low in NE region, the government needs to regulate output and input markets an effective manner as to increase the farmers' income level.

## Conclusion and Policy Suggestions

The North Eastern states are among the most backward in the country and agricultural development is much below its potential in this region. An attempt has been made in this paper to study the agricultural development in north eastern region/states of India. Agriculture is the main economic activity in the region and despite major impact of green revolution in the irrigated areas of the country; modernization of agriculture has escaped this region as evidenced by poor adoption of modern technologies, low consumption of fertilizers and other indicators of growth. The agricultural production system in the region is predominantly rainfed, mono-cropped at subsistence level. Pressure on rural land, hilly terrain, tribal population, economic backwardness of the area, low technological levels, low productivity coupled

**Table 14: Income, expenditure and Indebtedness of Farmers Households (July 2002- June 2003) (Rs/Annum)**

| States | Cultivation income | Total income | Cultivation Expenditure | Consumption Expenditure | 2-4 | 3-5 | Incidence of Indebtedness (%) | Extend of Indebtedness (Rs) |
|---|---|---|---|---|---|---|---|---|
| 1 | 2 | 3 | 4 | 5 | 6 | 7 | 8 | 9 |
| Arunachal P | 20856 | 89460 | 3564 | 35280 | 17292 | 54180 | 5.9 | 493 |
| Assam | 21504 | 37932 | 3370 | 32568 | 18134 | 5364 | 18.1 | 813 |
| Manipur | 11748 | 32892 | 4064 | 38988 | 7684 | -6096 | 24.8 | 2269 |
| Meghalaya | 39000 | 53952 | 8022 | 38388 | 30978 | 15564 | 4.1 | 72 |
| Mizoram | 33816 | 58344 | 1395 | 44856 | 32421 | 13488 | 23.6 | 1876 |
| Nagaland | 23004 | 43080 | 2308 | 53772 | 20696 | -10692 | 36.5 | 1030 |
| Sikkim | 12780 | 39096 | 3899 | 34500 | 8881 | 4596 | 32.8 | 2053 |
| Tripura | 10368 | 20904 | 4214 | 33504 | 6154 | -12600 | 49.2 | 2977 |
| NE Region | 21635 | 46958 | 3855 | 38982 | 17780 | 7976 | 19.9 | 1448 |
| All India | 11628 | 25380 | 8791 | 33240 | 2837 | -7860 | 48.6 | 12585 |

Sources: NSSO (2005, 2005a), Narayanamoorthy, 2006.

with low per capita income are the major features this area. The NE region continues to be a net importer of foodgrains even for its own consumption. The NE region has about 52 percent area (reporting area) under forest cover though there is large variation among the states. About 54 percent of rural population is directly dependent on agriculture as compared to 75 percent at national level. The average size of operational holdings is found to be lower in NE region as compared to the same at the all-India level. The cropping pattern in the region, with the exception of Sikkim, is characterised by predominance of rice as the lead crop, however, maize is the dominant crop in Sikkim. Food crops account for more than 66 percent of the gross cropped area in 2003-04, which is suggestive of prevalence of subsistence agriculture and lack of crop diversification. The cropping pattern however suggests that the known tendency of concentrate in foodgrains, rice as a food crop, mainly driven by urge for subsistence. The net irrigated area in the region does not constitute a significant proportion of net area sown and due to which irrigation coverage was also very low. Fertilser consumption in the region is very low and spread of HYVs has been very poor against the national average. It is observed that productivity of all the crops was lower than that of national level. About 20 percent of the households in NE region are in debt, as compared to about 49 percent at all India level. The incidence of indebtedness as well as extent of indebtedness is found to be higher where consumption expenditure exceeds the income of farmers' households. The policy suggestions are as follows:

1. The wide range of disparities among the states, the lack of location-specific and system-based technologies, limited availability of cultivable and farming technique bordering between primitive and traditional are the major reasons behind the slow agricultural development of NE region. Therefore, a promotional measure for HYV seed, fertilser use is required in these areas. Same time, efforts need to made for location specific research and putting more emphasis on agricultural extension services.
2. The population is growing fast and hilly lands are being diverted from production of foodgrains, through

slash and burn method of cultivation, which cause soil erosion, to horticultural and silvicultural plantations. Therefore, extensive soil protection measures are needed in the plains, along with the setting up of a well-managed food security system, particularly for the remote areas.

3. Also the development of different types of farming systems to irrigated and rainfed situation coupled with efficient management of soil and water should be worked out. Watershed approach of soil and water conservation and development of land based production programme in participatory mode need to be propagated.
4. To increase productivity of crops, development of location specific seeds and planting material is required. Adequate attention is necessity to increase use of fertilser and availability of credit.
5. Lack of infrastructure and markets is an important impediment in realising the potential of high-value agriculture in this region. Therefore, grater investment in roads and transportation, and development of innovative market institutions like cooperatives, self-help groups and contract farming that provide an assured market to the producers, quality inputs, technology and credit. Therefore, in order to increase the income level of farmers of NE region, substantial investment has to be made in agriculture and allied sector as well as in infrastructure development.
6. Irrigation intensity was low as the mountain terrain has somehow made it difficult to exploit the growing water potential an in most states sources other than canals and well supply most of all the irrigation water. In such areas techniques for harvesting rain water appears to be essential.
7. Though the incidence of indebtedness was low in NE region, the government needs to regulate output and input markets an effective manner as to increase the farmers' income level.
8. The North Eastern Region of India has high potential

to tap immense opportunities emerging in high-value sector provided the existing constraints are alleviated through appropriate policies and institutional arrangements (Birthal, et. al., 2006).

## REFERENCES


Banerjee, Amalesh and Minan Kar (1999), *Economic Planning and Development of North-Eastern States,* Kanishka Publishers, Distributors, New Delhi.

Bhalla, G. S. and Y. K. Alagh (1979), *Performance of Indian Agriculture: A District-wise Study,* Sterling Publisher, New Delhi.

Bhalla G. S. and Gurmail Singh (2001), *Indian Agriculture- Five Decades of Development,* Sage Publications, New Delhi.

Birthal, P. S.; A. K. Jha; P. K. Joshi and D. K. Singh (2006), "Agricultural Diversification in North Eastern Region of India: Implications for Growth and Equity, *Indian Journal of Agricultural Economics,* Vol. 61, No. 3, July September 2006, pp. 328-340.

Chand, Ramesh and Sonia Chauhan (1999), *Are Disparities in Indian Agriculture Growing,* Policy Brief No. 8, National Centre for Agricultural Economics and Policy research, New Delhi, October.

CMIE (2000), *Profiles of Districts,* Centre for Monitoring Indian Economy Pvt. Ltd., Mumbai, October.

CMIE (1998, 2006), *Agriculture,* Centre for Monitoring Indian Economy Pvt. Ltd., Mumbai, March.

CSO (2005), *Statewise Estimates of Value of Output from Agriculture and Livestock* (1990-91 to 2002-03), Central Statistical Organization, Ministry of Statistics and Programme Implementation, Government of India.

Dantwala, M.L. (1986), Strategies of Agricultural Development since Independence, *Indian Agricultural Development since Independence: A Collection of Essays,* Oxford and Publishing Co. Pvt. Ltd..

Deshpande, R.S.; M. J. Bhende; P. Trippaiah and M. Vivekananda (2004), *Crops and Cultivation, State of the Indian Farmer- A Millennium Study,* Vol. No. 9, Department of Agriculture and Cooperation, Ministry of Agriculture, Government of India, New Delhi.

EPWRF (2003), *Domestic Product of States of Indi: 1960-61 to 2000-01,* Economic and Political Weekly Research Foundation, Mumbai, June.

FAI (various years), *Fertiliser Statistics*, The Fertliser Association of India, New Delhi.

Ghosh, Nilabja (2003), "Organic Farming in North-East Hill Region in India", paper submitted for *INSEE's 3rd Biennial Conference on 'Biodiversity and Quality of Life'* held on 18-20 December, at Calcutta.

GOI (2001), *Report of the Working Group on Agricultural Development in Eastern and North Eastern India for the Formulation of the Tenth Five Year Plan*, Planning Commission, August, TFYP Working Group Sr. No, 40/2001.

GOI (2002), *Selected Socio-Economic Statistics India*, Central Statistical Organisation, Ministry of Statistics and Programme Implementation, Government of India.

GOI (2002a), *Indian Agriculture in Brief*, Agricultural Statistics Division, Directorate of Economics and Statistics, Department of Agriculture and Cooperation, Ministry of Agriculture, government of India, 27th edition, January.

GOI (2004), *Agricultural Statistics at a Glance*, Agricultural Statistics Division, Directorate of Economics and Statistics, Department of Agriculture and Cooperation, Ministry of Agriculture, Government of India, August.

GOI (2005), *Working Group Report on Agriculture and Allied Sector*, Planning Commission, Government of India.

GOI (2003, 2006), *North East Region Towards Economic Resurgence*, Annual Report, Ministry of Development of North Eastern Region, Government of India.

GOI (2006), *Annual Report 2005-06*, Ministry of Development of North-Eastern Region, New Delhi.

GOM (2005), *Economic Survey of Maharashtra*, Directorate of Economics and Statistics, Planning Department, Government of Maharashtra, Mumbai.

Joshi, P. K., and A. Gulati (2003), *From Plate to Plough: Agricultural Diversification in India*, paper presented for the Conference on Dragon and Elephant, IFPRI, Mimeo.

Joshi, P.K.; Ashok Gulati; Pratap Birthal and Laxmi Tewari (2004), Agricultural Diversification in South Africa: Pattern, Determinants and Policy Implications, *Economic and Political Weekly*, Vol. 39, No. 24, pp. 2457-2467.

NAAS (2001), *Strategies for Agricultural Research in the North-East*, National Academy of Agricultural Sciences, New Delhi, India, Policy Paper No. 9, June.

NABARD (2001), *Report of the Expert Committee on Rural Credit*, National Bank for Agriculture and Rural Development, Mumbai, July.

Narayanamoorthy, A. (2006), State of India's Farmers, *Economic and Political Weekly*, Vol. 41, No. 14, February 11, 2006, pp. 471-473.

NCAER (2004), *East India Human Development Report*, National Council of Applied Economic Research, Oxford University Press, New Delhi.

NHDR (2002), *National Human Development Report*, Planning Commission, Government of India, March.

NSSO (2005), *Situation Assessment Survey of Farmers, Indebtedness of Farmers Households*, Report No. 498, 59th Round (January-December 2003), National Sample Survey Organisation, Ministry of Statistics and Programme Implementation, Government of India, New Delhi, May.

NSSO (2005a), *Situation Assessment Survey of Farmers, Income Expenditure and Productive Assets of Farmers Households*, Report No. 497, 59th Round (January-December 2003), National Sample Survey Organisation, Ministry of Statistics and Programme Implementation, Government of India, New Delhi, December.

Sharma, R.K.; Brij Bala and H.R. Sharma (1997), "Regional Disparities in Agricultural Development – A Study of Mountains States", *Indian Journal of Agricultural Economics*, Vol. 53, No. 3, July-September, pp. 394-405.

World Bank (2005), *India Re-Energizing the Agricultural Sector to Sustain Growth and Reduce Poverty*, Agriculture and Rural Development Unit, South Asia Region, Oxford University Press, New Delhi.

http://www.pecad.fas.usda.gov; www.mdoner.gov.in; www.censusindia.com

# 4

# QUEST FOR RESOURCE SUSTAINABLE AND ECONOMICALLY HIGHLY REMUNERATIVE ALTERNATIVES TO AGRICULTURE DEVELOPMENT IN NORTH EASTERN REGION OF INDIA

*A. K. Mishra and J. P. Misra*

## Introducion

North Eastern Region of India is located between 21.5 degrees to 29.5 degrees North Latitude and 85.5 to 97.5 degrees E longitudes. The North East of India constitutes the states of Arunachal Pradesh, Assam, Manipur, Meghalaya, Mizoram, Nagaland, Sikkim and Tripura. Flanked by hills and with the mighty Brahmaputra River slashing a central path between its north and south, the North East is bounded by the states of Manipur, Meghalaya, Mizoram, Nagaland, Tripura with Assam comprising the heart. Arunachal Pradesh lies to its north and Sikkim a little away in the North West bordering China and Bhutan. Bangladesh and Myanmar lie to its southwest and east.

The western side of the North East is connected to the eastern part of the Indian subcontinent by a narrow land corridor, sometimes referred to as the Siliguri Neck or "Chicken's Neck". The hills of NE region contribute its major area with Assam and Sikkim states and Darjeeling on it periphery forming the North Eastern Himalayan zone in the

Indian territory while Bhutan and Chittagong Hills of Bangladesh lie outside the boundary of Indian republic. Hill states of India have a total geographical area of 5.63 per cent of the country (approximately 183,813 km$^2$) with a population of 1.13 per cent to be supported on its natural resources.

The NE states are quite different and peculiar in nature due to their distinct topography, (highly rugged steep mountains and hills to rolling topography of plateau region) terrain comprising mountainous region, hills and plateaus interspersed with valley lands in the foot hills of the hills, extreme climatic characteristics which results in above 11000 mm of average annual rainfall in Cherrapunjee and Mowsynram Ranges. Inaccessibility and remoteness of the places, dense forest cover and poor and underdeveloped infrastructure system are characteristics of the Indian North Eastern Region.

## Demography

The North Eastern region of India is sparsely populated as compared to other parts of the country. The total pupoulation in 2001 was 10286.10 millions. The population density in the region varies from 55-60 per sq. km as compared to 275-290 persons per sq. km in plains. The population growth rate in the region during the last decade has been 21.54 per cent and is expected to increase up to 25 per cent. Nagaland is the state with highest population growth rate during last decade.

## Physiology

The North Eastern region is characterized predominantly by hilly terrain, deep gorges and flat valleys. The average annual rainfall of the region varies from 2000 to 4000 mm. The climate varies from sub- tropical to extreme temperature and is bestowed with rich natural resources viz, water, petroleum, minerals, rare flora and fauna. The region is broadly divided into four major divisions: *(i)* Assam valleys *(ii)* Assam Himalayas *(iii)* Meghalaya hills and, *(iv)* Eastern high lands. Assam valley is 80-129 km wide and 725 km long. The river Bhahmaputra passes through the middle of the valley. The

Mishmi hills, Abor hills, Mikir hills, and Dafla hills are in Assam Himalayas in Arunachal Pradesh. The Meghalaya hills (Shillong Plateau) is in south of Assam valley. The Garo, Khasi and Jaintia hills, Mikir hills and North Cacher hills are in Shillong plateau. The Eastern high lands include the Patkai, Naga hills, Barail hills, high lands of Manipur and Mizo hills. Manipur lies on the east of the Barail range, whereas Tripura lies in the western flank of Mizo hills. Nagaland is situated between Manipur and Arunachal Pradesh.

## Climatic Resource and Agro–Climatic Sub–Division

The altitudinal differences coupled with varied Physiography contributes to climatic variations in the North Eastern region of India. The climate in the region varies from sub-tropical to alpine. Despite diverse physiographic characteristics of the region, sub-regional variations in the average seasonal temperature are not striking. Bulk of the annual rainfall in the region (64%) is received during June–September period. The North Eastern states of the region, through have 7.75% of the geographical area of the country receives about 12.7% of the total precipitation in the country. The area can be divided in to three climatic region viz.: *(i)* the cold humid monsoon climate of the frontier hilly region (above 200 m); *(ii)* wet sub-tropical monsoon climate and, *(iii)* humid mesothermal monsoon climate with heavy monsoon showers. Average annual rainfall in the region is about 2490 mm with highest rainfall of 12,500 mm in the world in the Cherrapunji Mawsynram state. The region, due to heavy rainfall, has very high humidity varying from 62% to 91% in rainy season and 42% to 65% in other seasons. The climatologic parameters are depicted in Figure 1. Accordingly the North Eastern India has been divided in the agro-climatic zones as: Alpine zone, Temperate sub-Alpine zone, Sub-tropical hill zones, Sub-tropical plain zone (valley areas) and Mild tropical hill zone.

## Status of Natural Resources in the Region

North Eastern region is endowed with rich natural resources of soil, water and vegetation. Management of resources of

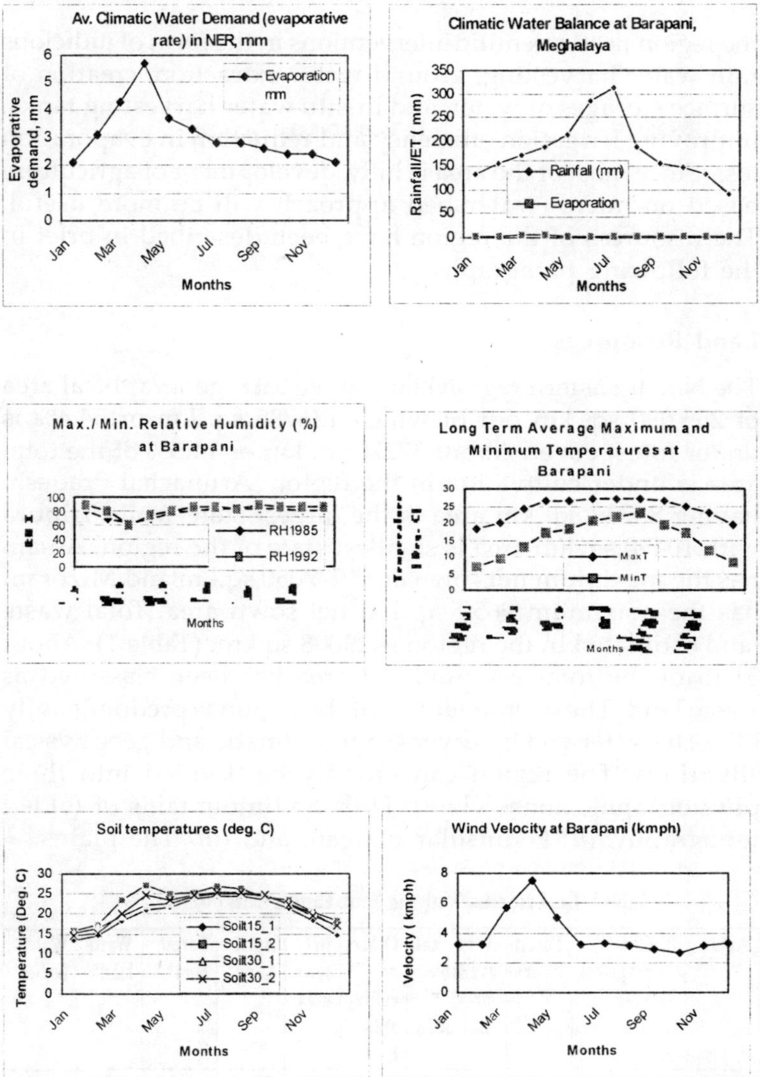


Fig. 1: Representative Agro-climatology of Barapani, Meghalaya, North Eastern India

the region need scientific interventions in the form of judicious rain water harvesting, ground water extraction, creation of surface storages of water, and in-situ water harvesting tanks, improving irrigation efficiency and reduction in evaporation loss. Since most of the area is hilly, development of agriculture based on micro- watershed approach will be more useful. The resources of the region have been described in brief in the following paragraphs.

## Land Resources

The North Eastern region states have total geographical area of 255,090 sq. km out of which 136785 sq. km or 54.4% is under forest cover. About 37220 sq. km or 14.5% of the total area is under cultivation in the region. Arunachal Pradesh, having 83749 sq. km area is the largest state while Tripura with 10490 sq. km area is smallest state of the region. Assam has the maximum net sown area 927060 sq.km and Mizoram has the maximum 9650 sq. km net sown area. Total waste land estimated in the region is 29008 sq.km. (Table 1). About 31.15 of the total geographical area has been classified as waste land. The overall terrain of the region is predominantly hilly characterized by diverse agro-climatic and geophysical situations. The region can broadly be divided into three physiographic zones viz. *(i)* Hills and mountains of folded topography, *(ii)* Peninsular plateau, and *(iii)* The plains.

**Table 1: Land use classification in North Eastern states ('000 ha).**

| State | Geo-graphical Area | Forest Area | Area not available for cultivation | Other un-cultiv-able land | Net sown area | Agril. Land* | Forest land* | Waste land* | Water bodies* |
|---|---|---|---|---|---|---|---|---|---|
| Ar. Pd. | 8374 | 5154 | 77 | 44 | 149 | 206.4 | 6935.1 | 114.7 | 80.4 |
| Assam | 7844 | 3071 | 2455 | 535 | 2706 | 424.9 | 1726.4 | NA | NA |
| Manipur | 2233 | 1515 | 1445 | 24 | 140 | 169.2 | 1414.1 | 29.0 | 10.7 |
| Meghalaya | 2243 | 851 | 226 | 646 | 202 | 731.6 | 1424.9 | 2.5 | 10.2 |
| Mizoram | 2108 | 1593 | 211 | 81 | 65 | 660.0 | 1956.5 | 0 | 0.06 |
| Nagaland | 1658 | 862 | 28 | 224 | 190 | 389.5 | 1142.2 | 0 | 0 |
| Tripura | 1049 | 631 | 131 | 40 | 270 | 154.8 | 3806.5 | 0 | 6.6 |
| Total | 25509 | 13677 | 4573 | 1594 | 3722 | 4897.8 | 14979.8 | 31.6 | 27.4 |

Source: NEC' 1995, * NRSA Estimates (NRSA, 1995)

Shifting cultivation, the predominantly form of agriculture prevalent in the region is practiced in about 3869 sq.km of area annually, affecting 14660 sq.km area at one time or the other. The system is uneconomical and has caused resource depletion and land and environmental degradation (Table 1). NRSA has given a modified LU/LC statistics of the region in 1995. According to it the Forest area in the region is 14979.8 thousand ha (Table 1). That has shown an increase in last decade.

## Agricultural Development in the North Eastern Region of India

The tribal population of the NE India is dependent on agriculture and related activities connected to land. Agriculture is practiced for sustenance and not for commercial activities except for some location specific commercial activities such as tea growing and arecanut planting. Sifting cultivation is the major form of agricultural practiced in the region. The pattern of agricultural growth has however remained uneven across regions and crops. The NER continues to be a net importer of food grains even for its own consumption. In spite of covering 7.7% of the country's total geographical area, NER produces only 1.5 % of the country's total food grain production. Agriculture provides livelihood support to 70 % of the population of NER The food grain production and productivity of some crops are given in the table 2. A comparison of the crop productivity of the North Eastern Region with the country showed that almost all the crops show a very poor performance as compared to the country level productivity. This is indicator of the production gap that may be abridged by adopting modern and scientific practices of commercial farming is the common understanding of the planners, policy makers, educationists and agricultural scientists. However, associated factors related to sustainability of the resources as well as economics need to be examined critically (Table 3).

**Table 2 : Area, production and productivity of various crops in North Eastern Region (1993-94)**

| *Crops* | *NER* | | | *All India* | | |
|---|---|---|---|---|---|---|
| | *A* | *P* | *Pr.* | *A* | *P* | *Pr.* |
| Rice | 3351.9 | 4680.0 | 1396.0 | 42033.6 | 78972.4 | 1879.0 |
| Maize | 109.4 | 192.2 | 1793.0 | 5989.4 | 9479.5 | 1583.0 |
| Wheat | 92.8 | 125.6 | 1353.0 | 24914.5 | 59139.3 | 2373.0 |
| Small millets | 46.1 | 38.4 | 833.0 | 192.9 | 933.3 | 475.0 |
| Total pulse grains | 146.8 | 92.1 | 627.0 | 22440.2 | 13099.6 | 584.0 |
| Total food grains | 3747.0 | 5068.9 | 1351.0 | 2557.7 | 20277.4 | 793.0 |
| Total Oil seeds | 391.8 | 215.6 | 550.0 | 11075.2 | 16387.9 | 15242.0 |
| Potato | 96.5 | 688.9 | 7139.0 | 58.0 | 189.4 | 3262.0 |
| Ginger | 11.6 | 63.4 | 5473.0 | 128.6 | 397.4 | 3090.0 |
| Turmeric | 11.5 | 12.0 | 1043.0 | 226.9 | 5021.9 | 22152.0 |
| Tapioca | 7.0 | 40.9 | 5843.0 | 139.7 | 1185.2 | 8484.0 |
| Sweet Potato | 16.5 | 65.9 | 3994.0 | - | - | - |

*Source:* Basic Statistics of North Eastern States, NEC, 2000. (A : Area ('000 ha), P: Production ('000 t), Pr: Productivity (kg/ha))

**Table 3: Trends in Rice Production and Productivity in North Eastern Region**

| *Year* | *Area ('000ha)* | *Production ('000 t)* | *Productivity (Kg/ha)* |
|---|---|---|---|
| 1990-91 | 103.36 | 119.8 | 115 |
| 1991-92 | 104.5 | 121.1 | 1146 |
| 1992-93 | 104.2 | 114.0 | 1094 |
| 1993-94 | 104.2 | 111.5 | 1094 |
| 1994-95 | 103.2 | 118.8 | 1080 |
| 1995-96 | 104.0 | 141.1 | 1075 |
| 1996-97 | 104.8 | 151.1 | 1346 |
| 1997-98 | 157.9 | 225.3 | 1427 |

*Source:* Basic Statistics of North Eastern States, NEC, 2000.

## Production and Productivity of Important Food Grains in NE Region

Rice is the major crop of the region accounting for about 89% of the area and 92% of the total food grains production. The other major crop is maize, wheat, millets, oilseeds and pulses.

Fig. 2 Annual CGR of Food Production and Population in NER ( 1980-95)

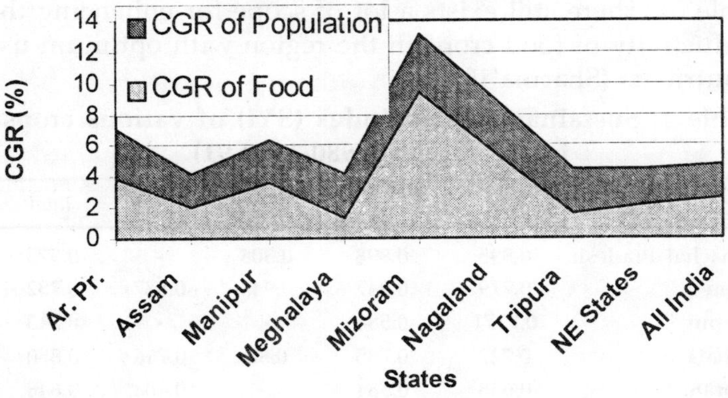


The region is deficient in food grains and the gap between demand and supply was widening further. During the years 1980-95, the annual Compound Growth Rate (CGR) of food grains production was a meager 2.11% as against 2.41% increase in the population. (Fig.2). The production in Meghalaya and Assam has lower annual; CGR than population. Sustainable Yield Index (SYI), which gives an idea of the overall yield sustainability of major crops, is shown in Table 4. The average productivity in the region is very poor. The annual CGR of productivity indicates that productivity of various food crops is declining in Meghalaya, while it is very poor in Assam and Nagaland states. This could be attributed to low use of fertilizer nutrients, traditional agricultural practices, mostly rainfed agriculture, lack of availability of good seed, suitable cropping patterns etc (Sharma, 1995).

The annual removal of major nutrients by various crops in the region is 294.08, 83.98 and 187.14 thousand tonnes of N, $P_2O_5$ and $K_2O$ whereas, replenishment is only 36.42 (12.3%), 10.92 (13.0%) and 7.91 (4.2%) thousand tonnes. This is a cause of concern as we are mining the soil at a faster rate that would affect the sustainability of food production system. The overall fertilizer nutrient response potential

(FNRP) achieved in the region is only 16%, with highest (34.4%) in Manipur and lowest (6.3%) in Arunachal Pradesh (Table 5). There still exists a lot of scope for enhancing the productivity of food crops in the region with optimum use of nutrients (Sharma'1992).

**Table 4: Sustainable Yield Index (SYI) of various crops in NE region (1980-1990-91)**

| *State* | *Rice* | *Maize* | *Wheat* | *Pulses* | *Total* |
|---|---|---|---|---|---|
| Arunachal Pradesh | 0.895 | 0.898 | 0.308 | - | 0.573 |
| Assam | 0.779 | 0.947 | 0.943 | 0.887 | 0.782 |
| Manipur | 0.7171 | 0.584 | - | - | 0.713 |
| Meghalaya | 0.727 | 0.735 | 0.88 | 0.716 | 0.850 |
| Mizoram | 0.658 | 0.584 | - | 0.408 | 0.648 |
| Nagaland | 0.647 | 0.564 | - | 0.426 | 0.647 |
| Tripura | 0.709 | - | - | 0.709 | 0.731 |
| NE State | 0.793 | 0.774 | 0.679 | 0.439 | 0.797 |

Source: Sharma, 1996.

**Table 5: Contribution of various inputs (%) in increasing food grain production in North Eastern states (1974-93)**

| *State* | *Irrigation* | *Fertilizer Nutrient* | *Area* | *HYV & improved practices etc* |
|---|---|---|---|---|
| Arunachal Pradesh | 6.4 | 17.4 | 65.9 | 10.3 |
| Assam | 2.0 | 44.4 | 33.0 | 20.6 |
| Manipur | 3.7 | 63.0 | 0.0 | 32.3 |
| Meghalaya | 3.1 | 38.0 | 26.6 | 32.3 |
| Mizoram | 1.2 | 48.0 | 14.0 | 36.4 |
| Nagaland | 6.2 | 39.9 | 30.7 | 23.2 |
| Tripura | 5.2 | 60.9 | 0.0 | 33.9 |
| NE States | 2.9 | 48.9 | 26.8 | 21.5 |

*Source:* Sharma, 1997.

## Natural Resources Degradation in Shifting Cultivation vis-a-vis Modern Agricultural Methods

The soil and runoff from *Jhuming* has been estimated to be very high (Table 6) in the shifting cultivation practice and hence ICAR has set up its research complex to study, understand, suggest alternatives to Shifting cultivation for its complete replacement with modern methods of agriculture (Jackson, 2001). The million dollar question is whether the above form of shifting cultivation that has been reported to have most degenerative effect on resources sustainability in the region and that resulted in to such high levels of resources degradation needing the immediate attention leading to creation of a full fledged Research Institution (ICARNEH) to replace Jhuming 25 years back has been able to achieve its goals? It is the time for introspection that whether after 25 years with all the expenditure in the form of man and money what we have been able to achieve is worth? To understand the intricacies of shifting cultivation practice in much better manner; one has to explore the Indigenous method of shifting cultivation and its intricacies, which took ages to evolve with the ingenuity of the farmers who have inhabited the area since long more scientifically. Their cultivation practices have evolved as a solution to the problems to a raw mind. The arguments can be put forth in favour of Jhuming over modern agriculture looking from the angle of resources sustainability. The modern agriculture is certainly not friendly to resources as a huge soil loss takes place as soon as the land is put under the modern tools and implements of primary and secondary tillage superimposed with high intensities and duration of continuous rainfall. Augmenting these vows are terrain characteristics. What is more daunting is the fact that as soon as the soil particles are detached and transported with water they are lost forever. As the siltladen water leaves the boundary of Indian territories these resources become the property of neighbouring countries which are never retrievable. The author has watched this phenomenon helplessly on the southern boundary of Meghalaya state. This has warranted the quest for new resources sustainable alternatives that are more remunerative.

Shall we allow such type of resources wastage in the name of developing modern agriculture or shall the scientific community feel concerned to preserve and protect the resources. The valid argument can be put forth in favour of other non-degenerative alternatives, which may provide economical security to the people of the region who can obtain food grains from other regions or countries if they have either money or material to exchange with. This concept has led to look for more economical vis-à-vis resources sustainable alternative tor the region.

### Per Capita Income

The Net Domestic Product of NER has increased from Rs 46162 crores in 2000-01 to Rs 50520 crores in 2001-02. Per Capita Income. NER as a whole is placed amongst the poorest regions in the country. The per capita income at Rs. 12407 is less than the national average of Rs. 17978/- (2001-02). There is a dire need to take provisions so that the per capita income of the region gets a boost. A vigorous policy of the government is therefore, inevitable which can not only boos the economic conditions of the poor tribals as well as ensure resources sustainability.

**Table 6: Table Runoff and Soil Loss from Jhum Cultivation Fields and Fallow in Meghalaya in North East India**

| *Particulars of Land Use* | *Runoff (% of Rainfall)* | *Soil Loss (t/ha/year)* | *Source* |
|---|---|---|---|
| Freshly burnt field under cropping (Steep slope, rainfall 2200 mm) | | | |
| 30 years cycle | 21 | 22.5 | Tokey and |
| 10 years cylce | 24 | 23.0 | Ramakrishna |
| 5 years cycle | 26 | 30.0 | (1981) |
| Fallows (regeneration) | | | |
| 5th Year | 19 | 1.1 | |
| 10th years | 13 | 0.8 | |
| Freshly burnt field under cropping (Rainfall 1785 mm on 40 % slope) | | | |
| 10 years cycle | 30 | 50.0 | Mishra and |
| 5 years cycle | 33 | 55.0 | Ramakrishna |
| Fallows (regeneration) | | | (1983) |
| 1st year | 25 | 7.1 | |
| 2nd year | 21 | 3.5 | |
| 10th year | 13 | 2.0 | |

*Source:* Jackson, 2001.

## Sustainability of Modern Methods of Agriculture on Steep Hill Slopes with Heavy Rainfall

Twenty five years before, the ICAR Research Complex for NEH region with its head quarters at Barapani near Shillong (Meghalaya) and regional stations in all the North Eastern states except Assam which already had an agricultural university came in to being. The major objective of the research institutions were to replace the Jhuming and to suggest more remunerative agricultural alternatives. It has now been concluded that Jhuming has become a way of life for the tribals of the North Eastern region that can not be replaced but modified. However, modern scientists, over a period of time have resorted to the sartorial research in their respective field of research. As for as the agriculture with modern tools and implements is concerned there has been a marked improvement in the ways the agriculture is being done in the North Eastern hills region. If carefully analyzed and compared it appears that the total soil losses would be far more than the traditional system of Jhuming which was using Dibbling as the method of sowing seeds and harvesting the grains leaving the stalks in the field when superimposed the intensity and duration of rainfall over the same plot which now is being subjected to intensive disturbance by the way of primary & secondary tillage using modern tools and implements. This warrants immediate attention of the planners and scientists to arrest the degradation due to modern methods of cultivation with or without soil conservation measures and to ensure the resources sustainability vis-a vis agricultutral development of the region.

## Quest for Resources Sustainable, More Economic Alternatives to Agriculture in North Eastern Region of India

In the North Eastern Region of India due to high rainfall of the order of 2000–400 mm per annum that is intense and for sort duration having high erosive energy water erosion is the major form of land degradation (Table 6). This problem gets aggravated with the cultural practices being done in the agricultural fields to realize economic production. All cares should be taken and the resources degradation should be checked by all means so that not only the soil but also the most important nutrients should be preserved. Since, agriculture and related activities are highly desired and cant be checked overnight, we may allow it to be practiced; but keep on looking for more remunerative alternatives as soon as possible with the help and support of the government agencies. Agro-based small-scale industries, semi permanent farming with perennial plants, animal based farming systems, aquaculture etc.; are some other alternatives that should be practiced in order to improve the socio-economic status of farming community with a wider national goal of preserving the national treasurer of Soil and Nutrients.

Few workable suggestions are made for adoption of other economically remunerative ventures for large-scale adoption in the NE region to ensure resources sustainability vis-à-vis economic well-being. State governments are already making some progress in tandem with the agricultural development but no one except researchers and scientist really seems to be concerned for resources conservation and management in true sense else the long term sustainability of natural resources should not have been put to jeopardy by modernizing the traditional agricultural practices on the pretax of making it more profitable and sustainable over traditional Jhum cultivation methods since last a quarter of the century. Some alternatives have been studied in detail (but not exhaustive) and recommended for wider adoption in the region in the light of above discussion.

### *1. Tea Plantation and Tea Industry*

North Eastern Region of India is one of the major tea growing areas of the country (Table 7). The tea industry of NER is the world's largest tea growing region having 16 % share, and is the largest producer and exporter of tea in India (share 55 %). Tea plants being perennial in nature that can withstand high intensity rainfall and provide the cover to the exposed land surface in turn helping to check the soil erosion to a large extent. Tea growing is being expanded to non traditional areas in other North Eastern states. Tea growing areas need a boost in the region for resources sustainability and profitability. Brazil having deforested its large tracts of rainforest has planted tea on the hills, which is providing both benefits of resources conservation with economic benefits.

**Table 7: Tea Industry of the NER (in million kg)**

| | *2000* | *2001* | *2002* | *2003* | *2004* |
|---|---|---|---|---|---|
| Production | 846 | 854 | 826 | 857 | 830 |
| Imports | 14 | 17 | 22 | 7 | 25 |
| Exports | 207 | 183 | 201 | 173 | 190 |
| Consumption | 653 | 673 | 685 | 697 | 710 |

*Source:* Report of the Meeting of the Committee on Financial Sector Plan for NER, March 2006.

### 2. *Other Plantation Crops*

The agro-climatologically and edaphically, conditions of North Eastern Region are highly favourable for development of plantation crops. On lower elevations of the plains, arecanut, banana and rubber have got very high potential. Rubber board is already experimenting with the rubber plantation in Guwhati. Suitable areas could be earmarked to expand the rubber plantation areas in non traditional areas. Areca nut is another economical plantation crop that can be propagated in the North Eastern Region big way looking to its huge market potential. In lower Brahmaputra valley region Banana plantation which has a highly diversified germplasm in the region can be commercialized.

### 3. *Spices Crops*

North Eastern Region offers great potentials for lager scale cultivation of Pepper, Bell pepper, Cinchona, Large and Small Cardamom. These crops require partial shade and moist ambient climatic conditions. Therefore, these spices can be commercially cultivated in mixed farming conditions in the region.

### 4. *Coffee and Cocoa Cultivation*

The agro climatic conditions of the region are quite favourable for cultivation of Coffee and Cocoa, which has got very high export potential. The government has to develop infrastructure for processing the coffee beans before taking up to popularize the coffee plantation.

*5. Cashewnut, Walnuts and Other Nuts*

In some areas of the region Cashewnut and Walnuts are having scope for commercial cultivation. In Garo hills region of Meghalaya farmers have already started growing cashew nut. Strategies should be formed for large scale cultivation if the conditions are favourable.

*6. Tropical and Temperate Horticulture*

High hills and plains of the region offer excellent conditions for development of Tropical and Temperate Horticulture is in unorganized primitive state of development as for as the area, production and post harvest processing is concerned. Area under commercial horticulture need to be enhanced. This will ease out the burden from the land for resources depletion. The North Eastern States produce varieties of tropical and sub-tropical fruits starting from Jackfruit, Pineapple to Orange, Pears, Peaches and Plums. From the data of productivity made available by National Horticulture Board, it appears that although North Eastern States produce good quality of fruits, their productivity per hectare is far below in comparison to other states of India. Amongst many reasons for low productivity per unit area, non-availability of quality planting material is one of the major causes. Conventional methods of propagation from uncertain Clonal mothers have become obsolete. The modern technology of Micropropagation through Tissue culture is gradually eliminating the conventional system of propagation. But in North Eastern States the conventional methods are still being practiced and till date infrastructure for making available high quality planting materials has not been created so far. If this bottleneck is overcome then the spread of horticulture to non-conventional areas can be ascertained.

*Food Processing Industries:* NER, by virtue of its diverse agro-climatic conditions, varied soil type and abundant rainfall has the promise of becoming an excellent sourcing point for high value horticultural produce and value added products unique to the region, for onward marketing both within the country and abroad.

### 7. *Ornamental Fish Culture and Composite Fish Culture*

There is a great potential in the NER in the fishery sector. Due to various reasons currently there is low fish production in the NER however, the majority of the population being non-vegetarian the demand is very high. At present the NER imports about 90,000 tonnes of fish from outside the region. This gap can be easily abridged by popularization of fish culture in the region. Due to very high rainfall in the region and undulating topography the region offers best potential for cultivation of fish for ornamental purposes for exports to foreign markets and composite culture for table purposes. Production of fishlings and hatcheries of inland fisheries is another suitable alternative. Innumerable number of rivulets can be converted into cesspools for fish rearing by artificial creation of dykes and bunds in these channels at regular intervals so that water can be retained round the year. The total fish production in the year 2003-2004 was expected to be 233. 46 tonnes.

### 8. *Dairy Farming*

Dairying is practiced in the region on a minor scale especially by the migrated labourers from the neighboring states. This is a highly remunerative venture associated with the agriculture. Sheep and goats are also seen but due to high humidity probably they are not able to withstand the climatic pressures. The region due to its high rainfall offers great potentials for fodder production round the year for dairy cattle. Having earned the much-needed foreign exchange the food grains can be imported form such areas, which have high productivity either in the country or from abroad.

### 9. *Rabbit farming*

Due to the predominantly non-vegetarian nature of the tribals of the region this species has found acceptance among farmers and locals for self consumption of rabbit for meat. Table size rabbit of the age of 3-4 months has been liked by the people and is in great demand. Rabbit farming also fetches wool and fur required for warm clothing essential for harsh winters in the North eastern Hills region.

*10. Pig Farming*

Pig meat (pork) is liked by almost everybody in the region. Backyard piggery needs a boost in form of large scale industry. Industrial houses can be invited to set up modern piggery farms and the meat processing units in the region. Possible bottlenecks such as high cots of feed can be overcome by promoting the trans border trade between neighboring countries and plains of Assam and Bangladesh.

*11. Rearing Cattle for Meat for Self Consumption and Export to Foreign Countries*

Almost all the North Eastern states are dependent for their meat supplies from the plains of Assam, Bihar, U.P. and W.B. The domestic demand for meat products as well a favourable climatic conditions for production of fodder and grasses provide very highly suitable atmosphere for commercil development of cattle farming for met purposes. Yak and Mithun Farming is already prevalent in Arunachal Prades and Manipur. The localized practice of this type can be popularized elsewhere in the region.

*12. Poultry and Duckery*

Poultry and Duckery are two promising ventures that have vast potential in th eregion. Tripura, Assam and plains region of Manipur and Meghalaya can be converted to the poultry production hubs in the North East. Duck can also be promoted for egg and meat purposes. Scientific studies are needed to adopt ducker and poultry, which has great liking of the native people of the area. The efforts for developing commercial farms not only for self-consumption and domestic markets but also an eye on export markets should be initiated.

*13. Cultivation of Medicinal and Aromatic Plants*

North Eastern Himalayan region is the mage biodiversity zone and the house of innumerable no of plant species having medicinal and aromatic properties. Opportunities can be explored to start the cultivation of such plant species which are perennial in nature and do not require disturbing of the land beneath due to weeding.

### *14. Floriculture/Ornamental Horticulture and Orchids Production for Export*

Agroclimatolgy of the North East is much suitable for commercial floriculture and Ornamental horticulture. The region has great potential to become one such hub of commercial floriculture that can be comparable to Holland due to its climate. The need for providing artificial cooling can be done away with due to its mild climate round the year. The international airport is the vital link required for the same. Orchid is one group of ornamental plants, which has large, export markets in Europe, America and Middle east. Almost all the hilly sates of the region have favourable climatic condition for commercialization of orchid cultivation and this fact needs to be explored more than any other thing. North Eastern State especially Meghalaya has 124 species of Orchids and amongst them there are high value species which can be grown for commercial flower production as well as commercial supply of planting material by micropropagation through Tissue culture technique. Commercial Orchids are complex hybrid which can not be propagated by seed and micropropagation through Tissue Culture is the only answer. Green house culture and production of commercial flower seedling in off season through tissue culture and supply at the onset of season is the widely practicised method of commercial flower production, which can also be adopted in North Eastern States.

### *15. Ramie and Jute Fibre*

Ramie and jute fibers are in great demand in the industrial sectors and this opportunity can be encashed by propagating their cultivation in suitable areas. The ramie is the strongest known vegetative fibre and is popularly known as "Reha" in Assam. The importance of remie fibre is increasingly felt in the textile industry because of its excellent quality compared to other fibres. The fibre of ramie is obtained from the bark of the plant which is composed of high celullose percent. This fibre is used in the textile industry for manufacturing of various type of superior quality of clothing material and it can also be mixed with rayon, terelyne, polyster, wool, cotton, jute and silk etc. The ramie fibre is in great demand for marine purposes.

### *16. Bamboo*

Bamboo is another commercial multipurpose tree specie that has tremendous potential of production in the region. Bamboo based industries and usage of bamboo products are innumerable. It is also very highly economic species if the cost of cultivation on pr ha basis is estimated. The problem with bamboo cultivation is that it is not done on large scale commercial basis. In recent past the bamboo has caught the attention of the authorities and efforts are being made to strengthen the bamboo cultivation in the region. The same need to be spread in almost whole of the North eastern Region of our Country. A bamboo groove of 10 sq. meter can fetch approximately Rs, 2000/- on maturity of approximately 100 bamboos @ Rs. 20/- local prices. This species alone can be more remunerative if cultivated as an industry. This means that the bamboo cultivation can fetch approximately 20 lakhs over a period of five years from a well organized and maintain bamboo area of one hectare.

### *17. Agro Forestry*

Agro forestry is already being practiced in the region at a limited scale. Several plantation companies have already established their forest plantations in many parts of the region. Agro forestry provides the poor farmers the much needed cash and the food items at a much lower input and labour. Various alternatives in agro forestry are already been tried and recommended. Novel alternatives and combinations of species can be experimented and tried so tat the farmers get the maximum benefits.

## Development of Agro-based Industry

Industrially, the NER continues to be the most backward region in the country, and the states in the region hardly have any industrial base, except perhaps Assam, because of its traditional tea, oil and wood based industries .To some extent Meghalaya has made some headway in setting up of small and medium industries. There are a number of factors contributing to the lack of industrial growth in the region like poor infrastructure, inadequate supply of electricity, violence and extortion etc. The growth of industries, large, medium and small has been uneven. The number of functioning large and medium industries in the NER has been negligible. The status of industries in the NER is as follows.

## Summary and Conclusions

Though there may be many other highly remunerative and economically viable options yet they might not have been sighted here yet it confirms the fact that there are ample number of other activities that can replace the need for cultivating the steep slopes permanently. Thus, from the above discussion it is evident that while the North Eastern Region has excellent scope for development of many economic alternatives to agriculture and farming they need to be provide with much needed help support and infrastructure facilities before establishing the lager scale projects. Also, the inherent built in nature of social-fabric of a large number of tribals of mutual cooperation can be utilized for establishing such ventures in which the farmers may be owners on a collective basis which will avoid confrontations. All the solutions may not be applicable for all regions but a suitable group of activities can be picked up and spread in the most favourable region so that farmers can learn to be independent financially and socially and the sustainability of the natural resources are not in jeopardy. The planners and policy makers should be aware that the highly degrading nature of activities in long run are going to create problems and more problems unforeseen so far of great dimensions until the corrective measures are taken well in advance with farsightedness and zeal.

## REFERENCES

N.E.C., (1995). *Basic Statistics of the North–East.*, Min. of Home affairs. Govt. of India, p. 40.

N.R.S.A., (1995). Report on area statistics of land cover generated using remote sensing techniques .NRSA, Hyderabad, 70 pp.

Jackson, M. G. (2001). Let us take another look at jhum Cultivation. *Asian Agri-history,* 5 (3) 197-223.

Sethi, R.M., Nair, G.K.M., Sheshagiri Rao and Singh, R.R., (1994) . Status of land degradation in India. *Jr. of Soil & Water Conservation.* 38 (3&4) : 163-172.

Sharma, U.C., (1992). Adequate use of fertilizer nutrients in North Eastern Hill region is the only viable alternative to achieve self sustainability in food grains in 2000 A.D., *Indian Fertilizer Scene,* Annual :113-122.

Sharma, U.C., (1995). Socio- economic aspects of acid soil management and alternative land use systems for North- Eastern states of India. In *Plant Soil Interaction at Low pH.* (R.A. Dates, et al (eds)): Kulwar Academic Press Netherlands, pp. 689-695.

Sharma, U.C., (1996). Yield sustainability & strategies in North Eastern states. *Productivity.* 36(4) :682-690.

Sharma, U.C., (1997). Balanced fertilization for improving productivity of north eastern region. *Fertilizer News,* 42:71-77.

Samra, S., Dhyani, S.K., and Sharma, A.R., (1999). Soil and Water Conservation strategies for sustainable agriculture in the north eastern hill region. Lead paper, Nat. Seminar on strategies for agriculture research in the North East. 10-12 Nov, ICAR Research Complex for NEH Region, *Abstracts* : XXIII-XXIV.

# 5

# TECH-ECONOMIC AND INSTITUTIONAL FACTORS IN THE DEVELOPMENT DYNAMICS OF AGRICULTURE IN NORTH-EAST INDIA

*M.P. Bezbaruah*

## Introduction

The North-East, comprised of the seven states of Arunachal Pradesh, Assam, Manipur, Meghalaya, Mizoram, Nagalnd and Tripura, constitutes a land surface of 255,083 square kilometres of hills and plains where, as per 2001 census, a population of 38.5 million belonging to different ethnic and cultural groups inhabited.[1] While the hills account for about 70% of area and accommodate about 30% of population of the region, the plains constituting the remaining 30% of area hold about 70% of its population. Though the overall density of population for the region works out to be 151 persons per square kilometre, it varies from 13 in mountainous Arunachal Pradesh to 340 in Assam, 80% of which is plain.

When India attained independence, the economy of the region was largely agrarian. Though a modern industrial sector existed, it was geared to colonial exploitation of the region's rich natural resource base and had very little linkage with the livelihood of the local population. Partition-inflicted isolation of the region coupled with some other factors prevented industrialization from taking off in the post independence period (Sarma 1966, Goswami 1981). Meanwhile the process of integration of 'excluded' and "partially excluded' areas—mostly the hill tracts inhabited predominantly by tribal population—into the modern administrative system of independent India resulted in emergence of small states and sub-states called autonomous/ district/regional councils within the region.[2] Establishment of new administrative machinery in these newly formed states induced a decade of service sector led growth from mid 1980s to mid 1990s. But that growth momentum has since petered away. Consequently the region today has remained more agrarian then the rest of the country in the sense that proportion of population dependent on agriculture for livelihood here exceeds the proportion for the country as a whole. Experts and policy makes are hence unanimous that the region's development depends critically on the development of its agriculture.

Due to the difference of topography, history and socio-economic organization, the evolution of agrarian structures and agricultural practices has taken different routes and shapes in the hills and the plains of the region. The problems of development of agriculture in the hills and the plains hence need to be discussed separately.

### The Issues Relating to Hill Agriculture

Traditionally shifting cultivation or *jhum* has been the dominant form of agriculture in the hills. When land used to be abundant and jhum cycles used to be long, the system was environmentally sustainable and it ensured reasonable food security in most times to the people living in the hills (Ramakrishnan, 1992). But over time, with growing population pressure and reduced availability of land for shifting, jhum cycles have become progressively shorter, and consequently the productivity of the system has declined and environmental degradation in the form of loss of forest cover and top soil has become increasingly pronounced (Dasgupta, 1996). Statistics show that area under *jhum* has decreased over the years and a gradual process of nomadic tribal village communities settling down by adopting cultivation of wet rice in valleys, various different crops in the terraces and other practices has been taking place. Yet the area affected by shifting cultivation still remains significant and interventions to win remaining shifting cultivators away to more sustainable ways of life need to be continued. Programmes and packages tried out as alternative to jhum cultivation has so far met with limited success (Borah 1993). The techno-economic viability of these programmes and their socio-economic acceptability to *jhum* cultivators is an area for debate and deliberation. Most of these suggested alternatives to *jhum* being of the nature of commercial farming, development of market connectivity, marketing organisation and management of risk associated with their practice are crucial factors for their success.

Traditionally land in the hills used to be held communally by a tribe or a clan and the user rights to land used to be distributed to families according to customary norms by the traditional authority, which may be the village council or the village chief. With the inroads of settled cultivation and urbanisation individual holdings started to emerge in many parts. But as these developments have not been followed up with cadastral survey and land settlement, evolution of land holding rights to full property rights has remained incomplete which limits the usefulness of land as collateral for financial transactions. While transition to alternatives of *jhum* such as plantation and horticulture requires infusion of investment, lack of progress toward document based property right regime can hinder the inflow of investment especially in the form of bank credit. Moreover as the old order of community ownership has decayed, absence of legal and institutional reforms has resulted in growing intra-tribal inequality in access to land and consequent emergence of landlessness (Ray and Kuri 2001). Though several scholars have raised and analysed these issues, these have not found adequate attention and treatment from policy makers.

## Agriculture in the Plains

In the plains, which consist of the Brahmaputra and the Barak Valleys of Assam, the Imphal Valley of Manipur and parts of Tripura, agriculture is settled in nature as it is in most parts of the country. Cropping pattern is dominated by foodgrains, mostly rice. Wheat, pulses and other cereals are of marginal importance. By all-India standards the average size of holding is small, but not smaller than in neighbouring West Bengal which has maintained a remarkable growth of crop production over the last two decades. Obviously the relatively lower yield of crops is due to other reasons such as inadequate penetration of productivity raising technology. Two obvious factors restraining the advancement of the HYV seed - fertiliser technology are high flood-proneness and low irrigation capacity. In the kharif season flood-proneness on the one hand renders shorter high yielding paddy varieties vulnerable and on the other hand investing in costly inputs like fertilisers a risky proposition. In the dry season the use of the HYV seed - fertiliser package is limited by small extent of irrigation coverage. In the Imphal valley, which is largely free from floods and has a better irrigation coverage, farmers apply inputs like fertilisers at much higher rate. Hence in contrast to the region as a whole the yield of rice in Manipur has been at levels comparable with the all-India levels.

Besides these physical constraints, there are several institutional factors holding back agriculture in the plains.

*(a)* In Assam plains tenancy is wide spread, mostly in the form of sharecropping, but it is almost completely informal. While at the national level incidence of tenancy has been declining, here, with the spread of education in farm families, a tendency to lease out land has been observed (Kuri 2003).

*(b)* Utilisation rate of public sector irrigation projects, which often suffer from lack of funds for maintenance, is abysmally low. Absence of the practice of collection of water charges and of effective water users' organisations does not help matters. The private tube well based irrigation systems are relatively better utilised. But holdings being typically of small size, even their capacity is not fully utilised. The water market has not come up extensively to facilitate better utilisation of tube well based irrigation systems (Dutta and Bezbaruah 2006). Following deployment of tube wells under RIDF and ARIASP (a World Bank aided

programme) in the 1990s there was a spurt in production of summer (*boro*) rice in the Brahmaputra Valley. But the failure to organise procurement resulted in a crash in the post harvest price and subsequently the production boom could not be sustained. Though farmers have increasingly taken to growing of vegetables, cold storage capacity has not come up concomitantly.

(*c*) There is urgent need of boosting both public and private investment. But delivery of institutional credit is very thin.

## Complementarities

Natural resource base wise, the farmers in the hills and the plains of the region can specialise in mutually complementary cropping patterns. In plain areas, which are better suited for the HYV seed—fertiliser technology, intensive use of better inputs with improved water management can greatly enhance production of foodgrains. Hills, being endowed with agro-climatic conditions for horticulture and other high value crops, can exploit the potentials of such commercial crops. The challenge therefore is provision of the infrastructure and the necessary organisational support, and putting in place the right institutions for converting such technical possibility to a socio-economic reality.

## Issues Requiring Further Probing

Some of the issues relating to agricultural development in the North East that call for further investigation and more intensive discussion can be listed as the following:

- Techno-economically viable alternatives to precarious hill agriculture systems currently in practice have been identified. At the same time it has also been recognized that successful transition to these alternatives will critically depend on efficient organization of the post harvest chain of storage, processing and marketing. But how this can be brought about in practice requires further discussion. In particular the desirability and feasibility of contract farming is worth more detailed probing.

- Institutional barrier in land and labour markets in hill areas have reportedly created hurdles for credit flow and resulted in potentially inefficient covert tenancy. These barriers are unlikely to be removed in the short run. The relevant question for the present time therefore is how best to tackle these issues within the existing set-up.
- Low penetration of technology and consequent slow growth of yield in the plains has been traced to weak rural infrastructure in general and poor road connectivity and paucity of irrigation in particular. Why these facilities have remained undeveloped in the North-East states in contrast to what has happened in similar physical conditions in neighbouring West Bengal deserves pondering. Is it merely an issue of governance and management or a deeper issue of institutional weakness and lack of enough grassroots empowerment?
- The issue of efficient organization of the post harvest chain is pertinent also for inducing diversification of rice centric agriculture in the plains to crops like potato and vegetables, which have shown some growth potential in recent years.
- Measures for addressing the problems of agriculture in the flood plains, such as enhancing access of farmers there to irrigation, storage and connectivity, deserve intensive discussion.
- Facilitation of expansion of the nascent water market with effective regulation with an eye to sustainable exploitation of ground water resource is another issue deserving attention.
- For growth in agricultural productivity in the longer run, the Approach Paper to the 11$^{th}$ Five Year Plan has called for prioritised research for extending the technological frontier (Government of India, 2006: p. 20). In that broad context, review of research priorities

for agriculture in the North-East region can be a worthwhile exercise.

## NOTES

1. In the 1990s Sikkim was added to these seven states in the denotation of the North-East of the Central Government and the state was included in the North-Eastern Council. However Sikkim shares neither geographical contiguity nor a common historical-economic evolution with the other seven states. Since the original seven states, form a more compact spatial-economic block, the North-East region here is viewed as comprised of these seven states only.
2. Under the British Raj hills of the region were classified as 'excluded' or 'partially excluded' areas depending on whether 'a tract of country was inhabited by a compact aboriginal population or the aboriginal population was mixed with other communities... Both these areas were excluded from the competence of the provincial and federal legislatures' (Agnihotri, 1996: 58) and the tribal communities living in such areas were allowed to continue with their traditional arrangements of self-governance.

## REFERENCES


Agnihotri, S.K. (1996), Constitutional Development in North-East India since 1947, B. Dutta Ray and S. P. Agarwal (ed) *Reorganisation of North-East India since 1947*, Concept Publishing Company, New Delhi.

Borah, D. (1993), Constraints of Agriculture Development in the Hilly Regions of North-East India, K. Alam (ed) *Agricultural Development in North-East India: Constraints and Prospects*, Deep and Deep Publications, New Delhi

Dasgupta, M. (1996), Crisis Facing *Jhumias* in Tripura, *Economic and Political Weekly*, January 6

Dutta, M.K. and Bezbaruah, M.P. (2006), Institutional Factors behind Variations in Utilisation Efficiency of Irrigation Projects in the Brahmaputra Valley of Assam, *Indian Journal of Regional Science* Vol. 38, No. 2

Goswami, Atul (1981), Assam's Industrial Development: Urgency of New Direction, *Economic and Political Weekly*, May 23

Government of India (2006), *Towards Faster and More Inclusive Growth: An approach to the 11th Five Year Plan*, Planning Commission

Kuri, P. K. (2003), Factor Market Imperfections and Explanation of Tenancy: Testing an Econometric Model using Evidence from Assam of North-East India, *Indian Journal of Agricultural Economics*, Vol. 58, No. 2, April-June.

Ramakrishnan, P.S. (1992), *Shifting Agriculture and Sustainable Development: an Interdisciplinary Study of North-East India* (Man and the Biosphere Series, Volume 10), UNESCO (Paris) and Parthenon Publishing Group.

Ray, N.C. and Kuri, P.K. (2001), *Land Reform in Arunachal Pradesh*, Classical Publishing Company, New Delhi

Sarma, J.N. (1966), Problems of Economic Development in Assam, *Economic and Political Weekly*, Vol. I, No. 7.

# 6

# INSTITUTIONAL FINANCE FOR DEVELOPMENT OF NORTH-EASTERN REGION: RETROSPECT AND PROSPECTS

*Gopakumaran Nair, G.*

The North Eastern Region (NER) is endowed with the flora and fauna, which are one among the most bio-diverse in the world. However, this worthy region of the country has been faced with various challenges, both natural and manmade, pulling back the developmental initiatives of the region. Institutional aid, albeit not in a coordinated way, were flowing to the region from time to time. However, there was no substantial improvement in the status of the region, which is still lagging behind on various counts when compared to other parts of the country. There is an imperative for planned efforts from various fronts to permeate these challenges being faced by this region. The region entail paramount attention for the development of agriculture and rural area since as high as 51 per cent in Mizoram to 89 per cent in Sikkim of the total population in the states stay in the rural areas and depend mainly on agriculture for their livelihood. The present paper, presented in 4 parts, analyses the relative backwardness of the area in terms of access to institutional credit for agriculture and offer plausible solution to mitigate the inequality that prevails.

## Relative Backwardness of the NER

NER is backward in terms of percapita income and people below poverty line when compared to the national average (Table 1). A comparison of column 3 with 4 exposes that, the states of Tripura, Nagaland and Sikkim although enjoy almost comparable percapita income of the country, share of people below poverty line is high in the range of 33.4 per cent to 37 per cent as against the national level of 26.1 per cent indicating high level of inequality. The region's population at 3.7 per cent of the total is increasing at faster rate - 25.29 per cent during 1991-2001 against national average of 21.54 per cent (*UA Shimray, 2004*).

**Table 1: Share of the NE States to the Country and Major Economic Parameters (2002-03)**

(%)

| *States* | *Share of agri. in SGDP* | *People below poverty line** | *Percapita Income (Rs. at 1993-94 prices)* | *State's share to All India* NSA | *Agri holding* | *Forest cover* | *Food-grain produ-ction* |
|---|---|---|---|---|---|---|---|
| 1 | 2 | 3 | 4 | 5 | 6 | 7 | 8 |
| Aruna. Pradesh | 26.2 | 16 | 8958 | 0.12 | 0.2 | 2.8 | 0.12 |
| Assam | 28.9 | 36 | 6520 | 1.94 | 1.9 | 9.0 | 1.92 |
| Manipur | 20.9 | 29 | 8751 | 0.10 | 0.1 | 2.4 | 0.18 |
| Meghalaya | 21.6 | 34 | 9727 | — | 0.1 | 1.4 | 0.11 |
| Mizoram | — | 19 | — | 0.07 | 0.1 | 2.2 | 0.07 |
| Nagaland | 33.3 | 33 | 11674$ | 0.21 | 0.4 | 1.4 | 0.15 |
| Sikkim | 21.0 | 37 | 11232$ | 0.07 | 0.0 | 0.9 | 0.05 |
| Tripura | 18.2 | 34 | 10255$ | 0.20 | 0.1 | 0.4 | 0.25 |
| All India | 24.0 | 26.1 | 11799 | 100 | 100 | 100 | 100 |

**Pertaining to 1999-2000, $ pertaining to 2001-02.*

*Sources: 1. RBI (2005), 2. CMIE, Various issues.*

Land available for agriculture purpose is limited, as forest cover constitutes more than 67 per cent against national average of 20.6 per cent. The region accounted for about 8 per cent of the total geographical area of the country but has only 3.4 per cent of total land put under agricultural purposes. Low echelon of the region is further enunciated by the fact that the region contributes only 2.8 per cent to total agricultural production in the nation as a whole. This confirms the low level of productivity of agriculture in the region as compared to the country. Various factors such as poor irrigational status of agriculture land, low level of input use, credit availability for agriculture, etc have heavily contributed to low level productivity in the region (Table 2). Investment in building irrigation capacity has been insufficient, ineffective and irrigation potential created has remained grossly underutilized down the years *(Mrinal Kanti Dutta, 2005)*. Small and marginal farmers dominate the land holding pattern in the region with more than 59 per cent of the holding below 1 ha. and 80 per cent in the category of below 2 ha *(CMIE, 2006)*. Use of electricity in agriculture (to total electricity used) is just less than 1 per cent in all the states except Tripura (18.7%) as against the national average of 25.3 per cent. One of the major upshots from Table 2 is the low level of credit availability to the area especially for agriculture.

**Table 2: Input Use in Agriculture in NER: A Comparison**

| *States* | *% of irrigated area (2000-01)* | *Fertilizer consumption (Kg/ha) (2003-04)* | *% share of agri. in electricity consumption* | *Credit availability 31 March 2003 KCC/1000 ha.* | *Credit (Rs/ Ha. GCA)* |
|---|---|---|---|---|---|
| 1 | 2 | 3 | 4 | 5 | 6 |
| Arunachal Pradesh | 16.4 | 2.9 | 0 | 10.0 | 152.9 |
| Assam | 5.6 | 47.5 | 0.52 | 15.1 | 482.6 |
| Manipur | 36 | 126.3 | 0.43 | 10.3 | 264.8 |
| Meghalaya | 22.6 | 19.5 | 0.7 | 22.0 | 2090.3 |
| Mizoram | 14 | 16.9 | 0 | 28.9 | 604.4 |
| Nagaland | 24.9 | 1.8 | 0.08 | 3.7 | 200.5 |
| Sikkim | 14.3 | 3.4 | 0 | 11.1 | 334.6 |
| Tripura | 13.9 | 37.1 | 18.74 | 24.6 | 867.1 |
| All India | 40.2 | 88.2 | 25.33 | 181.6 | 4942.4 |

*Sources :* 1.GoI (2004-05), 2. CMIE (2006), 3. NABARD (various issues)

## Financial Exclusion in the NER

Importance of adequate credit and its timely availability is of utmost importance in agriculture. In addition to providing control over key factors of production, credit also facilitate capital formation in agriculture and thereby improvement in productivity. In spite of directed credit supply in the region for a reasonably long period, credit available to the farming community is far from satisfactory. As may be noted from Table 2, availability Kisan Credit Cards (KCC)– an innovation in the farm credit products, per 1000 ha. of gross cropped area ranged between just 3.7 KCC per 1000 ha. in Nagaland and 28.9 KCC per ha. in Mizoram as against the national average of 181.6 KCCs (i.e., 2 % to 16 % of the national average only). Low level of credit access was also exemplified by the data contained in the NSS 59[th] round on indebtedness of farmer households. Level of indebtedness was lower than the national average in all the NE states except Tripura(Table 3). Infact, indebtedness in Meghalaya(4.1%) and Arunachal Pradesh (5.9%) were negligible, as against the all India situation of 48.6 per cent—a situation that warrant hectic inquisition.

**Table 3: Farm Household Indebtedness in NER**

*(Rs.)*

| *States* | *Farmhold indebted (%)* | *Avg. loan amount* | *Sources of borrowing (% to total)* | | | | | |
|---|---|---|---|---|---|---|---|---|
| | | | *Govt. sour-ces* | *Co-oper-atives* | *Banks* | *Money-lenders* | *Relatives /friends* | *Others** |
| 1 | 2 | 3 | 4 | 5 | 6 | 7 | 8 | 9 |
| Arun. Pradesh | 5.90 | 1493 | 6.1 | 0 | 20.8 | 0 | 50.7 | 22.4 |
| Assam | 18.1 | 813 | 7.0 | 2.7 | 27.8 | 15.5 | 24.7 | 22.3 |
| Manipur | 24.8 | 2269 | 1.5 | 0 | 16.7 | 32.9 | 40.1 | 8.8 |
| Meghalaya | 4.10 | 72 | 6.0 | 0 | 0 | 12.8 | 80.9 | 0.3 |
| Mizoram | 23.6 | 1876 | 24.3 | 3.1 | 49.9 | 0 | 19.3 | 3.4 |
| Nagaland | 36.5 | 1030 | 7.5 | 7.7 | 53.6 | 0.3 | 15.5 | 15.4 |
| Sikkim | 38.8 | 2053 | 34.8 | 0.0 | 23.0 | 7.3 | 6.7 | 28.2 |
| Tripura | 49.2 | 2977 | 16.4 | 2.8 | 60.5 | 2.0 | 11.9 | 6.4 |
| All India | 48.6 | 12585 | 2.5 | 19.6 | 35.6 | 25.7 | 8.5 | 8.1 |

*Include traders, doctors, lawyers, other professionals, etc.

*Source:* NSSO (2005).

Access to credit by the farmers with less than 0.4 ha land holding was almost negligible in states of Arunachal Pradesh, Mizoram, Nagaland and Meghalaya (less than 10 %). A close perusal of data *(NSSO, 2005)* reveals that except Tripura (46%) and Mizoram (33%), in all the NE states, less than 20 per cent of the indebted farmer households depended on institutional sources for credit. Except Manipur, where one-fourth of the total indebted farmer households depended on private money lenders, in all other states the dependence was less than 10 per cent. In majority of these states, farmer household's financial requirement was met through borrowings from friends, relatives and traders. The situation is perilous when we consider the average loan outstanding by the indebted farmers in these states. As against the national average of Rs.12,585 per indebted farmers household, average loan outstanding in NE states ranged between Rs.72 in Meghalaya to Rs. 2,977 in Tripura (varying from just 6 per cent to 24 per cent of the national average), inviting onerous attention for suitable rectification. The disparity in credit inclusion can be further exemplified by observing the trend in the state wise ground level credit (GLC) flow during the period from 1995-96 to 2003-04.

**Table 4: Ground Level Credit Flow for Agriculture and Allied Activities**

*(Rs. lakh)*

| *States/Regions* | *1995-96* | *% to All India* | *2002-03* | *2003-04* | *% to All India* | *CAGR* |
|---|---|---|---|---|---|---|
| A. Northern Region | 456645 | 20.73 | 1745488 | 2181869 | 28.7 | 21.6 |
| B. Eastern Region | 138178 | 6.27 | 386889 | 504740 | 6.7 | 17.6 |
| C. Central Region | 353968 | 16.07 | 1002957 | 1248734 | 16.4 | 17.1 |
| D. Western Region | 425381 | 19.31 | 914223 | 1012247 | 13.3 | 11.4 |
| E. Southern Region | 823957 | 37.40 | 2155794 | 2613740 | 34.4 | 15.5 |
| i. Arunachal Pradesh | 97 | 0.00 | 994 | 390 | 0.01 | 19.0 |
| ii. Assam | 3207 | 0.15 | 10127 | 19129 | 0.25 | 25.0 |
| iii. Manipur | 191 | 0.01 | 405 | 580 | 0.01 | 14.9 |
| iv. Meghalaya | 398 | 0.02 | 552 | 5184 | 0.07 | 37.8 |
| v. Mizoram | 115 | 0.01 | 684 | 544 | 0.01 | 21.4 |
| vi. Nagaland | 325 | 0.01 | 540 | 742 | 0.01 | 10.9 |
| vii. Sikkim | 98 | 0.00 | 325 | 425 | 0.01 | 20.1 |

| | | | | | | |
|---|---|---|---|---|---|---|
| viii.Tripura | 683 | 0.03 | 1759 | 3000 | 0.04 | 20.3 |
| F. North East Region | 5114 | 0.23 | 15386 | 29994 | 0.40 | 24.7 |
| All India* | 2203243 | 100 | 6955977 | 8698079 | 100 | 18.7 |

*Including other bonds, RIDF, disbursement by private sector commercial banks, etc.

*Source: NABARD & Rural Credit (various issues).*

A detailed data based analysis (Table 4) indicates that the share of different regions of total GLC in the year 2003-04 puts the Southern region at the top most position with a credit share of 34.4 per cent, followed by the Northern Region at 28.7 per cent, Central Region at 16.4 per cent, Western region at 13.3 per cent, Eastern Region at 6.6 per cent and lowest by NER at 0.4 per cent. The above positioning of the regions is on account of variations in the growth of GLC recorded by various states/regions. During the 9 years period under analysis, GLC in the country has registered annual compound growth rate of 18.7 per cent. Among various regions NE Region recorded the highest annual growth of 24.7 per cent followed by Northern Region (21.6%), Central Region (17.1%) and Southern Region (15.5%).

Though it is a welcome progress leading to the reduction in the inequality in the ground level credit flow, as compared to the share of area under cultivation (as explained earlier), the share of institutional credit flow in the region is still far from desirable. Reasons for the low level of credit flow in the region are further discussed in ensuing section.

## Factors Responsible for Low Level of Credit Flow in NER

Low level of GLC flow in the NER is the outcome of a variety of factors such as status of agricultural development, banking infrastructure, extent of repayment of loan, non performing assets of the banks, status of land reforms, low credit absorption capacity etc. Credit flow in the NER was adversely affected by weak banking infrastructure (Table 5) as well.

TABLE 5: BANKING INFRASTRUCTURE IN NER

| States | Bank branches (March 2000) | | | | Repayment$ | *Per branch performance (Rs. lakh) | | | |
|---|---|---|---|---|---|---|---|---|---|
| | CB | RRB | SCB | LDB | | Deposits | Advances | Business | Collection |
| 1 | 2 | 3 | 4 | 5 | 6 | 7 | 8 | 9 | 10 |
| Arunachal Pradesh | 49 | 19 | 31 | - | 30 | 647 | 466 | 1113 | 20 |
| Assam | 829 | 404 | 68 | 19 | 32 | 795 | 436 | 1231 | 25 |
| Manipur | 56 | 29 | 11 | 1 | 27 | 435 | 327 | 762 | 55 |
| Meghalaya | 128 | 51 | 37 | - | 31 | 1123 | 244 | 1367 | 21 |
| Mizoram | 26 | 54 | 9 | - | 54 | 1188 | 786 | 1975 | 129 |
| Nagaland | 64 | 8 | 22 | - | 55 | 677 | 211 | 888 | 77 |
| Sikkim | 42 | - | 1 | - | 76 | 512 | 350 | 863 | 29 |
| Tripura | 91 | 90 | 36 | 5 | 45 | 511 | 350 | 863 | 29 |
| All India | 51059 | 14461 | 742 | 1172 | 84 | 4976 | 4192 | 9168 | 2039 |

**for State cooperative banks for March 2005, $for State cooperative bank as percent of demand during 2004-05*

*Sources: NABARD (2001, &2004-05)*

The region with 8 per cent geographical area have only 2.5 per cent of the total number of commercial bank (CB) branches showing the extent of the area to be covered per branch. With the hilly terrain and sparse population, per branch business is also well short of national average. Weak cooperative structure and their non-viability has adversely affected the credit flow to agriculture in the region *(NABARD, 2001*). Efficiency parameters of state cooperative banks in the regions is well short of national averages on account of low level of repayment of loan, high level of NPA, lack of manpower in bank branches, unfamiliarity of the staff with local environment and culture, etc. Poor repayment of loans of state cooperative bank at less than 55 per cent of the demand (except Sikkim with 76%) when compared to the national average of 84 per cent not only reduced the fund available with banking system but also deterred the bankers.

One of successful innovations that has revolutionised the access of institutional credit among the less privileged section of the society is the micro credit innovation under the flagship of NABARD. One of the prominent features of the SHG Bank linkage programme in the country is the assertive role played by the Non Government Organization (NGO) as facilitator. However a major drawback in the NER is the lack of dominant and effective NGOs as they are still at budding stage and are facing resource crunch to operationalise the less attractive micro credit. Restricted operation by them also seal the chance of cross subsidizing micro credit operations with the surplus from other functions. The security oriented lending system followed by the banks in the region is not appropriate to the NER, where land is owned by the government or community and farmers are not equipped with land records to avail bank loan. Wide prevalence of tenancy system *(Roy, Niranjan and MP Bezbaruah(2002 and* shifting cultivation system(*jhum*) in the region were also not conducive to the banks for extending credit.

From the demand side, poor credit absorption capacity in the region has also adversely affected the GLC flow in the region *(NABARD, 2001).* Negligible share of area under assured irrigation (*Government of Assam, 2004*), small operational holding *(CMIE, 2006*), uneven terrain limiting the scope of farm mechanism, low level of input use, low level of technology adoption, etc had strong bearing on the demand for agricultural credit.

Overall infrastructure index in the region is well below 93 except Assam, which is having an index of 104 against the national average of 100(*CMIE, 2000*). Poor state of rural roads, which cater to the needs of transportation of agricultural inputs, poor connectivity with other part of the country *(Prabhakra, 2004),* weak infrastructure for post-harvest management and marketing linkages (*Dhar, 2002),* lack of ancillary units, cold storages, etc are factors responsible for generating enough demand for capital formation in the private sector. Social factors such as low level of literacy level, lack of awareness on the benefit and of modern high value agriculture, lack of proper training programmes for farmers, sensitive law and order situation (*Baruah Sanjib 2004*), increased corruption and criminalization of politics *(Hussain, 2003)* have also adversely affected credit flow.

## Credit Expansion in NER: Prospects

### *Building up Demand for Credit*

There are constraints on both demand side and supply side to the fostering up of institutional credit to agriculture in the region. Theses are to be tackled suitably for credit intensification. The agriculture in the region need to be modernized / diversified to improve the credit absorption capacity of the sector. Keeping in mind, the limited availability of land for cultivation and their constraints, increasing credit flow in the region is not possible without effective management of water and land resources. Considering the bio-diversity, soil and climatic conditions, which are congenial for growing high value crops like horticulture and medicinal plants, there is a need for devising appropriate regionally differentiated agricultural strategy to improve credit absorption capacity of the sector. As the growth of the sector, in future, will be on high value agriculture, improved agriculture extension by the government, agriculture universities, private entities, etc can go a long way in this area.

*Contract Farming*

Contract farming by involving private corporates has been proved to be an important tool for achieving crop diversification, introducing new/high-value crops, improved technology, promoting processing and exports, etc (*NABARD,2004*). The diverse agro-climatic condition conducive for variety, mostly horticulture, crops such as fruits, (banana, pineapple, orange, passion fruits, papaya, etc) spices, (ginger, garlic, black pepper, cardamom, chilies), sericulture, plantations like tea, rubber and bamboo, etc are suitable cultivation under contract farming arrangements. State specific advantages such as orchid cultivation in Arunachal Pradesh, mushroom cultivation in Manipur, spices in Nagaland, may be promoted under contract farming.

*Banking Infrastructure*

The region's dismal record of very low Credit Deposit Ratio, need amelioration through proper directives from the controllers. Resolute efforts are to be resorted to increase the outreach of credit by providing suitable banking infrastructure keeping in view the geographical constraints in the region. Suitable relaxation in the lending norms to match with the local tradition, land relations, land holding pattern, risk involved, etc are to be extended to enhance credit flow. Flow of bank credit to farmers, particularly to the small and marginal farmers in the NER need to be augmented through special schemes. Non existence/ inadequate land records which pose constraints for lending to the farming community especially for long term loans are to be overcome by efforts from governmental as well as banking side. Along with government's effort to improve land records, bankers may think of popularizing Joint Liability Groups, involving NGOs and community based organizations as banking correspondents to cover tenant farmers. As the role of farmers club(promoted by banks) in imbibing repayment ethics among the farmers and as a knowledge dissemination center has been well established elsewhere in the country, the same could be a replicated through initiative suitable for the region.

### *Alternative Credit Delivery*

There is a need to consider the way in which local community based organizations, NGOs could be used as banking correspondents and business facilitators in the wake of disturbing recovery performance in the region. Strengthening the capacity and the skills including training to enable the NGOs to be used by banks also deserve added importance. The concept of micro finance through SHG-Bank linkage programme has proved to be an effective tool in NER and has made enough inroads into the region. The recovery of loans under the programme is good, offering the potential of this concept. Number of SHGs credit linked in the region has shown tremendous growth from mere 93 in March 1999 to 12278 in March 2004(with a CAGR of 165.5% as against all India CAGR of 100.9 % during the period). Active involvement of government agencies is also warranted to overcome the shortage of suitable facilitating agencies in promotion of micro credit. As the NGO network in the region is very weak, grant based schemes available to the NGO, especially for skill and capacity building are to be implemented in the region on priority basis. Replicability of success of Farm Management Groups in Assam can also be deliberated upon.

## REFERENCES

Baruah, Sanjib (2003), Protective Discrimination and Crisis of Citizenship in North East India, *Economic and Political Weekly*, Vol. 38 No. 17 pp.1624-1626

Baruah Sanjib (2004), *Historising Ethnic Politics in NE India, Keynote address in the seminar on inter ethic conflict and ethenic crisis in NE India*, Dibrugarh University, 16-17 Feb 2004(unpublished)

CMIE (2000), *Economic Intelligence Service—Profile of districts*, Mumbai.

CMIE (2006), *Economic Intelligence Service—Agriculture*, Mumbai.

CMIE (various issues), *Economic Intelligence Service- Agriculture*, Mumbai.

Dhar, P.H. (2002), *The Economy of Assam*, Kalyani Publishers.

Government of Assam (2004), *Economic Survey, Directorate of Economics and Statistics*, Guwahati.

Governemnt of India (2004-05), *Economic Survey*, New Delhi.

Hussain, Monirul (2003), Governance and Electoral Process in India's North-East, *Economic and Political Weekly*, Vol 38, No. 10, Mar. 8-14, pp 981- 990.

Mrinal Kanti Dutta, and Madhunjya Prasad Bezbaurah (2005), Utilisation efficiency of irrigation Project in the Brahmaputra Valley: the Role of Market and Non Market Institutions, *7th NEEA Conference, held in Arunachal University* during 21–22 October 2005.

NABARD (2001), *NABARD's Initiatives for Development of NER, Mumbai.*

NABARD (2004), *Contract Farming as Means of Value added Agriculture (by Maharastra Economic Development Council), Mumbai.*

NABARD (2004), *Rural Credit and NABARD*, Mumbai.

NABARD (various issues), *Rural Credit and NABARD*, Mumbai.

NABARD (2004-05), *Key Statistics on Cooperative Banks*, Mumbai.

NSSO, 2005, Situation Assessment Survey of Farmers," Indebtedness of farmer households", NSS 59th Round, New Delhi.

Prabhakar, M.S. (2004), Is North East India Land Rocked, *Economic and Political Weekly*, Vol 39, No. 42, Oct. 16-22, pp. 4637-4643.

RBI (2004-05), *Handbook of Statistics on Indian Economy*, Mumbai.

Roy, Niranjan and M.P. Bezbaruah (2002), *Agricultural Growth and Regional Economic Development*, Mittal Publications, New Delhi.

U.A. Shimray (2004), Socio-political Unrest in the Region Called North-East India, *Economic and Political Weekly*, Vol 39, No. 42, Oct. 16-22, pp. 4637-4643.

# 7

# GROWTH AND DIVERSIFICATION OF AGRICULTURE IN THE NORTH-EAST INDIA

*K.C. Borah and D.K. Chakraborty*

## Introduction

North East Region (NER) of India comprises of seven states, viz., Arunachal Pradesh, Assam, Manipur, Meghalaya, Mizoram, Nagaland and Tripura. The region constitutes nearly 8.74 percent of geographical area of the country and possesses 3.8 percent population of the entire nation. Because of unique features of NER in respect of topography, about 75 percent of the geographical area of the region is hilly, which exhibits large-scale prevalence of shifting cultivation. In contrast, percentage of area under forest is as high as 51 percent in NER whereas the national average is only 22 percent. Though the overall land to man ratio for NER of 0.66 hectare per person is still favourable compared to the all India average of 0.32 hectare per person, but the percentage of cultivable area in the region (58.74 percent) is less than the national average (73.05 percent) and mono cropping is the predominant practice of cultivation. Unfortunately, despite having plenty of natural resources, NER could hardly achieve the desired economic development during the plan periods. As a result, the region continued to lag behind many other states of the country in the sphere of development even after completion of so many decades of planning. The economy of the region is generally characterized by the low level of per capita income, poor infrastructure, inadequate capital formation, inadequate exploitation of natural resources, slow industrial progress and prevalence of high unemployment. At present not even a single state of the region has a high per capita income than the national average. This assumes significance, because in 1997-98 Nagaland had a higher per capita income than the national average and in 1996-97 both Nagaland and Mizoram had higher per capita incomes than the national average.

The work participation rates for the rural areas of the region are not only meager but also shows widespread state wise variation. Whereas Mizoram has the highest rate of work participation, Assam has achieved the lowest position in total work participation rate in rural areas. Most of these workers are engaged in agriculture, some as agricultural labourers and others as the cultivators. Another feature of the workforce of the NER is noteworthy. The work participation rates of the rural areas show predominance of the female population, both in agriculture and household industry sectors. This makes it pertinent to examine the need for diversification of employment of female workers. Moreover the employment generation for peoples under SC, ST and OBC categories is not gearing up in the region.

Against this background, in this paper an attempt has been made to study the pattern of growth and diversification of agriculture in the NER. The present study is based on secondary data collected from various sources.

## Land Use Pattern

The pattern of land use of a region at any particular time is determined by the physical, economic and institutional framework taken together and above all by its pattern of cultivation. The pattern of cultivation depends mainly upon its physical characteristics like geographical location, topography and climate, its institutional framework like the social system and food habits, other resources available like capital and labour, etc.

**Table 1: Land Classification of the North East India, 1991**

| Categories | North East India | | All India |
|---|---|---|---|
| | Area ('000 ha.) | Percentage | (Percentage) |
| Reporting | 22536 | 100.00 | 100.00 |
| Cultivated | 5646 | 25.05 | 59.00 |
| Non Agricultural use | 1224 | 5.43 | 7.10 |
| Barren | 3351 | 14.77 | 6.40 |
| Forest | 11494 | 51.00 | 22.30 |
| Other Uncultivable | 821 | 3.65 | 5.20 |

*Source:* CMIE, September 1995.

It appears from Table 1 that percentage of cultivated area is 25.05 in the NER as compared to 59 per cent in the country as a whole. In contrast, area under forest cover is as high as 51 per cent in the NER, whereas the national average is only 22.30 per cent. However, the percentage distributions in all the states of the NER are not similar. Moreover the patterns of land holding systems in the NER are not same in all the states. The ownership system in the tribal societies of the hill areas is different from the ownership of land in the plains. So far as operational land holding is concerned, except Assam and Manipur, this has not yet been properly surveyed in other hill states.

**Table 2**

**Land Utilization Pattern in the North East India, 1998-99**

| States | Total Geographical area (000 ha) | Cultivable land (000 ha) | % of geographical area | Total crop area (000 ha) | Area sown more than once (000 ha) | Net sown area (000 ha) |
|---|---|---|---|---|---|---|
| Arun. Pradesh | 8374 | 293 | 3.5 | 250 | 65 | 185 |
| Assam | 7844 | 3410 | 43.47 | 3941 | 1239 | 2702 |
| Manipur | 2233 | 164 | 7.34 | 216 | 76 | 140 |
| Meghalaya | 2243 | 1082 | 48.24 | 263 | 45 | 218 |
| Mizoram | 2109 | 446 | 21.15 | 116 | 7 | 109 |
| Nagaland | 1658 | 619 | 37.33 | 286 | 25 | 261 |
| Tripura | 1049 | 309 | 29.46 | 488 | 207 | 281 |
| All NE | 25510 | 6043 | 23.69 | 5512 | 1657 | 3855 |
| All India | 328726 | 194715 | 59.23 | 192619 | 50019 | 142600 |

*Source: NEDFi Databank Quarterly*, Vol. 2, Issue 1, January 2003.

**Table 3: Average Size of Holding in the North East India (in ha), 1990-91 & 1995-96**

| States | Marginal | | Small | | Semi-Medium | | Medium | | Large | | Average Size of Holding | |
|---|---|---|---|---|---|---|---|---|---|---|---|---|
| | 90-91 | 95-96 | 90-91 | 95-96 | 90-91 | 95-96 | 90-91 | 95-96 | 90-91 | 95-96 | 90-91 | 95-96 |
| Arunachal Pradesh | 0.60 | 0.48 | 1.50 | 1.30 | 2.81 | 2.66 | 5.76 | 5.50 | 18.43 | 12.83 | 3.62 | 3.31 |
| Assam | 0.41 | 0.37 | 1.39 | 1.37 | 2.70 | 2.63 | 5.23 | 5.16 | 74.17 | 65.60 | 1.31 | 1.17 |
| Manipur | 0.55 | 0.57 | 1.37 | 1.37 | 2.56 | 2.57 | 5.01 | 4.67 | 12.16 | | 1.24 | 1.22 |
| Meghalaya | 0.54 | 0.49 | 1.38 | 1.25 | 2.54 | 2.41 | 5.46 | 4.50 | 13.00 | | 1.81 | 1.33 |
| Mizoram | 0.62 | 0.61 | 1.57 | 1.38 | 2.78 | 2.33 | 4.00 | | 0.00 | | 1.34 | 1.29 |
| Nagaland | 0.69 | 0.56 | 1.43 | 1.10 | 2.92 | 2.60 | 6.34 | 5.95 | 16.85 | 14.71 | 6.92 | 4.83 |
| Tripura | 0.40 | 0.33 | 1.53 | 1.40 | 2.69 | 2.50 | 5.14 | 6.00 | 121.57 | | 0.97 | 0.60 |
| All NE | 0.43 | | 1.41 | | 2.70 | | 5.56 | | 25.20 | | 1.60 | |
| All India | 0.40 | 0.40 | 1.44 | 1.42 | 2.76 | 2.73 | 5.90 | 5.84 | 17.30 | 17.21 | 1.57 | 1.41 |

Source: *NEDFi Databank Quarterly*, Vol. 2, Issue 1, January 2003 and *CMIE*, March 2005.

The proportion of cultivable area in the region is 63.79 per cent, which is lower than the national average of 73.23 per cent. States like Tripura, Assam and Manipur have higher percentage utilizations than the national average.

The following Table 3, gives information regarding operational holding and average size of holding in the state.

The average size of holding in the region is 1.60 ha. In 1990-91 compared to the all India average of 1.57 ha. The average size of holding declined in all the states of the region during 1990-91 and 1995-96. This indicates an increase in fragmentation of holdings in the region. It is evident from the Table that the average size of holding of most of the states is lower than the national average.

Table 4 reveals that except Nagaland and Manipur there has been an increase in the number of marginal holdings in other states of the region during 1990-91 and 1995-96. Arunachal Pradesh recorded a rise in the numbers of marginal, small, medium and large holdings during the same period; while in case of Assam, an increase may be noticed in all size groups except large holdings.

## Growth Rates

The NER recorded a growth in respect of Area, Production and Productivity of various agricultural crops. The compound growth rates in respect of area, production and productivity for important crops in the NER during 1984-85 to 1997-98 are presented in Table 5.

**Table 5: Compound Growth Rates (from 1984-85 to 1997-98)**

| *Crops* | *Compound Growth Rates* | | |
|---|---|---|---|
| | *Area* | *Production* | *Productivity* |
| Rice | 1.005 | 1.027 | 1.022 |
| Maize | 1.016 | 1.042 | 1.026 |
| Rapeseed & Mustard | 0.995 | 1.017 | 1.020 |
| Wheat | 0.974 | 0.990 | 1.017 |

*Source:* Calculated from *Basic Statistics of NER 2000*, NEC, Shillong.

**Table 4**

**No. of Holdings by Size Groups ('000 nos.) in the North East India (in ha), 1990-91 & 1995-96**

| States | *Marginal* | | *Small* | | *Semi-Medium* | | *Medium* | | *Large* | | *All Groups* | |
|---|---|---|---|---|---|---|---|---|---|---|---|---|
| | 90-91 | 95-96 | 90-91 | 95-96 | 90-91 | 95-96 | 90-91 | 95-96 | 90-91 | 95-96 | 90-91 | 95-96 |
| Arunachal Pradesh | 17 | 21 | 18 | 20 | 30 | 29 | 26 | 28 | 4 | 6 | 95 | 104 |
| Assam | 1451 | 1669 | 546 | 561 | 324 | 351 | 92 | 96 | 6 | 5 | 2419 | 2683 |
| Manipur | 69 | 69 | 49 | 49 | 21 | 21 | 3 | 3 | 0 | 0 | 142 | 142 |
| Meghalaya | 59 | 72 | 42 | 44 | 46 | 39 | 13 | 6 | 1 | 0 | 162 | 160 |
| Mizoram | 29 | 28 | 23 | 26 | 9 | 12 | 1 | 0 | 0 | 0 | 61 | 66 |
| Nagaland | 13 | 9 | 21 | 21 | 26 | 40 | 47 | 64 | 33 | 14 | 142 | 149 |
| Tripura | 217 | 247 | 69 | 40 | 28 | 12 | 4 | 1 | 0 | 0 | 318 | 301 |

Source: *CMIE*, March, 2005 and September, 1998.

A perusal of the estimated values indicates that rice productivity has recorded an increase of 1.022 per cent per annum in the region, whereas in case of maize, wheat and rapeseed & mustard the annual compound growth rates in productivity are 1.026, 1.017 and 1.020 respectively. Among the four crops considered, maize recorded the highest growth in productivity followed by rice. The same is the case with the growth of area under production and total production.

**Table 6: Area, Production and Yield of Foodgrain in North-East India, 1990-91**

| *States* | *Area under Food grain (000 ha)* | *Production (000 tonnes)* | *Yield (kg/ha)* |
|---|---|---|---|
| Arunachal Pradesh | 186.8 | 219.6 | 1176 |
| Assam | 2728.7 | 3532.1 | 1294 |
| Manipur | 172.5 | 390.7 | 265 |
| Meghalaya | 133.1 | 151.9 | 1141 |
| Mizoram | 77.5 | 133.8 | 726 |
| Nagaland | 199.5 | 211.9 | 1062 |
| Tripura | 272.7 | 556.1 | 2039 |
| All NE | 3770.8 | 5196.1 | 1378 |
| All India | 14509.5 | 199321.2 | 1601 |

*Source: NEDFi Databank Quarterly*, Vol. 2, Issue 1, January 2003.

The reliance on the production of food grains is seen in the region mainly for ensuring food security among the farmers. The area, production and productivity of food grain in the region are seen in Table 6. Tripura records the highest yield in food grain production whereas it is the lowest in Manipur.

## Cropping Pattern

Cultivation of rice is predominant in the entire region. About 63.18 per cent of the gross cropped area is used for the cultivation of rice alone. This indicates that, by and large, the cultivators of this region practice subsistence type of agriculture. Again, the crop intensity of the region is low, as double cropping and multiple cropping practices are not common. The area sown more than once worked out to be

16.38 lakh hectares, which is 37.30 per cent of the total net area. The rest of the 62.70 per cent of the cropped area used to raise one crop only. Besides rice, another important food crops grown in the region is maize. In the foot hills areas of the region, maize cultivation is suiTable and generally this crop is grown under shifting cultivation. Area under maize cultivation is little over 2 per cent of the total gross cropped area of the region. Except maize, cotton, sugarcane, spices, and the rest of the crops are grown in the plain areas of the region.

The area under food crops has increased marginally from 3729.30 thousand hectares in 1990-91 to 3756.30 hectares in 1997-98. Percentage shares of food grains and non-food grains in gross cropped area are reported in Table 7.

**Table 7: Percentage Share of Food Grains and Non-Food Grains in Gross Cropped Area**

| | *Food grains* | | | | *Non-food grains* | | | |
|---|---|---|---|---|---|---|---|---|
| *States* | *70-71* | *80-81* | *92-93* | *01-02* | *70-71* | *80-81* | *92-93* | *01-02* |
| A.P. | 43.18 | 88.65 | 71.88 | 69.81 | 56.82 | 11.35 | 28.12 | 30.19 |
| Assam | 74.90 | 74.73 | 71.25 | 71.03 | 25.10 | 25.27 | 28.75 | 28.97 |
| Manipur | 77.09 | 91.78 | 81.55 | 78.56 | 22.91 | 8.22 | 18.45 | 21.44 |
| Meghalaya | 59.08 | 55.78 | 55.15 | 51.24 | 40.92 | 44.22 | 44.85 | 48.76 |
| Mizoram | 4002 | 31.23 | 70.69 | 64.79 | 59.98 | 68.77 | 29.31 | 35.21 |
| Nagaland | 60.18 | 78.48 | 82.28 | 67.04 | 39.82 | 21.52 | 17.72 | 32.96 |
| Tripura | 78.55 | 73.58 | 58.87 | 59.32 | 21.45 | 26.42 | 41.13 | 40.68 |
| All India | 74.98 | 73.37 | 66.39 | 64.73 | 25.02 | 26.63 | 33.61 | 35.27 |

Source: *CMIE*, March 2005 and September 1998.

Except Tripura, other states of NER recorded a decline in the percentage share of food grains in gross cropped area during 1992-93 and 2001-02. Consequently, the share of non-food grains in gross cropped area recorded a marginal increase in these states.

During the same period area under sugarcane has declined from 44.00 thousand hectares to 35.80 thousand hectares. Area under vegeTables and tuber crops registered a decline from 260.03 thousand hectares to 189.40 thousand hectares.

The level of diversification of cropping pattern reflects the extent of economic development in the agricultural sector.

The level of crop diversification varies among region because of varied agro-climatic conditions and resource endowment of the farms. To examine the extent of crop diversification in the region, the Entropy Index (EI) is calculated. The EI for the NER for 1990-91 and 1997-98 are estimated as 1.172 and 1.083 respectively. The fall in the EI indicates concentration of agriculture in the region during 1990-91 to 1997-98. This downward trend diversification has got serious consequences on the sustenance of the rural economy.

## Yield Variability

A number of risks and uncertainties are faced by agriculture, which adversely affects investment and production decisions. Relative yield variability of major crops in Assam is presented in Table 8.

**Table 8**

**Mean Yield, Standard Deviation and Coefficient of Variation of Important Crops in Assam during 1972-73 to 2000-01**

| *Crops* | *Mean Yield (kg/ha)* | *Standard deviation* | *Coefficient of variation* |
|---|---|---|---|
| Rice | 1160.20 | 169.75 | 14.63 |
| Maize | 627.07 | 55.19 | 8.80 |
| Wheat | 1172.10 | 128.53 | 10.97 |
| Pulses | 445.03 | 66.67 | 14.98 |
| Rape & Mustard | 469.83 | 55.70 | 11.86 |

*Source:* Calculated from *CMIE*, December 2002.

The highest yield variability is found in Pulses (Coefficient of variation being 14.98), which can be attributed to the climatic conditions, damage caused by insects and late occurrence of flood among others.

## Conclusion

The northeast region has diversified agro-climatic and soil condition and offers congenial environment for the development of agriculture. But the region is lagging behind in this front from many states of the country. The reasons for low production and productivity of agriculture in the region are many and varied. Among them institutional, technical and natural factors are principal broad factors retarding the growth of agriculture in the region.

There is a necessity of increasing double-cropped area to increase the production of agricultural crops. With the introduction of micro irrigation through STWs the scope of increasing double and multiple cropping has raised in the region. Nevertheless, the productivity of most of the crops produced in the region is low in comparison to some other states of the country. Hence, efforts are needed for enhancing productivity of the crops through the introduction of HYV seeds, increase use of fertilizer and irrigation facilities.

The farmers of the region should go for a diversified cropping pattern. If carried out appropriately, diversification can be used as a tool to augment farm income, generate employment, alleviate poverty and conserve precious soil and water resources (Joshi, *et. al.*, 2004).

Marketing of agricultural produce is a gray area in the region's agriculture. The modern marketing infrastructure is somewhat non-existent in the region. Lack of infrastructure in the region makes the agricultural market 'middlemen dominated'. This results low returns and inadequate reinvestment in agriculture. Development of marketing infrastructure can go a long way in the development of this sector. With this end in view, the department of agriculture is trying to promote rural and terminal markets in the region. The Government of India's organizations like NAFED, TRIFED and NERAMAC are advised to intervene in the market during the peak arrival period of agricultural produce to prevent fall in prices. Setting up of necessary storage facilities in the form of grain storages and cold storages may give impetus to the growth of this sector.

Institutional credit is always viewed as an important instrument for generating output and enhancing productivity in the agricultural sector. But the institutional credit flow to the agricultural sector of NER is low in comparison to most of the states of India (Borah, *et.al.*, 2003). The credit deposit ratio for the region has been far from desirable over the years. The credit deposit ratio is not only low in comparison to the national average but also there is a sharp decline in recent years. This needs immediate attention of the banking sector.

Agricultural policies must take a holistic view in the above directions to enable the economy of the NER to realize its potential.

*Notes:* The Entropy Index (EI) of the following form is calculated in the present paper:,

$$EI = \sum_{i=1}^{N} P_i \log_e\left(\frac{1}{P_i}\right)$$

where $P_i$ = acreage proportion of the $i^{th}$ crop in total cropped area, N = total number of crops.

EI is regarded as an inverse measure of concentration having logarithmic character. The index increases with the increase in diversification and it approaches zero when there is perfect concentration (Kalamkar 2003).

## REFERENCES

*Basic Statistics of North Eastern Region 2000,* North Eastern Council, Shillong.

Borah, K.C. & Chakraborty, D.K. (2003): "Growth of Institutional Credit to Agricultural Sector in North East India", *Assam Economic Journal,* Vol. XVI.

Joshi, P.K., Gulati Ashok, Birthal Pratap & Tewari Laxmi (2004): "Agricultural Diversification in South Africa: pattern, Determinants and Policy Implications", *Economic and Political Weekly,* Vol. 39, No. 24.

Kalamkar, Shrikant S. (2003): "Agricultural Development and Sources of Output Growth in Maharashtra State", *Artha Vijnana,* Vol. XLV, Nos. 3-4.

*NEDFi Databank Quarterly,* Vol. 2, Issue 1, January 2003.

# 8

# AGRICULTURAL GROWTH AND DIVERSIFICATION IN EASTERN AND NORTH-EASTERN INDIA IN THE ERA OF ECONOMIC REFORMS

*P. M. Passah*

## Introduction

After the Independence of India in 1947, a planned attempt was made to bring about the agrarian transformation in the country with a view to improving the lot of the peasantry. The role of agriculture in the economy was then very vital as it contributed more than 55 percent to the GDP of the country and employed about 70 percent of the total workforce. As in most underdeveloped and developing countries, the agricultural sector the Indian economy was then most backward. Collin Clark (1982) attributes the cause of this backwardness of agriculture to the high density of agricultural population with low productivity. But it must be realized that immense improvements are possible in agricultural productivity in most parts of the world. What is needed is education and technological knowledge and course, capital. But this paper would not be directly confined to these requirements but to the agricultural growth and diversification of the agrarian economy of the Eastern and North-eastern India since Independence and more particularly since 1991 when economic reforms were launched in India. It must, however, be made clear that in the context of this Paper, Eastern India should include North-eastern India where it is not specifically mentioned.

## Agricultural Growth

### *Agrarian Reforms after Independence*

After Independence the most important reforms introduced in Indian agriculture dealt with *(i)* the abolition of zamindary (intermediaries) system, *(ii)* the introduction of tenancy reforms, *(iii)* the imposition of ceiling on land holdings, and, *(iv)* the distribution of the surplus lands among the landless and small landholders.

As a result of the abolition of intermediaries and the introduction of tenancy reforms, the area under tenancy declined fairly considerably in the four State of Eastern India during the 40 years from 1953 to 1993 as the following Table shows:

**Table 1**

**Area under Tenancy as % of the Total Area**

| *States* | *1953-54* | *1992-93* |
|---|---|---|
| Bihar | 12.40 | 3.91 |
| Orissa | 12.60 | 9.48 |
| West Bengal | 25.40 | 10.40 |
| Assam | 16.70* | 8.87 |
| All India | 20.55 | 8.28 |

*Assam figure relates to 1971-72(NSS 26th Round)

*Source: National Sample Survey on Landholdings*, NSS (8th Round 1953 and 48th Round 1992).

The percentage of area leased out varied from 11 to 26 percent in different parts of India, though the all- India average was 20 percent in 1953-54, which fell down to 8 percent by 1992-93 (Datt & Sundharam, 2004). Thus as the above Table shows the areas leased out in each State of Eastern India in 1992-93 was still above the all –India average.

The imposition of ceiling of existing landholdings appears to have failed in India due to mala fide and benami transfers, and proportion of small and marginal farmers obviously increased instead of decreasing. This is apparent from the trends of the average size of operational holdings shown in the following Table so far as the States in Eastern India are concerned.

**Table 2**
**Average Size of Operational Holdings in Eastern States**

| *States* | *1970-71 (ha.)* | *1990-91 (ha.)* |
|---|---|---|
| Bihar | 1.50 | 0.93 |
| Orissa | 1.89 | 1.34 |
| West Bengal | 1.20 | 0.90 |
| Assam | 1.47 | 1.31 |
| All India | 2.28 | 1.57 |

*Source:* Ministry of Agriculture, *Agricultural Situation in India* (1981) and *Agricultural Statistics at a Glance* (1995)

Although the extent of the area operated by the farmers with larger size of holdings in not readily available, it can be inferred from the above Table that a bulk of the peasantry subsists on marginal holdings and unless these peasants are either provided with alternative non- agricultural employment or are employed in medium and large farms by increasing their productivity, the chances of eradicating the rural poverty in Eastern India will remain bleak. Nevertheless, the position in West Bengal is different where security of tenure of the tenants and determination of fair rent has been successfully implemented (Datt & Sundharam, 2004). On the whole, the preponderance of small and marginal farmers with low production base continues to pose a great institutional challenge to the Eastern Region of India.

### *Technical Progress and Agricultural Growth*

Besides the agrarian reforms, the upgradation of production technology is another very important factor for agricultural development in India, which took place after Independence. This includes development of irrigation facilities, rural electrification, construction of roads and communication, establishment of agricultural universities, development of research infrastructure, seed- fertilizer technology, etc. It was these technological inputs that had led to the Green Revolution in 1970-71. But the revolution was limited to wheat, maize and bajra only and took place in Punjab, Haryana, western U.P. and in certain districts of Andhra, Maharashtra and Tamil Nadu.

The technological inputs were further upgraded. Hence in the second phase of the green revolution in the period from 1972-73 to 1982-83, the new technology had gradually extended to some parts of Eastern India and to the coastal areas of Andhra and Tamil Nadu in Southern India. The third phase from 1982-83 to 1990-93 marks a turning point in the agricultural development of India. During this Phase, agricultural growth had accelerated not only in the few irrigated regions but also extended to most regi0ons including the rainfed regions. The agricultural growth rate increased to an unprecedented level in the Eastern States particularly West Bengal which achieved the growth rate of 5.4 percent per annum (GOI). The following Table shows the agricultural output region wise from 1962 to 1999:

**Table 3**

**Region-wise Growth of Agricultural Output (1962-65 to 1990-93)**

*(in percent)*

| *Regions* | *1970-73 over 1962-65* | *1980-83 over 1970-73* | *1990-93 over 1980-83* | *1996-99 over 1990-93* |
|---|---|---|---|---|
| North West Region | 3.60 | 3.21 | 3.55 | 1.91 |
| Eastern Region | 1.57 | 1.09 | 3.63 | 1.25 |
| Centre Region | 0.73 | 3.13 | 3.33 | 4.50 |
| Southern Region | 2.40 | 1.38 | 3.43 | 2.19 |
| All India | 2.08 | 2.38 | 3.44 | 2.83 |

*Source:* G.O.I., Ministry of Agriculture, *Area and Production of Crops in India* (Various issues of the Report).

The above Table shows that Eastern Region had creditably performed among all regions during 1990-93 due to unprecedented growth in West Bengal.

The unprecedented growth in the output and yields of wheat and rice resulted to the self-sufficiency in foodgrains production in India during the 1980s. But the Table further shows the deceleration of agricultural growth rate during the 1990s, which hit the Eastern Region, the most.

### *Globalization and Deceleration of Growth During the 1990s*

The introduction of economic reforms failed initially to bring much buoyancy to the agricultural sector, which was somewhat, neglected. Therefore during the 1990s the trend of growth rate in agriculture has shown a significant deceleration as compared with the growth rate during the 1980s. The growth rate of the Gross Domestic Product from agriculture decelerated from 3.91 percent per annum during 1980-91 to only 1.94 percent per annum during 1991-99. The joining of WTO and the signing of GATT in December 1994 was expected that it would lead to increasing international trade in agriculture. The expected abolition or reduction of export subsidies by the developed countries was thought that it would bring about an increase in agricultural exports of developing countries. But that did not materialize and instead the agricultural sector in India is facing some serious problems. After high growth during early 1990s, deceleration had visibly appeared in the export at the end of 1996. This should have clearly affected the Eastern States whose agricultural growth had already slumped to the lowest level as shown in Table 3 above.

### *Productivity Improvement*

A major challenge before India is to reduce costs and raise productivity and quality of agricultural produce to global standards. In the long run, productivity increase has to be seriously considered, as the possibilities of expansion of farm area are minimal except the region stretching from north Bihar to Assam. The potentiality of expansion of farm area still exists in this region. (Passah, 2001) what is important is the enhancement of capital formation in agricultural productivity. It was very disheartening to observe that the Indian agriculture was starved of capital at the initial period of the economic reforms. That was the main reason for the subsequent deceleration of growth in agriculture. While the public sector investment declined, the private sector investment kept on fluctuating as can be seen from the following Table.

**Table 4**

**Public and Private Investment in Agriculture (1980-81 prices)**

| *Year* | *Public Sector (Rs. in crores)* | *Private Sector (Rs. in crores)* | *Total GCF in Agriculture* | *% Share of Agri. GCF in total GCF* |
|---|---|---|---|---|
| 1980-81 | 1892 | 2972 | 4864 | 18.9 |
| 1984-85 | 1822 | 3066 | 4888 | 14.8 |
| 1985-86 | 1631 | 3010 | 4641 | 12.1 |
| 1992-93 | 1179 | 4694 | 5873 | 11.2 |
| 1993-94 | 1269 | 4295 | 5564 | 10.8 |
| 1994-95 | 1426 | 5397 | 6823 | 9.6 |
| 1995-96 | 1394 | 6193 | 7587 | 9.9 |

*Source:* Ministry of Agriculture, *Agricultural Statistics at a Glance* (Various Issues of the Report).

From the above Table, it is clear that the percentage share of Gross Capital Formation (GCF) in agriculture keeps on decreasing until 1995-96 when it started increasing by a small margin. It must, however, be mentioned that public sector investment from 1980-81 till 1984-85 was always above Rs. 1,800 crore each year. Thereafter it declined while private sector investment kept on rising from 1987-88 onwards. The deceleration of agricultural growth has some serious repercussions. The region-wise data of capital formation are not readily available. But it could be surmised from the foregoing analysis that deceleration must be more serious in Eastern States with the exception of West Bengal. The main consequence of deceleration in agricultural growth is the falling income of the cultivators. Secondly, the expected gains from increased trade have not actually materialized because the growth rate of exports has decelerated. Further, there has been adverse impact on employment and trend reduction in poverty in rural areas.

## Agricultural Diversification

The main rationale of economic liberalization and globalization was the expected increase in agricultural exports of developing countries to the markets of developed countries because of increased market access. Since agriculture is labour-intensive, increased exports were expected to provide more productive employment to agricultural workers of developing countries not only in agriculture but also in the agro- processing industries. This in turn was likely to lead to gradual diversification of labour force from agricultural to non-agricultural occupations. But all this does not happen. The exports of major agricultural commodities from India did register and increase in the early 1990s, but this increase was short- lived. For instance, rice exports which had reached the level of 5(five) million tones in 1995-96 declined to 1.8 million tones by 1999-00. The slow-down in the exports of agricultural commodities from the country must have affected the exports from the Eastern region more since the deceleration in its agricultural growth had been the lowest among all regions as can be seen from Table 3 above.

## Crop Diversification

The deceleration in agricultural growth during the 1990s had not only resulted in the slow down in the exports of agricultural commodities but it had also painfully led to the slow process of workforce diversification away from low productivity agriculture in the country as a whole. The slow rate of agricultural diversification of rural workforce equally applies to crop diversification. The strategy in India initially was to raise foodgrains confining to two crops only, namely, wheat and rice, leaving aside other cereals and crops, the production of which remains stagnant for many years. India, the Eastern Region in particular, would do well to change its strategy in crop diversification. The entire region of Eastern India along the foothills from north Bihar to Assam and the Hills States in the North East is the potentially attractive area for crop diversification through horticulture development. The region could be converted into a National granary of horticultural produce with the potential to export these produce to other States in the country and abroad. The rest of Eastern India also has the potential of productivity increase if some more efforts and use of nutrients are adopted. With the brightening scenario of the plains in food crop production, the hilly regions in the north and northeast can go all out for horticultural development (Passah, 2001).

## Conclusion

It must be the aim of agricultural development to diversify the rural economy of Eastern Region, which is lagging behind the rest of India. The development of activities allied to agriculture would tremendously contribute to the diversification of agriculture in the region. There is also the need to develop rural industries based on agricultural outputs. The diversification would not only lead to higher incomes and employment of the rural population but would also result into even self-sufficiency and exporTable surplus of agricultural produce from this region.

## Rural Infrastructure

The entire rural infrastructure has to be refurnished to make rural activities viable. The Union Government has an ambitious plan to achieve Universal Rural Connectivity before 2007 through the National Rural Road Project. It is hoped the Eastern States are fully geared to fully implement the project. Besides rural transport facilities, minor irrigation works are to be built through public sector investment. Both these components of infrastructure will help the exploitation of extensive waste or fallow lands in Eastern India. The recent law passed on Rural Employment Guarantee Schemes and the Food- for- work Programme, if sincerely implemented and intensified, would not only lead to gradual diversification of labour force but also help in better food distribution and creation of durable rural assets.

## NABARD and RIDF

The Union government had created a corpus fund Known as the Rural Infrastructure Development Fund (RIDF) in NABARD with an initial amount of Rs. 2000 crores to extend loans to State governments to enable them to complete the implementation of the existing rural infrastructural projects relating to irrigation, rural roads, rural bridges, watershed development, flood development, inland waterways, etc. The RIDF has since grown to a massive Rs. 23,000 crores (Datt and Sundharam, 2004). It all depends on the capability of the Eastern States to fully draw or utilize funds from RIDF in order to gain maximum benefits from the Fund.

Both agricultural diversification and building of rural infrastructure would create demands in rural areas for more specialize agro-processing, more sophisticated trading and repair services and more professional education, health and personal services. They would also mitigate the excessive rural-urban migration that has been witnessed by the slums and shantytowns in various cities. A prosperous agriculture is therefore the only determining factor for jobs for rural poor. While the ECONOMIC Reforms Policy of the country has been directed towards Liberalization, Privatization and Globalization, the public sector policy has still to play a leading role in accelerating the pace of development in Eastern Region of India. And, for this to happen, a responsive government in each State of the Eastern Region has to keep in mind the welfare of the masses.

## REFERENCES

Collin Clark's Paper in A.N. Agarwal and S.P. Singh (ed.), *The Economics of Underdevelopment*.

Datt, R. and K.P.M. Sundaram (2004), *Indian Economy*, (S. Chand and Company Ltd.). p. 552.

G.O.I., Ministry of Agriculture, *Area and Production of Crops in India* (Various Issues of the Report).

Passah, P.M. (2001), "Agricultural Growth and Diversification in N.E. India: A Keynote Address" in the *Proceedings Volume of the 4th Conference of the North Eastern Economic Association* held on September 7-8, p. xix.

# 9

# ECONOMIC UPLIFTMENT THROUGH WATERSHED DEVELOPMENT IN NORTH EASTERN REGIONS OF INDIA

*R.P. Singh, Ranjit Kumar and N.P. Singh*

## Introduction

The North-East of India constitutes the seven 'sister' states of Arunachal Pradesh, Assam, Manipur, Meghalaya, Mizoram, Nagaland and Tripura. Flanked by hills and with the mighty Brahmaputra river slashing a central path between its north and south, the North East is bounded by the states of Manipur, Meghalaya, Mizoram, Nagaland, Tripura with Assam comprising the heart. This region of the country is mostly hilly and accounts for 7.8 % of the total land of the country. Steep slope in this region is being used for growing cereals, pineapples and tuber crops without using proper soil and water conservation measures. Sloping land is an important factor determining the feasibility, suitability and economics of land use options. As per the recommended soil conservation practices, land having more than 33 per cent slope is not suitable for arable farming (Dhyani *et. al.*, 1997).

The economy of the region is still primarily agrarian. The contribution of agriculture to state domestic income is much higher in this region, except for Meghalaya and Nagaland, where mining and forestry & logging, respectively are the important contributors to NSDP. Out of the 25.51 lakh hectare of the total geographical area of the region, net sown area accounts for 17.38 per cent and forests 52.38 per cent of the region's reporting area. Majority of area being hilly in the regions, even though population density is very low, the average size of operational holding is quite small and about 78 % of the farmers in the region belongs to small and marginal category. An important feature of land use in the region is that almost 16 % land area is barren and uncultivable. This could be partly brought under afforestation through suitable programmes (Government of India, 1983). The limited availability of land for cultivation thus suggests that for raising the agricultural production in the region, the emphasis have to be on raising the productivity by effective utilisation of water-seed-fertilizer technology.

Two distinct types of agriculture are practiced in the NE region: 'Settled' and 'Shifting' or '*Jhum*'. Settled agriculture is practiced in the plains, valleys, foothills and on terraced land on hill slopes. *Jhum* cultivation is practiced by the tribals in the hill areas of all the northeastern states. The area under *Jhum* cultivation in the NE is estimated at 30 % of the total area under settled agriculture. Among the individual states, Nagaland has the largest area under *Jhum* cultivation, where around 0.1 million hectare land or about 35 % of its forest land are under *Jhum*. With the Government's efforts, the total area under *Jhum* cultivation in the NE region has come down significantly from 2.7 million hectares in 1974 to 1.4 million hectares by mid-eighties. It has been estimated that nearly 18.6 million tones of soil, 0.6 million tones of organic carbon, 9.7 tonnes of available $P_2O_5$ and 5690 tonnes of available $K_2O$ are lost annually as a result of shifting cultivation in North Eastern Hills (NEH) region of India (Singh and Singh, 1981). Keeping the colossal loss accruing due to this, it was imperative to empirically demonstrate as to how newer soil conservation and irrigation measures can add to the increased agricultural productivity, thereby more income and restoring sustainability in the long run.

## Economy of North Eastern (N.E.) Region

The pace of development in the hilly areas and plains of northeastern region of India differ considerably. The valleys are economically active areas of the region and the

Brahmaputra valley the most active. Tribal population forms only one fourth of the population of the North East, despite the fact that in four states i.e., Mizoram, Meghalaya, Nagaland and Arunachal Pradesh, tribals are in majority and in Mizoram, they constitute as high as 95% of the population.

There are differences among the seven states in the North Eastern region with respect to their resource endowments, level of industrialisation as well as infrastructural facilities. The industrial sector has mainly grown around tea and timber in Assam and mining; saw mills and plywood factories in other parts of the region. The economy of the region is still primarily agrarian. The contribution of agriculture to state domestic product (SDP) is high in this region, except for Meghalaya (mining) and Nagaland (forestry and logging). On the other hand, growth in per capita income has almost stagnated in Assam and Arunachal Pradesh since the nineties, as the growth in per capita SDP in these two states during last decade had been far behind that of country as a whole as well as other sister states. The growth in per capita SDP has been appreciable to the tune of double digit in states of Tripura, Mizoram and Meghalaya (Table 1).

**Table 1: Per Capita State Domestic Product in N.E. Region (at current prices)**

*(in Rs.)*

| *Year* | *Arunachal* | *Assam* | *Manipur* | *Meghalaya* | *Mizoram* | *Nagaland* | *Tripura* | *All India** |
|---|---|---|---|---|---|---|---|---|
| 1990-91 | 5398 | 4281 | 3976 | 4375 | 4479 | 4990 | 3370 | 5365 |
| 1995-96 | 10956 | 7001 | 6894 | 8641 | 10953 | 11057 | 6828 | 10149 |
| 1996-97 | 10816 | 7394 | 7912 | 9243 | 12210 | 11805 | 8109 | 11564 |
| 1997-98 | 11642 | 7966 | 8804 | 10270 | 12393 | 13052 | 9658 | 12707 |
| 1998-99 | 12955 | 8826 | 9733 | 11914 | 13479 | 12408 | 11012 | 14396 |
| 1999-00 | 13669 | 10080 | 11059 | 13088 | 14909 | 12594 | 13195 | 15625 |
| 2000-01 | 14699 | 10718 | 11047 | 14632 | 18491 | 17629 | 15253 | 16729 |
| 2001-02 | 15701 | 11132 | 12683 | 15813 | 19704 | 18911 | 17383 | 17883 |
| 2002-03 | 15589 | 12074 | 13658 | 16803 | 22207 | NA | 18676 | 18988 |
| 2003-04 | 17393 | 13139 | 14766 | 18135 | NA | NA | NA | 21142 |
| CGR, % (1990-2003) | 9.43 | 9.58 | 10.89 | 13.24 | 14.99 | 11.95 | 17.82 | 12.46 |

* Per capita Net National Product, NA= Data not available

*Source:* Economic Survey, 2005-06, Government of India.

## Cropping Pattern in N.E. Region

The net sown area is hardly 14 per cent of the total reported area in the region. Agriculture is the main occupation of the people. The average size of holding in North East Region is more than the national average and it is 4.8 ha. The percentage of cultivable area over the geographical area in most of the NE State (22.09 %) is less than the national average (46.12%). It is highest in Assam (50.44%) and lowest in Arunachal Pradesh (3.20 %). Mono cropping is the predominant method of cultivation. In the absence of multi-cropping, little or no surplus is generated in the agricultural sector. Among all the NE States, Assam takes the first position both in terms of available cultivable area and net sown area. From a perusal of Table 2, it is evident that paddy is the most important crop and occupies more than 82 per cent of the total area under foodgrains, followed by pulses (5.7%), maize (3.5%) and wheat (2.6%).

**Table 2. Cropping Pattern in North Eastern Region, 2003-04**

*(per cent to GCA)*

| *States* | *Gross cropped area* | *Paddy* | *Maize* | *Small millets* | *Wheat* | *Food grains* | *Total pulses* | *Oil-seeds* | *Cropping intens-ity (%)* |
|---|---|---|---|---|---|---|---|---|---|
| Arun. Pra. | 268.00 | 44.48 | 14.40 | 8.32 | 1.57 | 71.49 | 2.72 | 9.89 | 163.40 |
| Assam | 3981.00 | 63.55 | 0.49 | 0.21 | 1.76 | 68.89 | 2.88 | 7.52 | 144.70 |
| Manipur | 216.00 | 72.96 | 2.04 | - | - | 77.69 | 2.69 | 0.88 | 154.30 |
| Mehgalaya | 277.00 | 39.60 | 6.10 | 0.97 | 0.32 | 48.70 | 1.70 | 3.54 | 120.40 |
| Mizoram | 118.00 | 50.17 | 8.90 | - | - | 63.22 | 4.15 | 6.36 | 100.00 |
| Nagaland | 378.00 | 39.74 | 9.47 | 2.65 | 1.46 | 60.00 | 6.69 | 12.49 | 113.50 |
| Tripura | 423.00 | 56.19 | 0.45 | - | 0.28 | 58.89 | 1.96 | 1.25 | 151.10 |
| N.E. Region | 5661.00 | 59.41 | 2.26 | 0.76 | 1.44 | 66.90 | 3.02 | 7.02 | 135.3 |

*Source:* Directorate of Economics & Statistics, Ministry of Agriculture, GOI.

Besides these main crops, the N.E. region has also been identified as potential area for coffee cultivation. However, not much headway has been achieved in this direction. Till date, around 8000 ha area is under coffee cultivation in this region with maximum in Nagaland state. Similarly, Arunachal

Pradesh, Assam, Manipur, lower reaches of hills of Meghalaya, Mizoram, Nagaland and Tripura have been identified as potential areas for rubber plantation. Therefore, with the help of Government and other public sector organisations, the area under rubber plantation has expanded to the tune of around 40 thousand hectares, with maximum area in Tripura followed by Assam state.

Tea is grown in almost all parts of Assam. The most important feature in the growing commercial importance of the State has been the remarkable expansion of the tea industry since the early part of the 20th century. Assam tea has been significantly contributing to India's Foreign Exchange earnings. Over the years, the North East including the seven North Eastern States, Darjeeling Dooars, Terai and now, Sikkim, have been responsible for the production of nearly 77 % of Indian tea, with Assam in itself producing more than a hefty 50 %. As of now, about 236 thousand hectare area is under tea plantation in total north east states.

## Water Resources and Irrigation Potential in NEH Regions

The irrigation potential of the NE states is enormous. According to the Central Water Commission, the total irrigation potential of the NE states is estimated at 36.65 lakh hectares. Among all the NE States, Assam has the highest irrigation potential of 2670 cu. km. However, both in Assam and Arunachal Pradesh, the ratio of gross irrigated area to gross cropped area has declined over the years. The ratio has remained constant for Meghalaya, while it has gone up in Nagaland and Mizoram, and improved marginally in Tripura and Manipur. Decline in the ratio of gross irrigated area to the gross cropped area in the NE region as a whole indicates that many secondary crops are no longer irrigated due to absence of adequate facilities. However, if the region's potential is suitably harnessed, adequate irrigation facilities can be provided for the entire cropped area of the region.

In Arunachal Pradesh, the ratio of net irrigated area to net sown area has not changed much over the last twenty years. In 2001-02, Manipur had an impressive 46.40% of its net sown area under irrigation, which was higher than the national average of 39.50%. In Mizoram, Manipur and Nagaland too there has been growth in irrigation over the years. Significantly, in the NE region, there is considerable difference between the ratio of gross irrigated area to the gross cropped area, and net irrigated area to the net sown area. This is because most of the agricultural land in the region receives irrigation water only once a year.

Singh (1993) studied the effect of land use and conservation measures on the soil loss and run off on the steep slope of NEH region of India. It was found that among all the land uses like agri-horticulture, agriculture with bench terrace, agriculture with contour bunds and secondary bamboo forest, except shifting cultivation, all were effective from soil conservation point of view. With appropriate measures of soil and water conservation, about 80 -100 per cent of annual rainfall may be retained or conserved in the deep soil profile of the watershed and/or allowed to percolate up to the groundwater or streams or ponds situated below, even if the slope is very steep and rainfall is high. This practice will greatly add to the irrigation requirement and will drastically reduce the soil erosion in the hilly as well as plain regions.

## Land Use System and Watershed Management in N.E. Regions

Major land use systems practiced in NEH region are shifting cultivation for growing cereals and other food crops, 'Bun' for tuber crops agriculture on bench terraces, agriculture on contour bedded slopes, agriculture in valley lands, agro-forestry, animal husbandry and horticulture. Most of the land use systems *viz.* shifting cultivation, Bun, unrestricted grazing in case of animal husbandry and cultivation of slopes with contour bunding results in heavy soil erosion, loss of fertility, environmental and ecological imbalance. However, there are pockets where effective indigenous land and water management techniques have been developed and used. But such systems are confined to limited area only. Bench terrace cultivation in Nagaland, valley land cultivation in Aptani plateau in Arunachal Pradesh and bamboo drip irrigation system in Jowei district of Meghalaya are examples of some indigenously developed technologies for land and water management.

Various schemes were launched by the development agencies of State as well as Central Governments and North East Council to replace the shifting cultivation in the form of *Jhum*ia Colonization, regrouping of villages, introduction of terrace cultivation on the hill slopes, introduction of horticultural and/or plantation based economy in place of agro-based economy, etc. But, most of them got very little success due to lack of need -based components in the technologies.

Watershed based land and water management mainly involves *in-situ* retention of rainfall for crop production and safe disposal of excess rainfall without causing the soil erosion and affecting the crop production. In the hilly region, micro-watersheds are appropriate basic unit for managing the land and water resources.

## Various Farming System Under Watershed Development—Case Study

The experiment was conducted in the farming system research block at ICAR Research Complex for North Eastern Hill Region (Umiam) in the Meghalaya State. It is situated in the central part of Meghalaya in the East Khasi hill districts near Shillong. Before the treatments, all the micro-watersheds were under natural vegetation with top portion under pine forest and lower portion was under tall grasses and bushes providing 100 per cent coverage to the soil surface.

Considering potentiality of various land uses, the experimental watershed was divided into eight micro-watersheds having the top of hill slope under pine forest and lower portion of the hill slope under planned land use system as shown in Table 3. Various land use system selected for the study were livestock, forestry, silvi-pastoral, agri-horti-silvi-pastoral, horticulture, natural fallow and timber farming system. Treatments were imposed in 1984. Contour bunds, bench terraces, half moon terraces, contour trenches and grassed waterways were the major mechanical soil and water conservation measures adopted in various watersheds.

Details of farming systems and conservation measures adopted in various micro-watersheds are given in Table 3. Bench terraces, contour trenches and bunds were planned in such a way that the vertical interval did not exceed 1.5 metre and all the terraces drained their excess run-off into a common grassed water way. Contour bunds in silvi-pastoral and horticultural systems were planned at 5 metres vertical intervals with a view to cut off the path of surface water flow and divert the same to grassed waterways. Small water storage pools were created with the use of gully plugs in main watercourses so as to act as silting basin.

*Livestock based farming system (FS-W1):* Out of total 1.39 ha land under this micro-watershed, 0.80 ha was put under cultivation of perennial grasses on the steep slope of 32.02 per cent. Milch cows were maintained on the biomass produced. Continuous type contour trenches of 0.5 x 0.5 $m^2$ average cross section at 1.5 m vertical interval have been used as soil and water conservation measures.

*Forest farming system (FS-W2)*: Total area was 3.89 ha with an average slope of 38 per cent. No conservation measures were used in this micro-watershed to allow natural retention of rainwater with in the forest plantation.

*Silvi-pastoral farming system (FS-W3):* Out of total area of 2.94 ha, 2.04 ha was under pine forest and 0.90 ha under planned silvi-pastoral system at the slope of 32 per cent without any conservation measures in the system.

*Agro-pastoral farming system (FS- W4):* An area of 0.41 ha under terraces was put under main crops while area of 0.33 ha was under perennial fodder grasses. Bench terraces in lower reaches and contour bunds in upper reaches in the micro-watershed at 1m vertical interval have been used as soil and water conservation measures.

**Table 3: Crops and Conservation Measures Under Various Watershed Based Farming Systems**

| *Farming System* | *Crops* | *Conservation Measures* |
|---|---|---|
| Livestock (FS-W1) | Guinea grass, broom grass, maize, field beans and oats | Contour trench |
| Forestry (FS- W2) | *Michelia obloga, Symangtonia populnea* and *Alnus nepalensis*-Timber & fuel trees | Nil |
| Silvi-pastoral (FS- W3) | *Symangtonia populnea, Bauhinia purpurea, Thysaolaena maxima* (Broom grass), and *Brachiaria ruziensis* (Congo-sinal) | Nil |
| Agro-pastoral (FS- W4) | Maize, paddy, groundnut, soybean, French bean, linseed, ginger, oats, radish, mustard, Broom grass and guinea grass | Bench terrace & Contour bunds |
| Agro-horti-silvi-pastoral (FS-W5) | French bean, maize, ginger, mustard, radish, chilli, sweet potato, Orange, lemon, pear, Pine trees, *Symangtonia populnea, Alnus nepalensis,* Broom grass and guinea grass | Contour bunds bench terraces & small water pools |
| Horticulture (FS-W6) | Orange, lemon, guava, radish, chilli, sweet potato and French bean | Contour bunds, bench terraces & small water pools |
| Natural Fallow (FS-W7) | Natural vegetation dominated by pine trees | Nil |
| Timber Forestry (FS-W8) | *Michelia champaca* | Nil |

*Source:* Annual Report, 2001-02, ICAR Research Complex for NEH Regions.

*Agro-horti-silvi-pastoral farming system (FS-W5):* The farming system on a 41.7 per cent hill slope in an area of 0.93 ha had 3 components, crop production (0.2 ha) on lower terraced land, horticulture (0.29 ha) on middle portion with half moon terraces and silvi-pastoral system (0.44 ha) on upper portion. Bench terraces at 1.0 m vertical interval and contour bunds at 0.5 m vertical intervals were used as conservation measures.

*Horticulture farming system (FS- W6):* Out of total area (3.13 ha) of this micro-watershed, 2.17 ha was under pine forest and 0.96 ha was under planned land use with average slope of 53.2 per cent. Bench terraces at 1.0 m vertical interval in the lower portion of planned land use were used as a conservation measure.

*Natural fallow system (FS –W7):* Average slope of this micro-watershed (1.03 ha) is 45.9 per cent. No soil and water conservation measures were used. This was fenced to check the biotic interference and allow the natural regeneration of vegetation.

*Timber forestry farming system (FS–W8):* The shifting cultivation was planned in this watershed. Total area and average slope of micro-watershed were 0.52 ha and 41.4 per cent, respectively. No conservation measures were used.

*Farm Pond:* A pond with total catchment area of 5.74 ha was constructed at the foot hills to harvest the water from agri-horti-silvi-pasture, horticulture and, natural farming systems. The capacity of the pond is one hectare meter.

**Table 4. Details of Experimental Watersheds Under Various Farming Systems**

*2001-02*

| *Particulars* | *Live-Stock* | *Fore-stry* | *Silvi-pasture* | *Agri-pasture* | *Agri-horti-silvi-culture* | *Horti-culture* | *Natural fallow* | *Timber farming* |
|---|---|---|---|---|---|---|---|---|
| *Total area (ha)* | 1.39 | 3.89 | 2.94 | 0.64 | 1.58 | 3.13 | 1.03 | 0.52 |
| 1. Forest land (ha) | 0.45 | 3.05 | 2.05 | 0.06 | 0.55 | 2.17 | 0.08 | 0.02 |
| 2. Not suitable for cultivation (ha) | 0.14 | - | - | - | 0.10 | - | - | - |
| 3. Net cultivated area (ha) | 0.80 | 0.84 | 0.90 | 0.74 | 0.93 | 0.96 | 0.95 | 0.50 |
| Average slope (%) | 32.02 | 38.6 | 32.18 | 32.42 | 41.77 | 53.18 | 45.87 | 41.35 |
| Max. length (m) | 301 | 320 | 295 | 240 | 260 | 515 | 250 | 185 |
| Max. width (m) | 65 | 230 | 175 | 65 | 85 | 85 | 70 | 48 |

*Source:* Annual Report, 2001-02, ICAR Research Complex for NEH Regions

## Run-off and Soil Loss Under Various Farming Systems

Table 5 clearly reveals that all the farming systems except agri-pasture generated less soil loss than the tolerance limit of 11 t/ha/annum. The agri-pasture generated 13.85 t/ha/

annum soil losses due to erosive crops such as maize and paddy occupying more area. The farming systems like forestry, natural fallow and timber farming generated least soil loss due to conservation friendly forest trees. As far as run-off is concerned, all the farming systems except livestock based silvi-pasture and agri-pasture based systems generated fewer run-offs than tolerance limit of 88 mm/ha/annum. The greater runoff in livestock based, silvi-pasture and agri-pasture based systems was due to the occupation of area under more runoff generating crops such as paddy, napier, guinea and congo-signal crops. While due to forest trees, the runoff is less in farming systems forestry, natural fallow and timber farming.

**Table 5. Comparison of Run-Off and Soil Loss of Different Farming Systems**

*(per hectare per annum)*

| *Particulars* | *Live-Stock* | *Fore-stry* | *Silvi-pasture* | *Agri-pasture* | *Agri-horti-silvi-culture* | *Horti-culture* | *Natural fallow* | *Timber farming* |
|---|---|---|---|---|---|---|---|---|
| Run-off (mm) | 158.59 | 32.00 | 106.79 | 174.01 | 94.68 | 37.12 | 8.10 | 32.00 |
| Soil loss (ton) | 9.08 | 0.80 | 2.70 | 13.85 | 5.33 | 5.04 | 0.02 | 0.45 |

*Source:* Thimmappa (2001).

## Cost and Returns from Different Farming Systems

The study conducted showed that the total returns varied in different farming systems ranging from Rs. 2853 per hectare in natural fallow to Rs. 94326 per hectare in livestock farming system (Table 6). The total cost was least in natural fallow (Rs 194/ha) and the highest (Rs. 58905/ha) in livestock based farming system. Net income was highest in horticulture (Rs. 40115/ha) and lowest in silvi-pasture farming system (Rs. 6746/ha).

In contrast to this, a study conducted for the same micro-watersheds by Singh (1995) revealed that only two farming systems namely Agriculture land use (FS-W4) and Agri-horti-silvi-pastoral land use (FS-W5) produced positive net return of Rs. 5460.00 and Rs. 9091.76, respectively. Even horticulture based land use system (Fs-W6) suffered loss. This might be because the horticultural crops had not reached to full maturity and full bearing level. Similarly, livestock system (FS- W1) with one milch cow was not profitable.

Similarly, it has been estimated that in Meghalaya state, by adopting eight watershed based farming systems, it would be possible to generate a net income of Rs. 4222 million and an annual employment of 19.68 million man days. The estimated soil loss is 0.942 mt/ha per annum and the investment needed to treat 265 thousand hectares is Rs. 2635 million (Table 7).

**Table 6. Comparison of Cost and Returns of Different Farming Systems**

*(per hectare)*

| *Particulars* | *Live-Stock* | *Fore-stry* | *Silvi-pasture* | *Agri-pasture* | *Agri-horti-silvi-culture* | *Horti-culture* | *Natural fallow* | *Timber farming* |
|---|---|---|---|---|---|---|---|---|
| Total Return (Rs.) | 94326 | 9391 | 11236 | 65401 | 33883 | 62650 | 2853 | 15598 |
| Total Cost (Rs.) | 58905 | 2520 | 4490 | 39170 | 12107 | 22535 | 194 | 2861 |
| Net Return (Rs.) | 35421 | 6871 | 6746 | 26231 | 21776 | 40115 | 2660 | 12736 |
| B-C Ratio | 1.60 | 3.73 | 2.50 | 1.67 | 2.80 | 2.66 | 14.74 | 5.45 |

*Source:* Thimmappa (2001).

**Table 7. Estimation of Income, Employment and Soil Loss for Meghalaya State**

| *Farming system* | *Net return (Rs. Million)* | *Employment (million mandays)* | *Soil loss (mt/ha/ annum)* | *Invest. Cost (Rs. Million)* | *Area ('000 ha)* |
|---|---|---|---|---|---|
| Livestock | 1336 | 5.8 | 0.118 | 424 | 32 |
| Forestry | 247 | 0.3 | 0.004 | 455 | 33 |
| Silvi-pasture | 238 | 0.5 | 0.059 | 337 | 35 |
| Agri-pasture | 802 | 4.9 | 0.261 | 460 | 29 |
| Agri-horti-silvi-pasture | 951 | 3.3 | 0.278 | 304 | 41 |
| Horticulture | 1163 | 4.6 | 0.251 | 163 | 38 |
| Natural fallow | 100 | 0.04 | 0.004 | 60 | 37 |
| Timber | 251 | 0.24 | 0.004 | 271 | 20 |
| Total | 4222 | 19.68 | 0.942 | 2635 | 265 |

*Source:* Thimmappa (2001).

From income and employment generation point of view, it seems that the livestock based farming system in North-Eastern regions looks most suitable. However, soil loss is least in those systems having less intervention and where natural vegetation like forestry and timber trees have been allowed to prosper.

## Conclusions and Policy Options

North East India is a region as interesting and varied as India itself. It has hilly as well as plain regions on both sides of the mighty Brahmaputra. The pace of development in the hilly areas and plains differ considerably. Steep slope in this region is being used for growing cereals, pineapples and other tuber crops without using proper soil and water conservation measures. The economy of the region is still primarily agrarian. The contribution of agriculture to state domestic income is high in this region. The net sown area account for 17.38 per cent and forests 52.38 per cent of the region's reporting area. An important feature of land use in the region is that almost 16 % land area is barren and uncultivable. *Jhum* cultivation practiced by the tribals in the hill areas is estimated at 30 % of the total area under settled agriculture, which is taking heavy toll on the environment in terms of loss of soil and other nutrients from the local land.

Watershed development is becoming popular in this region because it promises a win-win solution in which natural resource conservation and agricultural productivity are complementary. It is also assumed that it has favorable implications for poverty alleviation. The aforesaid watershed based farming systems by and large proved to be a worth mentioning effort. With technically sound soil and conservation measures, the livestock based or horticulture based farming systems or agri-horti-silvi-pasture land use system may help to increase the income as well as provide round the year employment to the local population across the different income groups. Besides, such systems will also help in checking the soil loss occurring due to ongoing '*Jhum*' cultivation. But, the policy maker should be cautious that for watershed development to succeed on a large scale, projects must find a way for all stakeholders to share in the net gains generated.

## REFERENCES

Dhyani, B.L., J.S. Samra, G.P. Juyal, Ram Babu and V.S. Katiyar (1997), 'Socio-economic Analysis of a Participatory Integrated Watershed Management in Garhwal Himalaya (Fakot Watershed)', Bulletin No. T-35/D-24. CSWCRTI, Dehradun.

Government of India (1983), *Task force on shifting cultivation*, Ministry of Agriculture, New Delhi.

Government of India (1990), 'Watershed atlas of India', All India Soil and Land Use Survey, Department of Agriculture and Cooperation, Ministry of Agriculture, New Delhi.

Singh, A. and M.D. Singh (1981), □Soil Erosion Hazard in NEH Region'. Research Bulletin No- 10, ICAR Research Complex for NEH Region, Barapani, Shillong, India.

Singh, A. (1993). Watershed management in India: A Retrospect and future possibilities, *In* D.L. Deb (ed.), *Natural resource management for sustainable agriculture and environment*, Angkor Publishers, New Delhi, 202-217pp.

Thimmappa, K. 2001. Economic Evaluation of Watershed Based Farming Systems for Sustainable Production in North Eastern Hill Region. Ph.D. Thesis (unpublished) submitted to P.G. School, IARI, New Delhi.

# 10

# STATUS OF COFFEE DEVELOPMENT PROGRAMMES IN NORTH-EASTERN STATES OF INDIA—A PERFORMANCE ANALYSIS*

*K.B. Umesh, B.M. Shashidhara and Venu, B.N.*

## Introduction

The North Eastern Region of India with a landmass of 2,62,500 sq.kms comprises seven States Viz., Arunachal Pradesh, Assam, Manipur, Megalaya, Mizoram, Nagaland, Tripura, Sikkim covering about 255.09 lakh hectares, which is 7.76 per cent of total geographical area of the country. The region has 3.73 per cent of the total population of the country and contributes 2.6 per cent to the Net Domestic Product. The region has about 4500 Kms of International Border with Bhutan, China, Tibet, Myanmar and Bangladesh. Most of the land has difficult terrain and mountainous periphery. Relatively left behind in the race of modern development, the entire northeastern region is a virtual treasure trove with its one hundred languages and dialects and distinct culture and faiths. About 13 per cent of the population of Assam, 29 per cent of Tripura, 31 per cent of Manipur, 81 per cent

---

* This paper is part of a larger study on "Coffee Development Programmes/Schemes Implemented by Coffee Board in Non Traditional Area and North Eastern Region during X plan: An Impact study", supported by the Coffee Board (Bangalore), Ministry of Commerce and Industry, GOI.

of Meghalaya and almost the entire population of Arunachal Pradesh, Mizoram and Nagaland are tribals who are considered as most vulnerable in the country. The region has unique distinction of having diverse hill ecosystems covering more than two-third of total geographical area. The hill areas have wide range of altitude up to 5,000 meters. The riverine plains, swamps, *tilla* land and char areas are the other agro-ecological situations. The region receives abundant rainfall with the world's highest rainfall of 12,000 mm at Mawsynram in Megalaya, in one hand and the rain shadow belt in Nagaon district of Assam with ± 1,200 mm rainfall, on the other hand.

Agriculture is the dominant economic activity providing employment to 64.28 per cent of total workers. The region is bountifully endowed with bio diversity and is ideally suited to cultivate whole range of plantation crops, spices, fruit and vegetables, flowers and herbs much of which could be processed and exported to the rest of the country and worldwide. The development of plantation crops has been given a special place in the development strategy for the northeastern region; particularly the hilly areas to prevent denudation of forest cover and conserve depletion of soil through Jhum cultivation practiced by tribals in the region. In this direction, the Government of India, through Coffee Board introduced coffee expansion and consolidation programes in these regions during ninth (IX) and tenth (X) plan periods.

## Shifting Cultivation in North-Eastern States

In North Eastern States (NES) shifting cultivation (jhum) is a common land use practice. There are varying estimates of areas affected by shifting cultivation by different agencies. The area affected by shifting cultivation in the region was reported to be 2.80 million Ha. by North-Eastern Council (1975), 7.40 million Ha. by FAO (1975) and 3.81 million Ha. by a Task Force on Shifting Cultivation (1983). The area under jhum cultivation in the NES is estimated at 30 per cent of the total area under settled agriculture. Over the last few years, the total area under jhum cultivation (1.73 million hectare) in the NES has come down significantly due to sincere efforts by the government. For centuries, tribals and forests have coexisted in harmony. The lives of the tribal communities are intricately linked to the forests in which they live. Most tribal families subsist below the poverty line. Each tribal family has very small land holdings and with primitive methods of cultivation, income from rainfed agriculture is not sufficient to meet their basic needs. Though labeled as inefficient, shifting cultivation is a major mode of production in many areas. As a result, 15-20 year cycle of shifting cultivation on a particular land has now been reduced to 2-3 years. This has resulted in large-scale deforestation, soil and nutrient loss and invasion by weeds and other species.

**Scope of the Study**

Coffee development programme in the NES was undertaken primarily to wean away the local inhabitants from the hazardous practice of Jhum/ shifting cultivation and to provide permanent source of income to the tribals. Accordingly, Coffee Board proposed a perspective plan for northeast during 1970s in association with NCAER. The Coffee Board has identified 44000 hectares of land as potential area for coffee cultivation in 322 villages of 39 districts of NES.

The Coffee Board during Ninth Five Year Plan implemented a Special Area Programme for overall development of coffee in the NES with the active participation of the concerned state Governments and other agencies. The main focus of the programme was to promote coffee cultivation as suitable commercial and eco friendly activity in the tribal sector. The Coffee Board has further strengthened its efforts through "Coffee Development in North-Eastern States" during the Tenth Five Year Plan. The primary objectives of the scheme were to promote coffee cultivation as an alternative to Jhuming practice and thereby save forest from denudation, provide permanent source of livelihood to the tribal growers, consolidate the existing area and improve production, productivity and quality, rejuvenate the existing coffee area in North Eastern States with a view to increase the existing productivity level of below 100 Kgs/Ha to a modest level of 250 Kgs/Ha, expand coffee area of 3000 Ha.

under tribal sector, provide market support and promotion scheme, create market development network, provide back up support at the post harvest stage, promote coffee consumption through promotional campaigns, setup roasting and grinding units, quality control laboratory to cater to the need of tribal farmers, enhance the learning levels of the tribal farmers in coffee technology by participatory method, upgrade the technological skills of growers, NGOs, traders and other related groups. Against this backdrop, the present study was taken up to assess the status of coffee development programmes in NES and its performance.

## Methodology

The study was taken up during March-April, 2006 covering ninth (1997-98 to 2001-02) and tenth plans periods (2002-03 to 2006-07). The information was collected through review of reports, interaction with Coffee Board extension personnel's, scientists, officers, tribal coffee growers, ethnic groups, growers associations, corporations, state government agencies and other stakeholders. The information was elicited through personal interview, group interaction and focus group meetings. Besides secondary information relating to area, production and productivity was collected from the Coffee Board. The primary data from various stakeholders' was also collected through personal interview method. The information collected was analyzed (up to 2005-06 based on the availability of data) using simple statistical tools to draw meaningful conclusions.

## Results and Discussion

In North Eastern States (NES), the Coffee Board has identified nearly 44000 Ha of land suitable for cultivation of coffee spread across seven states, of which 10748 Ha (24.43 %) of land is brought under coffee. Of the planted area, only 60 per cent of the area is under bearing (Table 1). Across seven states, the distribution of coffee is uneven. Except in Nagaland (98.65%) and to some extent in Mizoram (29.46%) in other states the coverage of area is very poor considering area suitable for coffee cultivation. Barring Manipur, in other states both Arabica and Robusta variety of coffee is grown. Among seven states, only three states (Meghalaya 74.78%, Nagaland 70.07 %, and Assam 54.39%) the coffee bearing area is somewhat satisfactory.

In NES, various agencies have been involved in the promotion of coffee (Table 2). The agencies involved for the promotion of coffee in Assam, Meghalaya and Nagaland had the objective of establishing coffee plantations and later handing them over to the tribal beneficiaries for their sustenance. In other states viz., Arunachal Pradesh, Manipur, Mizoram and Tripura the objective was to cultivate coffee as a commercial venture and also as a part of afforestation programme aimed at maintaining ecological balance.

The state-wise targets and achievements relating to expansion (bringing new area under cultivation) and consolidation (strengthening the existing plantations to increase productivity) programmes revealed that, in Assam, Arunachal Pradesh, Manipur, Meghalaya, Mizoram and Tripura the targets have not been achieved during the tenth plan period both in physical and financial terms (Tables 3 and 4) and there is a mismatch between target set and performance achieved. The performance relating to area expansion programme has exceeded the target set during the tenth plan period in Nagaland (Table 3). However, in respect of consolidation programme the achievement is far from satisfactory.

Similar trend was observed even during IX plan period both in physical and financial achievements (Table 3) except in Arunachal Pradesh and Mizoram where physical achievements in respect of consolidation and expansion have shown good progress. North Eastern States as a whole, during IX plan, under expansion programme, the area brought under coffee was hardly 36 per cent of the target and under consolidation programme, the achievements (10%) were far from satisfactory (Table 5). During X plan, the achievements in respect of expansion programme was satisfactory (95%) compared to IX plan achievements. However, in respect of consolidation programme, the achievements are not impressive

(23%). Similar trends were observed in respect of financial targets during IX plan. During X plan period, the financial achievements have far exceeded the targets envisaged (125%) indicating excess utilization of budget under expansion. In respect of consolidation programme, the achievements are far from satisfactory (only 15 %). During Xth plan (up to 2005-06), the Board has envisaged expansion of coffee area by 3000 Ha in NES (Table 6). The physical achievements are satisfactory and with respect to utilization of funds, hardly about 50 per cent is being used. With respect to consolidation programme during X plan (up to 2005-06), the performance both physical (23%) and financial (14%) is very poor and in Manipur no attempt has been made at all (Table 7) though the target set during the plan period is 3000 ha.

## Distribution of Coffee Growers and Productivity in North Eastern States

The productivity of coffee in NES clearly indicates that, the maintenance of coffee gardens by the tribal growers is far from satisfactory. During cropping season, particularly in Kharif tribal growers were involved only in Jhum cultivation (slash and burn of forest area, cleaning of the area, preparing the land for cultivation) for paddy, corn, ginger etc., neglecting the management of coffee plantations. The objective of increasing productivity level to a modest level of 250 Kgs/Ha (objective of Coffee Board during tenth plan) from the present level of below 100 Kgs/Ha has not been realized (Table 8). Poor maintenance and low plant population per unit area, lack of remunerative price are the principal reasons for the decrease in productivity. Among the coffee growers vast majority of growers (99.89%) belong to small category. The average expenditure per hectare incurred by the tribal growers is found to be about Rs. 1500 to Rs.2000, which is mainly expended towards labour charges i.e., weeding and harvesting. In the entire NES, majority of tribal growers are not using chemicals (fertilizers and pesticides) for cultivation of coffee, thus, the coffee produced in this region is very close to organic coffee.

## Market Development

In NES, Coffee Board is procuring coffee directly from the growers or through agencies of the state governments. The procured coffee (parchment/cherry) will be sent to the only curing unit established by the Board at Guwahati. After curing it is being transported to Bangalore for marketing through Auction. The Board while procuring coffee from the growers makes an initial payment of about 20 to 30 per cent of the prevailing market price. The final payment to the growers will be made after the sale of coffee in Auction Centre at Bangalore, which normally takes a year or so. Hence, the present system of marketing is not well received by large number of growers since they expect full payment at the time of procuring coffee by the Board. In order to overcome this problem, the Mizoram government (Soil Conservation Department) has introduced a novel scheme of payment of full money at the time of procurement of coffee from the tribal growers. With respect to quality upgradation of coffee, the Boards efforts seems to be not encouraging since only few Hullers and Pulpers have been provided (Table 9). The branding of specialty coffee is totally absent except the little effort made by Mizoram government. During 2005-06, the Board has procured 163.71 tonnes of coffee in the region. Among state, Assam (39.89%), Arunachal Pradesh (23.02%), Meghalaya (19.86%) and Mizoram (11.91%) are the major coffee contributing states. In NES, Robusta is the major coffee variety produced particularly in Assam, Arunachal Pradesh, Meghalaya and Tripura with a total of 99.21tonnes contributing to 60.61 per cent of the total procurement. However, in Manipur Arabica is the only variety produced and marketed.

## Bottlenecks for Expansion and Consolidation of Coffee in North Eastern States

The programmes initiated by the Coffee Board have not taken the cognizance of local ethnic social structures and the ethnic institutions either in the planning, or implementing the programmes. There are no plantations having coffee as an entire crop and coffee is being treated as subsidiary crop along with other crops like, arecanut, mango, ginger, tamarind, banana, pepper, cashew, orange and indigenous lime species, squash, butter fruit and others. The gestation period of coffee has become really a barrier in expansion programme, since the growers look for quick income generating enterprises. There is lack of suitable crop diversification programme in coffee, which can generate quick income.

In NES, coffee is grown without developing shade trees and there are coffee plantations having no shade trees at all there by affecting productivity. Lack of location specific suitable technologies is hindering the expansion of coffee in the region. Added to this, the involvement from the state governments is very marginal. Further, concerted efforts were not made in establishing functional linkages with respective state line departments and other Commodity Boards to work cohesively for the holistic development of tribals except in Mizoram. The branding of specialty coffee is totally absent except the little effort made by Mizoram government.

Majority of growers expressed that marketing of coffee is a serious constraint owing to lack of infrastructure facilities, poor quality, high transportation cost, and delayed payment by the Board, and low prices. The growers are realizing Rs. 40 per kg for parchment and Rs. 20 per kg for cherry coffee. Recently few traders are also procuring coffee directly from growers including traders from Myanmar. However, this kind of trade is constrained by legal hurdles. In NES, majority of population do not drink coffee. There are no promotional measures undertaken by any agency including Coffee Board to promote the consumption of coffee among the tribals. Lack of quality upgradation of coffee is also hindering production and development of coffee. In few areas, coffee growers have already abandoned coffee plantations due to low price and lack of market facility. Majority of tribals prefers to go for Jhum cultivation irrespective of whether they own coffee plantation or not in order to meet their food requirement. The opinion of the growers revealed that, nearly 78 per cent have 'favourable' to 'very favourable' attitude and 22 per cent are possessing still unfavourable attitude towards the Coffee Board's programme. This calls for the organisation of more number of extension educational activities to change the attitude of tribals from unfavourable to favourable status.

## Conclusions

In NES, the coffee development programmes have created a marginal impact on the income, employment and the livelihood on the tribal households. The impact of coffee development programmes on socio-economic development of tribals is slow as coffee is taken up as one of the components in different cropping systems followed. The promotional programmes may take some more time to alter the mindset among the tribal coffee growers with respect to discontinuation of Jhum cultivation. The Jhum cultivation according to tribal coffee growers ensures food security with no cost to them. Through the efforts of Coffee Board, a total of about 10000 hectares of land was brought under coffee producing about 164 tonnes rehabilitating around 5000 tribal families in NES on gainful employment.

The north-eastern region is probably the most suitable area in India for organic farming because of its location specific advantages. The farmers of this region are de facto organic producers, as they are not using fertilizers and pesticides. Transforming the region in to an organic food producing zone and simultaneously connecting it with the domestic as well as international market can certainly lead to higher return from coffee.

However, the Coffee development Programmes in NES could not give the expected results because of lack of proper market support and quality up gradation programmes resulting in neglect and poor management of coffee plantations by tribal growers, non involvement of ethnic groups (lack of community/group approach) both in planning and implementing the programmes and lack of interest and poor participation by various agencies related to promotion of coffee.

Though the contribution of coffee in socio cultural as well as economic change cannot be isolated in the absence of a focused social research, the Coffee Board has succeeded slowly in transforming the mindset of the tribals on the harmful effect of shifting cultivation. The programmes initiated by Coffee Board have greater impact especially among socially secluded tribals. This indicates that Coffee Board has succeeded partially in motivating and making them to go for coffee cultivation, especially among those who believed in 'shifting cultivation' for their livelihood. However, majority of the tribals are still undecided with respect to shifting cultivation. The mental outlook, prejudiced ideas, social exclusion, insurgency, lack of market facility etc. might have contributed for their un-decidedness.

**Policy Implications**

1. Considering the efforts made by the Coffee Board and the level of productivity and the production in NES, the programmes may not be sustainable in the long run. Hence, the Board has to seriously reorient current strategy of coffee development in NES. The Group/ community approach needs to be perused seriously to overcome several constraints faced by individual growers.
2. There is need to focus on a concerted manner on coffee promotion and development confining to specific and suitable areas to make expansion programme more meaningful.
3. Coffee has vast socio economic as well as ecological benefits. The coffee plantation is one of the measures for afforestation, which can help to provide a green cover to the denuded hillocks and thereby improving the entire eco system. This crop holds out the promise to accelerate the rural development by opening up vast avenues for the generation of employment and income. In view of this, serious efforts are very much desirable for expansion \ consolidation of coffee in these states
4. The future programmes should focus on farming system approach in order to validate and fine-tune coffee based farming system models for the NES. Such an Integrated farming system models should involve crop diversification, to generate quick income.

5. Though coffee development programmes brought significant improvement in socio-economic and cultural traits of tribal coffee growers, by generating supplementary family income, the Coffee Board has to take more interest in educational activities in mixed cropping especially with other compatible horticultural crops like pepper, mango, jack, orange, etc.
6. As majority of tribal coffee growers possess small land holdings, there is a need to introduce group/ community approach to take up processing, value addition and marketing.
7. There is a need to vigorously pursue and educate the tribals on the impact of Jhum cultivation on the forest coverage and its consequent effect on environment by involving various groups viz., NGOs, community organizations (church), corporations, tribal councils, community leaders, development departments, etc.
8. The Coffee Board should pursue an integrated approach involving other commodity boards like Spices, Tea, Cashew and Coconut to increase the income, employment and livelihood security among tribals.
9. Market development and support is the need of the hour for the expansion and consolidation of coffee in NES. Hence, the Coffee Board needs to facilitate in procurement, transportation, processing, and marketing of coffee in collaboration with various agencies in order to sustain coffee promotion and development.
10. In order to sustain coffee cultivation in the region, there is a need to initiate programmes to promote and enhance consumption of coffee among the population.

## REFERENCES

1. Coffee in the North Eastern Region – An Overview, May 2005, Research Department, Coffee Board, Bangalore
2. Office of the Joint Director, Coffee Board, North Eastern Region, Guwahati, Assam.

**Table 1. Current Status of Coffee in North Eastern States (2005-06)**

| *State* | *No. of Districts* | *Area Found suitable for Coffee (Ha)* | *Total Planted Area (Ha)* | *Total Bearing Area (Ha)* | *% of bearing area to planted area* |
|---|---|---|---|---|---|
| Assam | 11 | 16000 | 2031.29 (12.70) | 1104.83 | 54.39 |
| Meghalaya | 5 | 8000 | 2391.00 (29.89) | 1788.00 | 74.78 |
| Nagaland | 6 | 4000 | 3946.00 (98.65) | 2765.00 | 70.07 |
| Mizoram | 2 | 4000 | 1178.40 (29.46) | 462.65 | 39.27 |
| Manipur | 5 | 4000 | 102.00 (2.55) | 20.00 | 19.61 |
| Tripura | 3 | 4000 | 493.65 (12.34) | 130.00 | 26.33 |
| Arunachal Pradesh | 7 | 4000 | 606.00 (15.15) | 141.50 | 23.35 |
| Total | 39 | 44000 | 10748.34 (24.43) | 6411.98 | 59.66 |

*Source:* Office of the Joint Director, Coffee Board, North Eastern Region, Guwahati.

**Table 2. Agencies Involved in Coffee Expansion in North Eastern States**

| *Sl. No.* | *State* | *Agency Involved* |
|---|---|---|
| 1. | Assam | Assam Plantations Crops Development Corporation |
| 2. | Arunachal Pradesh | Soil Conservation Department |
| 3. | Manipur | Forest Department/Plantations Crops' Development Corporation |
| 4. | Meghalaya | Soil Conservation Department |
| 5. | Mizoram | Soil Conservation Department |

| | | |
|---|---|---|
| 6. | Nagaland | Nagaland Plantations Crops Development Corporation |
| 7. | Tripura | Forest Department/Plantations Crops Development Corporation |

*Source:* 1. Office of the Joint Director, Coffee Board, North Eastern Region, Guwahati.

**Table 3. Physical Performance During IX and X Plan in North Eastern States**

| Programmes / Schemes | Area in Hectares | | | | | | | | | | | | | |
|---|---|---|---|---|---|---|---|---|---|---|---|---|---|---|
| | Assam | | Arunachal Pradesh | | Manipur | | Meghalaya | | Mizoram | | Nagaland | | Tripura | |
| | T | A | T | A | T | A | T | A | T | A | T | A | T | A |
| Expansion- IX plan | 1500 | 400.20 | 500 | 204.50 | 1750 | 344.20 | 2000 | 1045.75 | 150 | 241.70 | 150 | 35.5 | 600 | 98.9 |
| Expansion- X plan | 300 | 238.30 | 250 | 233 | 100 | 87.90 | 200 | 164.30 | 700 | 644.75 | 1200 | 1261 | 250 | 211.9 |
| Consolidation-IX plan | 2150 | 291 | Nil | 2.5 | 750 | 39 | 100 | 43.50 | 2750 | 76.50 | 2650 | 360 | - | - |
| Consolidation –X plan | 700 | 247 | Nil | 10 | 100 | Nil | 500 | 88 | 250 | 1 | 1200 | 338.5 | 250 | 3.5 |

Note: T – Target, A - Achievement

Source: 1.Office of the Joint Director, Coffee Board, North Eastern Region, Guwahati.

2. Achievements in respect of X plan are up to 2005-06.

**Table 4. Historical Description of Sugar Industries in NER**

| Sl. No. | Sugar Mill/Factory | Location | Ownership sector | Establishment | Installed capacity (TCD) | Mills | Status |
|---|---|---|---|---|---|---|---|
| 1. | Assam Cooperative Sugar Mills Limited | Baruabamangaon Sibsagar (Assam) | Coop | 1958 | 1000 | 4 mills (736x1450 mm) | Closed in 2000-01 |
| 2. | Nowgong Cooperative Sugar Mill Ltd. | Kamrop, Nowgong (Assam) | Coop | 1986 | 1250 | - | Closed in 1998-99 |
| 3. | Cachar Sugar Mills Limited* | Chargola, Cachar (Assam) | Public | 1971 | 1250 | 4 mills isgec (712x1524) | Became sick and closed in initial stage |
| 4. | Manipur Sugar Mills Company Limited | Kabowakhing Central Manipur (Manipur) | Public | | 1250 | - | |
| 5. | Nagaland Sugar Mills Company Limited | Dimapur, Kohima (Nagaland) | Public | | 1000 | 4 mills (762x1524 mm) | closed in 1995-96 |

*The year when the actual production commenced was1976

Source: Bhartiya Sugar Directory, 1983-84; Indian Sugar, 2005

## Table 5: Physical and Financial Performance during 9th and 10th Plan in North Eastern States (aggregate)

| Sl. No. | Programmes/ schemes | Physical (Ha) Target | Achievements | Achievements (%) | Financial (Rs. in lakhs) Target | Achievements | Achievements (%) |
|---|---|---|---|---|---|---|---|
| 1. | Expansion- IX plan | 6650 | 2370.75 | 35.65 | 997.52 | 265.03 | 26.56 |
| 2. | Expansion- X plan | 3000 | 2836.15 | 94.60 | 450.00 | 290.44 | 64.54 |
| 3. | Consolidation -IX plan | 8400 | 812.50 | 9.67 | 1380.06 | 81.64 | 5.91 |
| 4. | Consolidation –X plan | 3000 | 688.00 | 22.93 | 450.00 | 66.65 | 14.81 |

*Source:* 1. Coffee in North Eastern Region-An Overview, May 2005, Research Department, Coffee Board, Bangalore.

2. Office of the Joint Director, Coffee Board, North Eastern Region, Guwahati.

3. Achievements in respect of X plan is up to 2005-06.

## Table 6. Expansion of Coffee during X Plan Period in North Eastern States – Targets and Achievements

| Sl. No. | State | Target(Ha) Physical (ha) | Target(Ha) Financial (Rs. in lakhs) | Achievement up to 2005/06 Physical (ha) | Achievement up to 2005/06 Financial (Rs. in lakhs) | Target 2006/07 Physical (ha) | Target 2006/07 Financial (Rs. in lakhs) |
|---|---|---|---|---|---|---|---|
| 1 | Assam | 300 | 45 | 238.30 (79.43) | 23.49 (52.20) | 82 | 7.38 |
| 2 | Arunachal Pradesh | 250 | 37.5 | 233 (93.20) | 20.97 (55.92) | 40 | 3.6 |
| 3 | Meghalaya | 200 | 30 | 159.30 (79.65) | 14.85 (49.50) | 50.70 | 4.56 |
| 4 | Mizoram | 700 | 105 | 644.75 (92.11) | 61.03 (58.12) | 150 | 13.50 |
| 5 | Manipur | 100 | 15 | 87.90 (87.9) | 9.25 (61.67) | 500 | 45 |
| 6 | Nagaland | 1200 | 180 | 1261 (105.08) | 140.37 (77.98) | 311.50 | 28.04 |
| 7 | Tripura | 250 | 37.50 | 211.90 (84.76) | 20.48 (54.61) | - | - |
| | Total | 3000 | 450 | 2836.15 (94.54) | 290.44 (64.54) | 1133.57 | 102.02 |

*Note:* Figures in parenthesis indicates percentage.

*Source:* Office of the Joint Director, Coffee Board, North Eastern Region, Guwahati.

## Table 7: Consolidation of Coffee during X Plan Period- Targets and Achievements in North Eastern States

| Sl. No. | State | Target(Ha) | | Achievement up to 2005/06 | | Target 2006/07 | |
|---|---|---|---|---|---|---|---|
| | | Physical (ha) | Financial (Rs. in lakhs) | Physical (ha) | Financial (Rs. in lakhs) | Physical (ha) | Financial (Rs. in lakhs) |
| 1 | Assam | 700 | 105 | 247 (35.28) | 22.23 (21.17) | 55 | 4.95 |
| 2 | Arunachal Pradesh | - | - | 10 | 0.9 | - | - |
| 3 | Meghalaya | 500 | 75 | 88 (17.60) | 10.02 (13.36) | 10 | 0.9 |
| 4 | Mizoram | 250 | 37.5 | 1 (0.40) | 0.09 (0.24) | - | - |
| 5 | Manipur | 100 | 15 | Nil | Nil | - | - |
| 6 | Nagaland | 1200 | 180 | 338.5 (28.20) | 33.09 (16.92) | 50 | 4.5 |
| 7 | Tripura | 250 | 37.5 | 3.5 (1.40) | 0.32 (0.85) | - | - |
| | Total | 3000 | 450 | 688 (22.93) | 66.65 (14.81) | 115 | 10.35 |

*Source:* Office of the Joint Director, Coffee Board, North Eastern Region, Guwahati.

*Note:* Figures in parenthesis indicates percentage achievement against targets.

## Table 8: Number of Growers and Average Productivity of Coffee in North Eastern States (2005-06)

| States | Number of growers | | |
|---|---|---|---|
| | Small | Large | Productivity (kg/ha) |
| Assam | 1292 | 4 | 60 |
| Andhra Pradesh | 315 | 2 | 156.3 |
| Manipur (Arabica) | 104 | - | 27 |
| Meghalaya | 2415 | - | 30 |
| Mizoram | 1120 | 1 | 59.45 |
| Nagaland | 1452 | 1 | 12 |
| Tripura | 577 | - | 31 |
| Total | 7275(99.89) | 8(0.11) | 53.68 |

Note: Small: Less than 10 Ha.; Large : More than 10 Ha.

## Table 9. Market Development Programmes during X Plan in North Eastern Region

| Sl. No. | Particulars | Assam | Arun. Pradesh | Manipur | Meghalaya | Mizoram | Nagaland | Tripura |
|---|---|---|---|---|---|---|---|---|
| 1. | System of procurement | By Board | By Board | By Board | By Board | By Board | By Board | By Board |
| 2. | Market infrastructure | | | | | | | |
| | a. Traders | At Guwahati | | | | | | |
| | b. Processors owned by APCBL | | | - | - | - | - | - - |
| 3. | Quality upgradation | | | | | | | |
| | a. Hullers | 3 | - | - | 1 | 2 | 2 | 1 |
| | b. Pulpers | 2 | - | - | - | 6 | 2 | - |
| 4. | Brand name | - | - | - | - | ZOCAFE | - | - |

*Source:* Office of the Joint Director, Coffee Board, North Eastern Region, Guwahati.

# 11

# ISSUES IN GROWTH AND SUSTAINABILITY OF SUGAR SECTOR DEVELOPMENT IN NORTH-EAST REGION OF INDIA

*Ashwani Kumar Sharma and D. V. Yadav*

## Introduction

The North East Region (NER) of India, comprising eight States, namely, Arunanchal Pradesh, Assam, Manipur, Meghalaya, Mizoram, Nagaland, Sikkim and Tripura covering 26219 thousand ha, accounts for about 8 percent of the country's geographical area, 3.02% of gross cropped area (GCA) in 2002-03, 16.72% of country's forests and 4.4% of India's rural population. Land is a critical resource in many of the NE States. The net cropped area per rural person is 0.12 ha compared to 0.19 ha at the all India level. The extent of cultivable land and its percentage utilisation in the region varies from state to state. The percentage of cultivable area to its geographical area and the percentage utilisation of cultivable area in NER is 24.58 % and 63.52 %, respectively (is less than the national average levels) and mono cropping is the predominant method of cultivation. The region has a wide variety of agricultural crops because of its diversities in topography, altitude and climatic conditions. Foodgrain crops continue to predominate the agriculture sector despite slight drop in area due to diversification towards horticultural crops. During the last 14 years, the area under foodgrains

**Table 1. Changes in the Percentage Share of Major Crops/Crop Groups in NER**

| *States* | *Rice* | | *Wheat* | | *Foodgrains* | | *Fruits* | | *Vegetables* | | *Sugarcane* | |
|---|---|---|---|---|---|---|---|---|---|---|---|---|
| | *1990-91* | *2003-04* | *1990-91* | *2003-04* | *1990-91* | *2003-04* | *1990-91* | *2003-04* | *1990-91* | *2003-04* | *1990-91* | *2003-04* |
| Arunachal Pradesh | 50.23 | 47.11 | 1.46 | 1.58 | 76.19 | 68.71 | 8.18 | 15.82 | 6.92 | 7.91 | 0.00 | 0.09 |
| Assam | 64.36 | 65.05 | 2.52 | 1.80 | 71.3 | 65.61 | 1.9 | 2.73 | 5.83 | 5.84 | 1.04 | 0.65 |
| Manipur | 87.89 | 73.87 | 0.00 | 0.00 | 81 | 55.74 | 9.4 | 8.1 | 5.9 | 34.75 | 1.06 | 0.25 |
| Meghalaya | 43.82 | 41.10 | 1.90 | 1.20 | 55.46 | 49.7 | 10.08 | 9.06 | 10.79 | 13.42 | 0.04 | 0.04 |
| Mizoram | 71.06 | 59.37 | 0.00 | 0.00 | 78.35 | 75.27 | 11.77 | 17.27 | 7.59 | 6.18 | 0.73 | 0.89 |
| Nagaland | 62.30 | 57.39 | 0.00 | 1.74 | 83.14 | 66.97 | 2.48 | 7.96 | 3.9 | 8.38 | 1.83 | 0.43 |
| Sikkim | 8.81 | 11.74 | 8.40 | 3.59 | 64.54 | 60.32 | 5.07 | 9.76 | 5 | 11.27 | 0.00 | 0.00 |
| Tripura | 61.50 | 66.03 | 0.78 | 0.30 | 65.54 | 62.24 | 10.11 | 5.44 | 6.82 | 7.31 | 0.44 | 0.28 |
| NER | 61.84 | 61.86 | 2.25 | 1.60 | 71.08 | 64.41 | 3.76 | 4.78 | 6.12 | 8.14 | 0.89 | 0.54 |
| India | 21.96 | 21.83 | 12.76 | 14.28 | 68.82 | 65.94 | 1.55 | 2.14 | 3.01 | 3.29 | 1.90 | 2.31 |

*Source:* CMIE Agriculture; National Horticulture Board, Agricultural Statistics at a Glance

in NER has reduced from 71% to about 64.4 percent (Table 1). Rice is the staple food crop in NER occupying 62% of GCA. In non-rice fields, diverse mixture of 8 to 10 crops is grown in a mixed farming system. Cash crops like fruits, and vegetables including potato are also being grown to a sizable extent. The share of fruits has increased from 3.76% to 4.78% and that of vegetables from 6.12% to 8.14% over the last 15 years. The yield levels of almost all the crops are lower compared to national average. The lack of development of irrigation infrastructure is the main reason for under development of NER agriculture. The percentage of NIA to NSA is very less in NER states, varying from 10% to 32%. About 87% of NIA is irrigated by canals in Assam and 59% in Tripura. Tubewell irrigation is limited to only some pockets in Assam and in Tripura. The irrigation potential is much larger than the actual achievement and only a small percentage of the ultimate capacity for providing water to the land has been used. The share of area sown more than once is very limited highlighting the absence of double/multi-cropping as a result of either poor irrigation infrastructure or flood prone conditions that are not suitable for the cultivation of crops other than rice. Sugarcane, though, grown in about 0.54% of the GCA of NER (Table 1), its cultivation to the extent of about 30 thousand ha in NER is still popular and is the main cash crop in some areas/pockets in the region such as flood-prone/valley areas of Assam, Nagaland and Tripura.

The Government of India is emphasizing for the overall development of NER region and had allocated sizable funds for the agricultural development in the region. Since NER has vast area as flood-prone areas not suitable for the cultivation of crops other than sugarcane or rice, the overall development of NER requires that sugar industry be revived as the welfare of sugar industry is linked to the welfare of the farmers. For the revival of the sugar industry and the development of sugar sector in the region, it is important that main issues concerning growth and sustainability are better understood at the site, local and regional level so that a sound strategy could be generated. This chapter provides an in-depth analysis of the issues concerning growth and sustainability of the sugar sector in NER. Both aspects of sugar sector development viz., the sugarcane cultivation as well as sugarcane processing have been covered to provide a thorough understanding for the development of an holistic approach for the sugar sector in NER.

## Sugarcane Cultivation Scenario in NER

Sugarcane in NER is mostly grown in Assam in 0.65% of total cropped area (TCA) in 2003-04, in Nagaland (0.43% of TCA), and for NER as whole in 0.54% of TCA. Sugarcane is cultivated in almost all NER states, except Sikkim. While the percentage share of sugarcane in cropping pattern over the last 14 years has increased at all India level, it has decreased in NER cropping pattern to about half per cent (Table 1). Sugarcane crop is normally sown in March-April and is harvested in November to January in NER while it is shown in November to February in other sub-tropical states. NER receives good amount of rainfall compared to other regions of the country so most of the irrigation requirements of sugarcane are met through rainfall and the crop is grown mostly as the rain fed crop in region in sharp contrast to 6 to 8 irrigations given to the crop in UP and 20 to 36 irrigations in tropical region leading to a substantial rise in the per ha cost of sugarcane cultivation. It also highlights that the cost of cultivation of sugarcane in NER is low compared to other states because sizable expenditure is incurred on meeting irrigation requirements of the crop in other sub-tropical and tropical states. Since the crop is raised as a rainfed crop in the area, it also infers that it is subjected more to the vagaries of the monsoon in the absence of well-developed and reliable irrigation/ drainage infrastructure, and adequate cane development activities. The sugar recovery in NER has been less (less than 9%) compared to that in the rest of India and accordingly the region has been grouped (along with WL and Orissa state) as a low sugar recovery zone of India. In Assam, sugarcane cultivation is practised by the farmers in the flood-affected areas of Kamrup, Sibsagar, Nagaon and Lakhimpur districts as the crop sustains the floods.

## Trends in Sugarcane Area and Productivity

The status of sugarcane cultivation over the last 35 years with respect to its area, production and productivity and their growth behavior in terms of compound growth rates for 4 different periods with $P_4$ as the post WTO period (Table 2) highlights that sugarcane cultivation in NER has declined over the years. Sugarcane was grown in about 45000 ha in early seventies and its area rose to about 58000 ha by early eighties (during QE year 1981-85) but the sugarcane area depressed thereafter and by the advent of the new millennium, sugarcane was being grown only on 30 thousand ha in NER. Even such an extent of area devoted to sugarcane in NER is more than the sugarcane acreage in states like Rajasthan, W.B. and in Orissa.

Assam is the main state, producing sugarcane to a sizeable extent in about 26000 ha area (87% of NER) followed by Nagaland (3.67%), Tripura (3.44%) and Mizoram (1.77%). Manipur had more area under sugarcane compared to Mizoram, ranging from 1.5 to 1.8 thousand ha upto early nineties but subsequently the area under sugarcane in Manipur declined considerably at the rate of 14.75 per cent per year during the post WTO era. Its contribution has now been relegated to 5th position. Sugarcane cultivation saw a spurt in acreage upto mid-eighties (1971-72 to 1984-85 period) particularly in Assam, Nagaland and Manipur as the area under sugarcane increased at an annual compound growth rate of 3.07%, 8.61% and 5.02%, respectively, resulting in an annual growth rate of 3.06% for the whole NER compared to only 1.63% growth rate at the country level. This may be attributed to the establishment of sugar factories in Assam and Nagaland during this period i.e., in late seventies. The acreage under sugarcane in Assam decreased thereafter resulting in an overall decrease in the area at the rate of 1.49% per year for the period from 1971 to 2003.

**Table 2. Trends in Area and Productivity of Sugarcane in NER**

*Area (000 ha)*

| *States* | *Growth Rates (%)* | | | | *Quenquennium Ending year* | | | |
|---|---|---|---|---|---|---|---|---|
| | *P1* | *P2* | *P3* | *P4* | *1975-76* | *1985-86* | *1995-96* | *2003-04* |
| Assam | -1.49 | 3.07 | -3.31 | -4.27 | 38.00 | 49.56 | 37.02 | 26.13 |
| Manipur | -2.87 | 5.02$^{ns}$ | -9.35 | -14.75 | 1.68 | 1.82 | 1.52 | 0.53 |
| Meghalaya | -2.99 | -1.66$^{ns}$ | -1.09 | - | 0.20 | 0.18 | 0.10 | 0.10 |
| Mizoram | -2.14 | -0.10 | -0.34 | -7.99 | 1.30 | 0.90 | 0.86 | 0.80 |
| Nagaland | -2.29 | 8.61 | -9.19 | -12.28 | 1.74 | 3.46 | 3.56 | 1.10 |
| Tripura | -2.54 | -1.57 | -4.20 | -4.45 | 2.22 | 1.92 | 1.58 | 1.03 |
| Arun. Pradesh | n.a. | n.a. | n.a. | n.a. | - | - | - | - |
| All NEH Region | -1.60 | 3.06 | -3.67 | -4.80 | 45.14 | 57.84 | 44.64 | 29.93 |
| All India | 1.74 | 1.63 | 2.01 | 0.76$^{ns}$ | 2650.00 | 3095.24 | 3788.66 | 4256.00 |
| Productivity (t/ha) | | | | | | | | |
| Assam | 0.15 | 1.48 | -0.91 | -1.71 | 38.43 | 44.46 | 39.86 | 37.10 |
| Manipur | 0.37$^{ns}$ | -0.13$^{ns}$ | 2.50 | 7.47 | 35.50 | 35.01 | 32.37 | 80.93 |
| Meghalaya | -8.58 | 2.29 | -16.80 | -29.75 | 36.00 | 41.90 | 22.00 | 2.00 |
| Mizoram | -4.36 | -21.01 | -0.38 | -2.43 | 42.50 | 4.45 | 7.68 | 5.59 |
| Nagaland | 1.82 | 0.12 | 2.94 | 2.53 | 30.88 | 31.16 | 43.87 | 49.27 |
| Tripura | 1.19 | 0.98 | 1.38 | 0.12 | 35.31 | 38.00 | 46.57 | 48.03 |
| Arun. Pradesh | n.a. | n.a. | n.a. | n.a. | - | - | - | - |
| All NEH Region | 0.19$^{ns}$ | 1.12$^{ns}$ | -0.57 | -1.16 | 38.03 | 42.55 | 39.47 | 37.56 |
| All India | 1.14 | 1.29 | 0.53 | -1.25 | 50.08 | 57.61 | 67.18 | 63.68 |

*Note:* The time periods for which the growth rates have been worked out are as follows: P1 = 1971-72 to 2002-03; P2 = 1971-72 to 1984-85; P3 = 1985-86 to 2003-04; and P4 = 1995-96 to 2003-04.; ns means non-significant. The units for area and productivity are thousand ha and t $ha^{-1}$, respectively.

The advent of WTO regime and the economic reforms in the country does not seem to have any influence in reversing the trends. On the contrary, during post WTO period characterized by low sugar prices by the turn of the century, the area under sugarcane declined at a much faster rate in NER. The sugarcane acreage at present is about 52% to that of mid eighties level. Sugar factories were also closed and the sugar industry of NER came to a complete halt in this period. The region was already having limited sugarcane crushing capacity in sugar factories varying from 1000-1250 tonnes of cane per day (TCD) and that too closed by the end of the century. As a result the sugarcane cultivation decreased and relegated to a much smaller scale for catering to the demands of gur and khandsari units.

So far as the productivity is concerned, sugarcane yields in the NER are highly depressed. The productivity is quite low in NER at 37.56 t/ha against the national average of 63.68 t/ha in TE year 2001-03 highlighting large gaps between the national and NER averages. The yield levels for sugarcane in NER are low because most of the sugarcane is grown under rainfed, flood-prone and under jhum conditions, and is thus subjected to variability due to small aberrations in rainfall. NER as a whole has experienced a mixed trend in productivity growth in different periods but in post WTO period, the sugarcane productivity has decreased significantly with an annual rate of -1.16%. Also, in Assam (and Mizoram), the productivity has declined in P3 (period 1985-2003) and in P4 (post WTO period). However, states like Manipur has experienced compound annual growth rate of 7.47% in productivity in post WTO period followed by Nagaland (2.53%) and Tripura (0.12%). Meghalaya has experienced a drastic reduction in productivity in the period 1985-2003 as well as in post WTO period due to high growth of fruit acreage and growing of sugarcane only as intercrop in fruit plantation.

## Stability in Acreage and Productivity of Sugarcane

The value of index of instability of area, production and productivity has found to be less in post WTO period in Assam, Nagaland, and Tripura as well as for the entire NER (Table 3) exhibiting more stable productivity during the period. Mizoram followed by Manipur and Meghalaya exhibited higher instability in sugarcane acreage and production. However in Manipur, the instability indices were very high. The index of instability in productivity was lowest in Assam amongst NER states highlighting lack of fluctuations due to same climatic and soil conditions over the years. The index of instability is low at all India level too. Yield and area both contributed towards fluctuations in sugarcane production. The instability in area in P2 was higher in Manipur, Mizoram and Assam and lower in P3 while in case of Maghalaya, Nagaland and Tripura, it was higher in later period P3 compared to P2. In major sugarcane producing states of Assam, Nagaland, and Tripura, the instability in productivity was lower than the instability in production in P3 and P4 highlighting the importance of area instability in these states while in other states the instability in productivity is higher than that of production. A deep peep into the association between productivity and variability of the crop highlights that the performance of the sugarcane crop was better in Nagaland followed by Tripura for its positive growth, high mean productivity and low instability indices. In case of major sugarcane producing state of Assam, the area and yield are more stable compared to any other NER state. It highlights that with the reduction in area under sugarcane, well-defined soils/areas where the cultivation of other crops was not so successful, had remained exclusively for sugarcane cultivation.

**Table 3: Instability Index of Sugarcane Area, Production and Productivity in NER in Different Periods**

| *Periods* | *Assam* | *Manipur* | *Meghalaya* | *Mizoram* | *Nagaland* | *Tripura* | *NER* | *All India* |
|---|---|---|---|---|---|---|---|---|
| Area | | | | | | | | |
| P1 | 14.76 | 44.61 | 19.23 | 58.36 | 42.36 | 11.49 | 15.49 | 6.40 |
| P2 | 5.40 | 47.81 | 13.35 | 64.05 | 23.16 | 8.73 | 6.41 | 7.73 |
| P3 | 4.83 | 19.28 | 20.68 | 44.63 | 31.54 | 10.39 | 5.59 | 5.27 |
| P4 | 3.79 | 13.92 | 0.00 | 48.59 | 52.79 | 13.74 | 5.79 | 4.07 |
| Production | | | | | | | | |
| P1 | 21.60 | 44.12 | 35.62 | 130.04 | 46.54 | 13.69 | 19.98 | 9.43 |
| P2 | 9.81 | 51.26 | 17.09 | 67.15 | 31.15 | 12.36 | 8.26 | 9.81 |
| P3 | 6.69 | 14.56 | 35.32 | 42.48 | 36.71 | 12.85 | 6.60 | 9.37 |
| P4 | 5.59 | 26.01 | 48.75 | 40.07 | 43.36 | 10.00 | 6.27 | 6.64 |
| Yield | | | | | | | | |
| P1 | 9.60 | 51.74 | 34.87 | 111.75 | 15.83 | 7.78 | 7.82 | 5.85 |
| P2 | 11.20 | 5.36 | 13.40 | 52.19 | 21.97 | 8.29 | 10.24 | 4.85 |
| P3 | 4.54 | 53.25 | 35.60 | 143.07 | 10.66 | 7.46 | 3.18 | 5.69 |
| P4 | 3.55 | 50.91 | 48.75 | 169.80 | 10.48 | 7.33 | 1.47 | 4.57 |

The sugarcane crop suffered due to a general lack of adequate cane development activities in the area. Since the sugar factories, the only and the ultimate market for sugarcane in NER were not providing adequate price and non-price incentives to the farmers, the sugarcane cultivation became less remunerative in the region compared to other crops like summer rice, banana and tea in the region. Owing to the labour intensiveness of cultivation, low yields and low sugar recovery resulting in comparatively higher cost of production, and low output prices resulting in low returns to the farmers in comparison to those obtaining from summer rice and banana, the sugarcane area depleted in major sugarcane producing areas of the region. As a result, the sugarcane area has also been substituted partly by these crops. Even the sugarcane growing flood-prone areas of Assam where the irrigation infrastructure in the form of state tube-wells (STW) has been developed by the state government have been substituted by summer rice. Its cultivation increased from 0.69% to 6.19% in Kamrup and from 0.72% to 19.39% in Nagaon registering a growth rate of 7.46% and 2.97%, respectively, in nineties (Talukdar & Beka, 2005). To compete with summer rice, sugarcane crop needs to have substantial increase in its productivity as well as an assured market for the easy disposal of the produce. The barriers to raise the mean productivity of sugarcane in such areas need to be broken by developing site-specific technology and flood-resistant varieties. This could be achieved only under cane development activities of the sugar factories as most of the extension work for the crop including the introduction of the new varieties in their command area is to be carried out by the sugar mills as per the mill requirement. The only hope in this context, thus, lies in the establishment of efficient sugar mills in the area and their proper functioning. The private sector is venturing to set up new sugar mills even in small states like Himachal Pradesh which have very less sugarcane area compared to Assam. Hence, it is very likely that the private sector may also generate adequate incentives to the farmers in NER in general and Assam in particular.

## Sugarcane Processing Scenario in NER

During early eighties, NEH Region had 4 "vacuum pan sugar factories", 3 in Assam and 1 in Nagaland (Table 4). Manipur has also one sugar-manufacturing unit. In Assam, two sugar factories were in co-operative sector and one was in public sector whereas in Nagaland and Manipur, the only sugar factory is in public sector. The sugar industry of NER has gone defunct with one industry in Assam remaining closed since short after its setting up, second in 1998-99 and the third sugar factories closed in 2000-01. In Nagaland, it closed way back in 1995-96. At present not a single sugar factory in NER/ Assam is in operation contrary to a fact that Assam Government had once planned to set up 6-8 sugar factories in the state.

**Table 4. Historical Description of Sugar Industries in NER**

| *Sl. No.* | *Sugar Mill/Factory* | *Location* | *Ownership sector* | *Establishment* | *Installed capacity (TCD)* | *Mills* | *Status* |
|---|---|---|---|---|---|---|---|
| 1. | Assam Cooperative Sugar Mills Limited | Baruabamangaon Sibsagar (Assam) | Coop | 1958 | 1000 | 4 mills (736x 1450mm) | Closed in 2000-01 |
| 2. | Nowgong Cooperative Sugar Mill Ltd. | Kamrup, Nowgong (Assam) | Coop | 1986 | 1250 | - | Closed in 1998-99 |
| 3. | Cachar Sugar Mills Limited* | Chargola, Cachar (Assam) | Public | 1971 | 1250 | 4 mills isgec (712x 1524) | Became sick and closed in initial stage |
| 4. | Manipur Sugar Mills Company Limited | Kabowakhing Central Manipur (Manipur) | Public | | 1250 | - | |
| 5. | Nagaland Sugar Mills Company Limited | Dimapur, Kohima (Nagaland) | Public | | 1000 | 4 mills (762x1524 mm) | Closed in 1995-96 |

*The year when the actual production commenced was 1976.

*Source:* Bhartiya Sugar Directory, 1983-84; Indian Sugar, 2005.

The sugar mills in NER were of limited installed capacity varying form 1000 to 1250 TCD with 4 mill units. Some mills were very old with obsolete technology and machinery. Assam had annual installed sugar capacity of 0.20 lakh tonnes while Nagaland had 0.12 lakh tones in the initial phase and it reduced to 184 thousand tones in Assam and 64 thousand tones in Nagaland by early ninties. The limited installed capacity in Assam was not utilized for more than 40% in the entire period and its utilization varied from 30–38%, it was only 22% by the closure time, exhibiting an annual decline with a compound rate of –0.964%. By the turn of the century, Assam utilized only 14.6 % of the installed capacity, though its utilization for corporative sector sugar factory was a little higher. The detailed analysis of operational scenario of sugarcane processing and sugar production scenario (Table 5) reveals that the mill operational conditions had not remained favorable in the region compared to the rest of India. The average duration of the sugarcane crushing season in Assam was maximun only upto 86 days (3 months approximately) during seventies, when more sugar factories were being established, which gradually declined to mere 36 days with an annual compound decline rate of –2.80 per cent per year (Table 6), leading to ultimate closure of the sugar factories. As a norm, the sugar factory may function upto 180 days to run to its full capacity. In other parts of the country, the sugar factories run at least for 4 to 5 months but in NER, the average duration of the crushing season was very less and it gradually decreased with the passage of time. The cane crushed by sugar industry in Assam varied a lot from 71 thousand tones in early seventies to as high as 108 thousand tones in late eighties and ultimately reducing to 50 thousand tones by quenquennium ending (QE) year 2000-01. For the entire period, the cane crushed declined at a rate of –0.908 per cent per year. The sugar production was fluctuating between 6 to 8.4 thousand tones in Assam upto early nineties and came down to mere 4.4 thousand tones by QE year 2000-01. Whereas all other states have experienced increase in sugar recovery, the percentage of sugar recovery in NER has also gone down. The average sugar recovery in the region varied from 7.9% to 8.71% and declined at an annual compound growth rate of –0.307% during the period under reference. At all India level, the sugar recovery was about 10.17% in 2004-05. Since 1% loss in sugar recovery causes loss to the factory in lakhs of rupees per day, the overall loss to the sugar mills due to low sugar recovery may be quite substantial, affecting the very viability of the sugar mill. Since the payment of sugarcane to the farmers is linked to the sugar recovery in the mills, the farmers also got lower prices for their cane and their profits dwindled over the years. Lower prices for cane also resulted in making the crop less remunerative compared to other crops such as summer rice.

Sugar industry in Assam was predominantly a co-operative sector entity as 2 mills out of 3 mills were in the co-operative sector and the co-operative sector sugar production constituted about 97% of the total industry sugar production in early seventies (Table 7) and about 86% during later period. A look into the cooperative sector reveals that the duration of the crushing season as well as the sugar recovery was higher in co-op sector compared to the public/ total industry scenario. The capacity utilization was more than 50% for most part of the period and was higher compared to the total industry. Its installed capacity was 62.5% of the total installed capacity during eighties as well as in nineties. The scenario existing in Nagaland was worse than that existed in Assam as the only factory was a public sector sugar mill. The analysis reveals that co-operative sugar factories were a bit better compared to public sector sugar mills.

**Table 5. Quenquennium Average of Sugar I ndustry Performance Indicators in NER**

| *Particulars* | *Quenquennium ending year* | | | | | |
|---|---|---|---|---|---|---|
| | *1975-76* | *1980-81* | *1985-86* | *1990-91* | *1994-95* | *2000-01* |
| *Assam (Total Industry)* | | | | | | |
| Sugar factories (average no.) | 1 | 2 | 1.6 | 2 | 2 | 1.4 |
| Total cane crushed (000 t) | 71 | 89.2 | 78 | 108.2 | 87.4 | 49.8 |
| Duration of crushing season (days) | 86.2 | 70.8 | 63.6 | 73.8 | 54.4 | 36.4 |
| Average sugar recovery (%) | 8.716 | 8.33 | 7.784 | 7.9 | 8.014 | 8.02 |
| Total sugar production (000 t) | 6.4 | 7.4 | 6 | 8.4 | 7 | 4.4 |
| Installed capacity | | 0.2 | 0.2 | 0.2656 | 0.184 | 0.184 |
| Installed capacity utilised (%) | | 30 | 30.14 | 35.62 | 38.72 | 21.86 |
| *Assam (Cooperative sector)* | | | | | | |
| Sugar factories (average no.) | 1 | 1 | 1 | 1 | 1 | 1 |
| Total cane crushed (000 t) | 71 | 74 | 56.4 | 108.2 | 71.2 | 46.8 |

| | | | | | | |
|---|---|---|---|---|---|---|
| Duration of crushing season (days) | 86.2 | 97.6 | 77.2 | 73.8 | 60.2 | 48.4 |
| Average sugar recovery (%) | | | | | 8.33 | 8.108 |
| Total sugar production (000 t) | 6.2 | 5.8 | 4.2 | 8.4 | 6 | 3.8 |
| Installed capacity | | 0.08 | 0.08 | 0.17 | 0.12 | 0.12 |
| Installed capacity utilised (%) | | 50 | 52.88 | 56.98 | 49.1 | 33.24 |
| *Nagaland* | | | | | | |
| | *1975-76** | *1980-81* | *1985-86* | *1990-91* | *1994-95** | *2000-01* |
| Sugar factories (average no.) | 1 | 1 | 1 | 1 | 1 | |
| Total cane crushed (000 t) | 38 | 79.2 | 49.6 | 56.2 | 29.5 | |
| Duration of crushing season (days) | 54.25 | 97.6 | 62.4 | 70.4 | 39.25 | |
| Average sugar recovery (%) | 6.32 | 8.49 | | | | |
| Total sugar production (000 t) | 2.75 | 6.8 | 4 | 4.6 | 2.25 | |
| Installed capacity | n.a. | 0.12 | 0.12 | 0.0976 | 0.064 | |
| Installed capacity utilised (%) | n.a. | 58.3 | 30.42 | 53 | 35.05 | |

## Table 6
## Growth Rates of Sugar Industry Performance Indicators

| *Particulars* | *Growth rates* |
|---|---|
| Assam | |
| Total cane crushed (000 t) | -0.908 |
| Sugar production (000 t) | -1.047 |
| Duration of the sugarcane crushing season (days) | -2.80 |
| Average sugar recovery (%) | -0.307 |
| Installed Capacity utilised (%) | -0.964 |
| Nagaland | |
| Total cane crushed (000 t) | -1.491 |
| Sugar production (000 t) | -4.060 |
| Duration of the sugarcane crushing season (days) | -1.612 |

## Table 7. Role of Co-operative Sector in Sugar Industry in NER

| *Particulars* | *Quinquennium ending year* | | | | | |
|---|---|---|---|---|---|---|
| | *1975-76* | *1980-81* | *1985-86* | *1990-91* | *1994-95* | *2000-01* |
| Co-op sector sugar production (% of total production) | 96.88 | 78.38 | 70.00 | 100.00 | 85.71 | 86.36 |
| Co-op sector installed capacity (% of total industry installed capacity) | n.a | 40 | 40 | 62.5 | 66.85 | 62.50 |

The average sugar production was highest to 6.8 thousand tones in QE year 1980-81 which afterwards reduced to 2.25 thousand tones showing an overall negative growth rate of –0.406%. Since all the growth factors were moving in reverse direction, it led to sharp increase in the cost of sugar production compared to the rest of the country. The cost per unit of production from the region including W.B and Orissa was higher by 32 % from all India level unit cost of production of sugar highlighting that the sugar industry in the region was not efficiently working compared to other sugar mills in the rest of the country. The cost of production grew at an annual compound growth rate of 8.27% per over the period (Table 8).

**Table 8. Trends in Unit Cost of Sugar Production in NER and India**

| *Particulars* | *Quinquennium ending year* | | | | | |
|---|---|---|---|---|---|---|
| | *1975-76* | *1980-81* | *1985-86* | *1990-91* | *1994-95* | *2000-01* |
| | | | Cost of production | | | |
| Unit cost of sugar production (Rs/qtl) | | | | | | |
| Assam including WB, Orissa | 199.89 | 253.838 | 367.178 | 514.028 | 823.742 | 1293.528 |
| India | 178.995 | 232.632 | 329.526 | 458.914 | 622.69 | |
| Unit cost of sugar production(% of all India figures) | 111.67 | 109.12 | 111.43 | 112.01 | 132.29 | - |
| Growth rate (%) in unit cost of sugar production (Rs./qtl) | | | **8.269 | | | |
| Sugar consumption (000 t) | 81.6 | 114 | 142 | 185.2 | 219 | |

**Refers to the period 1971-2000.

It would be appropriate to mention here that the shortage of the cane in Assam was not the sole reason for the closure of the industry in the state. Insurgency in Assam and some other states may have led to overall industrial sickness in the region as the region is predominently characterized by overall industrial sickness. Out of a total of 49 large industries in Assam, 16 industries are in severe financial crises, 13 are in operation but loosing, 11 are un-operative whereas only 6 are performing reasonably well and 3 are improving. So far as the sugar industry was concerned, it also suffered on technical grounds. The health of the sugar industry in NER was not good because the sugar factories were about 30 years old with obsolete machinery. Mechanical breakdowns were more frequent due to obsolescence of technology and machinery, and in some years the sugar factory remained closed for almost the entire season. Poor management and lack of technical expertise was also one of the major factor for very low efficiency in sugar factories in Assam. The sugar mills found it difficult to pay on time for the sugarcane supplied by the farmers. On an average about 1000 persons are employed in a sugar factory and the average number of employees in sugar factories in Assam was 946. The percentage distribution of employees in Assam and in some of the leading sugar producing states reveals (Table 9) that the proportion of skilled and highly skilled workers in Assam sugar mills was very less (7.6%) and that of unskilled workers was very high (61.4%) compared to states such as UP, Maharashtra, Tamil Nadu and All India average levels. Had the sugar industry in NER worked on professional lines and extended its capacity with new machinery and modern plants as well as taken up adequate cane development work and strengthened road network/transport facilities in the area, the situation might have been different.

**Table 9: Composition (in %) of Different Categories of Employees in Sugar Factories**

| *State* | *Unskilled* | *Semi skilled* | *Skilled* | *Highly skilled* | *Clark* | *Super-visory* | *Tech & Prof.* | *Admin & exec-utives* | *Other* |
|---|---|---|---|---|---|---|---|---|---|
| Assam | 61.4 | 14.9 | 4.8 | 2.8 | 15 | 0.9 | - | 0.2 | - |
| UP | 57.9 | 19.1 | 7.5 | 1.6 | 9.5 | 0.7 | - | 0.3 | 2.9 |
| Mahar-ashtra | 46.2 | 21.0 | 12.8 | 1.7 | 12.3 | 1.6 | 1.1 | 0.5 | 2.8 |
| T.N | 46.7 | 16.9 | 12.5 | 2.7 | 14.3 | 1.9 | 1.2 | 0.8 | 2.7 |
| All India | 53.1 | 18.0 | 9.3 | 1.7 | 11.6 | 1.2 | 0.6 | 0.5 | 3.1 |

*Source:* Table adapted from Rao, 1983-84.

The above analysis reveals that the most discerning feature with the sugarcane cultivation in NER is that its area and productivity both are declining for lack of incentives generated to the farmers in the region with the sick functioning of sugar factories, and very limited crushing capacity of gur and khandsari units as an alternative to process the sizable quantum of sugarcane in the absence of sugar mills. Since the sugarcane is the raw material for sugar mills, the closure of sugar mills was the major factor responsible for shrinkage of area under sugarcane in the area. This highlights that the public sector or the co-operative sector mills of the region were not operating as efficient units, may be due to limited crushing capacity of 1000-1250 TCD of the mills, frequent breakdowns with the obsolete machinery, the general industrial sickness (most of the industries in the region were running losses) or the backwardness of the region. This also highlights that adequate cane development activities by the sugar mills for either raising the productivity of the crop (and making it a profitable venture for the farmers) were not carried out or the assured market for the disposable of the produce was not developed to the desired extent in the region. The analysis further reveals that the growth rates in area under cane, its yield rate and total production had declined at a much faster rate than in terms of the cane crushing capacity, quantity of cane crushed and sugar production highlighting a deleterious and a multiplier effect on sugarcane cultivation which may have been initiated at the factory level resulting in low mill operational efficiency and inordinate delays in payment of cane prices to the sugarcane growers.

## *Khandsari* and *Gur* Manufacturing in NER

Existing in unorganized sector, Gur and Khandsari industry (cottage industry) is another potential market for sugarcane. Khandsari along with gur accounted for more than 60% in 1970-71 and 20% in 2002-03 utilization of total cane grown in the country. In NER region, reverse may be the case as only 3.53% to 5.93% of the total cane produced has been crushed in factories in Assam and about 84 to 86% i.e. about 1-2 million ton cane has been used for gur and khandsari manufacturing/ other uses to the extent of 130 to 185 thousand tones (Table

10) from about 1283 gur and khandsari manufacturing units in the state

**Table 10. Estimate of Sugarcane Utilization for *Gur* and Khandsari in Assam**

| *Particulars* | *QE 1975-76* | *QE 1985-86* | *QE 1995-96* | *QE2000-01* |
|---|---|---|---|---|
| Sugarcane production 000t | 1460 | 2209 | 1474 | 969 |
| Cane crushed in Sugarcane factories 000 t | 71 | 78 | 87.4 | 49.8 |
| Per cent cane crushed in sugar factories | 4.86 | 3.53 | 5.93 | 5.14 |
| Assumed %age for seed and chewing | 10 | 10 | 10 | 10 |
| Sugarcane utilized for ( %) gur and Khandsari | 85.14 | 86.47 | 84.07 | 84.86 |
| Sugarcane utilized for gur and Khandsari 000 t | 1243 | 1910 | 1239 | 822 |
| Estimate of gur production in Assam including khandsari (000 t) | 130.5 | 185.4 | 120.8 | 102.8 |

Since the sugar factories in NER are closed, it is the gur and khandsari units that are taking care of the sugarcane being grown in NER at present. The annual survey of the industries has shown that raw sugar as well as and refined sugar including sugar cubes is being produced in Assam, Meghalaya, Nagaland and Tripura. The ex-factory quantity and value of products and by-products is given in Table 11. Sugar worth Rs.9.29 crores was produced during 2002-03 (when the only one industry functioning in Assam was closed in 2000-01). It is interesting to note here that sugar is being produced in Assam and Nagaland even after the closure of the large sugar factories. Sugar has also been produced from other NER states like Tripura and Meghalaya also where no big sugar factory is existing. It thus highlights that crystal sugar is being manufactured from mini plants such as Khandsari plants having sulphur units.

**Table 11. Ex-factory Quantity and Value of Sugar Products in NER (Year 2002-03)**

| *Product* | *State* | *Qty (tonnes)* | *Value (in 000 rupees)* |
|---|---|---|---|
| Sugar, raw | Assam | 678 | 9615 |
| | Nagaland | 4 | 63 |
| | Tripura | 161 | 2657 |
| | Sub-total | 843 | 12335 |
| Sugar, refined including sugar cubes | Assam | 3866 | 55485 |
| | Meghalaya | 1455 | 21032 |
| | Nagaland | 201 | 1952 |
| | Tripura | 134 | 2118 |
| | Sub-total | 5656 | 80587 |

*Source:* Govt of India, Annual Survey of Industries, 2003-04, Vol-II-Part I.

The above analysis highlights that though khandsari production in NER has helped the industry in catering to the emerging needs of the local markets by producing refined sugar and sugar cubes, it has not provided adequate incentives to sugarcane growers despite the fact that about 84-86% of cane was at the disposal of gur and khandsari units. Medium-sized Khandsari mills existing in Nagaland and Manipur have experienced some teething troubles like low-recovery due to indifferent quality of cane, inclement weather etc. and as a result the performance was not up to the mark. Gur and khandsari production has decreased over the years at a higher rate of -1.46% per annum (Table 12) against a growth rate of 0.803% per annum at the country level. This may be due to limited crushing capacity of these units. Though separate estimate for gur as well as khandsari are not available, the gur making is still continuing in NER, particularly in the hilly tracts. Sugarcane cultivation for gur production is popular in NER in almost all the states. The gur produced is of better quality or we may say organically produced, as the sugarcane cultivation is carried out with no use of fertilizers in the region. Some organic sugar farms have also come up in NER for gur manufacturing. It is very popular in some specialized pockets such as Wokha district of Nagaland in the jhum fields and there is also good demand for it in the rural areas.

Adequate value addition and market support is needed for this product for better marketing outside the region.

**Table 12. Growth Rates in *Gur* and *Khandsari* Production**

| *Particulars* | *Growth rate (%)* |
|---|---|
| India | 0.803 |
| Assam | -1.46 |

The present policies governing khandsari production are not in the interest of farmers nor these policies have encouraged the growth of khandsari industry either. There are controls which hamper the growth of the industry. Though government gives rebate to sugar mills but no such rebate is available to khandsari units. These units are to be encouraged particularly in areas where sugar factories have failed to make any significant impact so that the interests of sugarcane growers are also protected. Hilly tracts in NER states have districts/small pockets of sugarcane growing 100ha to 300 ha which may be the ideal sites for these industrial units to begin with if the sugar industry is to be revived in the region. Khandsari, a SSI unit with 100 to 250 tonnes crushing capacity per day is comparatively easy to set up in about six months time, even in backward rural areas of NER. Once the farmers may get confidence in the sugarcane business and are well equipped with the suitable vars, then only any big sugar factory is to be taken up. It could be taken if the area under sugarcane is at least about 7000 ha. The setting up of a sugar factory with private partnership in the state could be carried out if sound contract arrangements for sugarcane farming are also made by the industry to ensure proper raw material supply for the mills.

## Status of Research and Development Support for Sugarcane

Most of the research efforts in NER are concentrated on soil conservation aspect, for rice crop, and recently on horticultural crops. The sugarcane crop of NER has not been properly covered and is least researched upon crop in the NER. The investment for sugarcane development is also inadequate.

There is no cane development activity by the sugar mills as all mills are closed, the state and central efforts were also inadequate. The budgetary allocation is also meager by the respective states. It is very strange that the allocation for Assam state, the major sugarcane producing state of NER, is less than that for Manipur (Table 13). Whatever allocation was made for sugarcane development in Assam only a small percentage of it i.e. 2.5% have been expended in sharp contrast to Nagaland, Mizoram and Manipur States where 62 to 73% of the outlay was expended. Even Tripura also expended about 23 % of its budget allocation. During the last 10 years from 1995-96 to 2005-06, about 474 lacs were expended for sugarcane development in NER out of which about 51% were expended in Manipur, 25% in Mizoram, and 18% in Tripura and only 1.16% in Assam. It is thus clear that, in the absence of a sugar mill/khandsari unit in operation in Assam, sugarcane development efforts have also become a least priority in the region.

**Table 13. Investment for Sugarcane Development (Rs. lacs) in NER**

| *Particulars* | *Assam* | *Manipur* | *Mizoram* | *Nagaland* | *Tripura* | *NER* |
|---|---|---|---|---|---|---|
| Total Allocation in 10 yrs | 211.6 | 384.641 | 172.36 | 115.46 | 95.99 | 980.051 |
| Total expenditure in 10 years (1995-96 to 2005-06) | 5.49 | 240.988 | 120.75 | 84.45 | 21.87 | 473.548 |
| Expenditure as % of outlay (10 yrs total) | 2.59 | 62.66 | 70.06 | 73.14 | 22.78 | 48.32 |
| Share of states in NER total Expenditure (%) | 1.16 | 50.89 | 25.5 | 17.83 | 4.62 | 100 (473.55) |

*Source:* Directorate of Sugarcane Development (GOI), Lucknow.

Low quality cane has remained one of the main constraints of sugar industry in NER. Un-cleaned cane with green leaves at the top is supplied to the industry resulting in low sugar recovery. The average quality of cane has improved in UP, Punjab and TN with the release of improved varieties of cane, it is still low in Assam for want of better focused research work. Manipur Government has recently emphasized to introduce suitable varieties of cane which would lead to higher output and help in rendering their units viable, hence enhanced its expenditure on sugarcane development. Better performing early varieties having high sugar content are of immediate need of the sugar industry so their suitable proportion in the total sugarcane plantation could be ensured which may result in the early start of the industry in the season and get the desired duration in the crushing season. Industrial oriented research efforts in NER are immediately required to ensure sugar industry revival. The research programmes also need to be redrawn for meeting new challenges of the industry such as cogeneration and ethanol production in the sugar factories. Though, some research work at the national level has been done on the suitability of varieties of sugarcane for gur and also on its quality, the organic product from this region need support in terms of marketing research.

## Policy Implications

Major sugarcane growing states like UP, Maharashtra and Tamil Nadu are also facing the challenge of the closure of many sugar factories in the public and the co-operative sector (Table 14). However, in these states, under the new industrial promotion policies of respective states, in place of obsolete machinery and old plants of limited crushing capacity of 1000-2000 TCD, very huge sugar complexes of 5000 TCD to 11000 TCD along with the facility for ethanol production and co-generation have been set up by the private sector. Even the sugar industry of Bihar where almost all the sugar factories in the public and co-operative sector were closed is being revived with the active participation of the private sector. In sharp contrast, the closure of sugar mills in co-operative and public sector in NER/Assam has not been compensated by the establishment of new and modern sugar mills in the private sector.

**Table 14. State-wise Numbers of Non-Functional Sugar-Mills (Year 2004-05)**

| *State* | *Maharashtra* | *Bihar* | *UP* | *Punjab* | *A.P.* | *TN* | *Assam* |
|---|---|---|---|---|---|---|---|
| Sugar Mills Closed (No.) | 36 | 18 | 16 | 4 | 5 | 3 | 3 |

Source: ISMA, New Delhi.

Sugarcane in new setting has more prospects to offer to the sugar factories/khandsari units and to the regional economy. The profit centre of sugar industry is gradually shifting towards effective use of by-products. It has more value in cogeneration and alcohol production in areas where sugar recovery is less. The possibilities for ethanol production could also be explored. Since the recovery in khandsari units remains less at about 6.5%, the molasses could be used for the manufacture of by-products like alcohol and feeds etc. It is evident from the existence of distilleries in Assam and Nagaland, one of which with annual licensed capacity of 1620 kilolitres is attached to the sugar factory in Assam.

The efforts for developing NER are gaining momentum and the sugar industry of NER requires a fresh re-look as the fortunes of the sugarcane farmers are closely linked with the sugar factory/processing unit in the area. Assam has sugarcane acreage of about 26 thousand ha which is constrained by poor irrigation infrastructure, lack of transport facilities, low remunerative prices for sugarcane and the overall sugar industry sickness need to be redressed and revived by ensuring better cane development activities. The issues of poor quality of cane, its inadequate quantity and poor management of sugar factories in NER need to be addressed to the root cause. Under the new setting in free trade regime, private sector participation and healthy competition among the sugar mills in NER is to be encouraged. The incentives to each sugar factory need to be provided and linked with the extent of its cane development activities. The problem of insecurity due to insurgency which has resulted into lack of private investment (private sugar factory) in the region also needs to be tackled on priority to generate adequate business environment by means of a suitable industry promotion policy.

Since the prospects of expansion in sugarcane area in the country are now extremely limited due to the high water requirements and the claims of the other crops, NER offers some hope if its area of seventies is restored. As per existing norms of cane area reservation, a sugar factory has to operate in an area of a radius of 7.5 km if there is a competition between the mills. If 40% of the command area is supposed to be under sugarcane, the cane area reserved per sugar factory works out to be 7 thousand ha. There are no competing sugar mills and the fear of cane reservation in a small zone is not existing and thus at least 2 sugar factories or one sugar complex could be run satisfactorily as Assam is still having 26 thousand ha under sugarcane. Sugar factories/complexes in the private sector would be the answer under the new setting.

## Concluding Remarks

Sugarcane is a commercial crop in the country and is grown for cash earnings by the farmers. Sugarcane, though a more profitable crop relative to other crops in the area where it is planted (National Commission on Farmers, 2005), it did not flourished much in the North East Region (NER) of India which comprises Assam as a major and Nagaland and Tripura as the other important sugarcane growing states of NER. The sugar factories in NER are closed and the sugarcane cultivation has declined over the years. With the improvement in the policy environment by the central and state govts and the revival package from NABARD in recent years, sugar industry is now de-licensed (though still continues with lot of regulations and controls). The scenario of late nineties has changed. The domestic prices of sugar increased for the first half of this decade and the exports of sugar also risen. The era of chronic shortage of sugar in the country is over and free trade regime under WTO regime has emerged. The number of sugar factories in other regions of the country has increased and the new sugar mills being established are modern and much bigger sugar complexes having facilities for co-generation also. These new sugar complexes are mostly in the private sector and with their establishment, the farmers have also got remunerative prices for the crop in recent years. Despite such favourable change in major sugarcane growing states of the country, the sugar scenario of NER has not changed. The closed sugar mills in co-operative and public sector in NER/Assam affecting the welfare of the sugarcane growers in the region need to be compensated by the establishment of new and modern sugar mills in the private sector. An adequate business environment by means of a suitable industry promotion policy linked to the cane development activities and the welfare of the farmers also need to be generated.

## REFERENCES

Bhartiya Sugar Directory, 1983-84.

CMIE Report on Agriculture, 2003, 2004, 2005 and 2006.

Cuddy, J.D.A and Della Valle, P.A (1978) Measuring of instability of time series data, *Oxford Bulletin of Economics and Statistics*, Vol.40, No.1 pp. 79-85.

Deshpande, V.V. (1984) Role of Khandsari Units in Sugar Industry in particular and in our Economy in General, *Bhartiya Sugar Directory* 1983-84 edition, pp. 55-63.

Government of Assam (Various Issues). Economic Survey, Assam, 1981-82, 1985-86 and 1989-90, 2003-04.

GOI (2002) Basic Statistics of North Eastern Region 2002.

Jayanta, Madhab (1999) North East : Crisis of Identity, Security and Under Development, *Econ. & Politic. Wkly.* 34 (6), pp. 320-322.

National Commission on Farmers (NCF) (2005) Saving Farmers and Saving Farming crisis to confidence, Second Report, National Commission on Farmers, Ministry of Agriculture, GOI, New Delhi.

Pramod Kumar and Badal, P.S. (2004) Growth and Instability of horticultural crops in North eastern India. *Agricultural Situation in India.* 41(7). pp. 499-504.

Purkyastha, Gautam (2001) Rural Indebtedness in Assam, Changing Scenario. *Econ & Politic. Wkly.* 36(25), pp. 2239-2241.

Ramswamy, C. (2004) Constraints to growth in Indian Agriculture: Needed Technology, Resource Management and Trade Strategies, Presidential Address. *Indian Jn. Agric. Econ.*, 59(1):1-41.

Rao, P.J.Manohar (1984) Development of Indian Sugar Industry – Historical, *Bhartiya Sugar Directory* 1983-84 edition, pp.1-54.

Sahi, H.N (2001). Improved practices of sugarcane, Indian Institute of Sugarcane Research, Lucknow.

Saikia, Anura (2000) Employment Patterns of Rural Women and there Involvement in Decision–making: A study of Jorhat District of Assam. *Proc. Women in Agriculture and Rural Development*, Indian Society of Agricultural Economics, Bombay, pp. 44-53.

Salunkha, D.K. and Degade, V.G. (1984) Gur Industry in India, *Bhartiya Sugar Directory* 1983-84 edition, pp.65-76.

Sarma, B.K. (2004) Content Analysis of Indian Journal of Hill farming. (short communication), *Indian Journal of Hill Farming* 17 (1&2), pp. 153-154.

Sharma, Ashwani Kumar and Yadav, D.V. (2006) Trends and Growth Analysis of Sugarcane Cultivation and Processing in North East Region of India. Summary of paper presented in 66th Annual Conference of The India Society of Agricultural Economics. Indian. Jn. Agric. Econ.61(3):357.

Singh A.J and Byerlee, D. (1990) Relative variability in wheat yields across countries and over time. *Journal of Agricultural Economics,* vol 14, no. 1, pp. 21-32.

Singh, Kisan and Singh, Bakhtawar (1985) The National Research grid for sugarcane in India, Indian Institute of Sugarcane Research, Lucknow

Singh, S.P. (2006) Technical and Scale Efficiencies in the Indian Sugar Mill: An Inter-state comparison. The Asian Economic Review. Vol. 48 (1), pp 87-99.

*Statistical Hand Book, Assam,* 1993 and 1995

Talukdar, K.C and Beka, B.C. (2005) Cultivation of summer Rice in flood Plains of Assam-- an Assessment of Economic Potential on Marginal & Small Farms. *Agricultural Economics Research Review.* Vol. 18(1), pp. 21-38.

The Fertilizer Association of India (2005). *Fertilizer Statistics 2004-05.*

# 12

# INDIAN TEA INDUSTRY: GROWTH AND TRADE PROSPECTS IN CONTEXT OF GLOBAL COMPETITION

*K.C. Talukdar*

## Introduction

Tea is the most popular of all beverages in the world and is manufactured from the young tender shoots of the plant, *Camellia sinensis* (L.). It has the cultivation history of over 5000 years and it is cultivated in more than 45 countries. Annually 3300 million kilograms of tea are produced from around 2668 thousand hectares of land and are consumed in 65 countries across the globe.

Tea industry in India is one of the oldest and well organised, labour intensive, agro-based industry and plays an important role in the national economy. India stands first in production of tea in 19.00 per cent of total area under tea and which stands second in the world. The industry contributes more than Rs 1200 crores per annum to the Gross National Products creating employment to three milli ›n people directly and indirectly. Traditionally, tea is an important foreign exchange earner and makes a significant contribution to the central exchequer. In the recent years it has been loosing its position in the world tea market. Due to stiff global marketing competition, India is loosing its traditional markets, which are being overcome by Sri Lanka, Kenya, China and also by Indonesia. During 1951 to 2003, total production of tea in

India declined from 41.87 to 27.31 percent and share of export declined from 44.96 to 12.51 percent during the same period.

**Table 1. Production of Tea in India**

| *Year* | *Assam* | *West Bengal* | *South India* | *All India* |
|---|---|---|---|---|
| 1951 | 150 | 78 | 52 | 285 |
| 1961 | 182 | 86 | 81 | 354 |
| 1971 | 224 | 104 | 103 | 435 |
| 1981 | 305 | 128 | 123 | 560 |
| 1991 | 401 | 155 | 191 | 754 |
| 2002 | 433 | 188 | 207 | 838 |
| Annual rate of growth (%) | 3.70 | 2.76 | 5.84 | 3.80 |

*Source:* Different issues of Tea Statistics, Tea Board of India.

Tea in India is grown in 5.16 lakh hectares of land in 2002, and the production of tea has increased from 285 mkg in 1951 to 838 mkgs in 2002 at the rate of 3.80 per cent. Production of tea is increasing faster in south India followed by Assam because of entry of small growers in the tea sector (Table 1)

## Data and Methodology

The study was conducted for India and other competing countries of tea based on secondary data collected from different published sources of Tea Board of India. International Tea Committee, Brokers at Kolkata, Russel Tea Company at Kolkata, Tea Auction Centre at Guwahati, Toklai Tea Research Station, Jorhat, Assam and Small Tea Grower's Advisory Cell, Assam Agricultural University, Jorhat. Both Tabular and functional analyses were carried out to derive the parameters for their interpretations.

### *Export and Sales Growth*

Exponential growth models were fitted to time series data of tea exported from India and through different methods of sale from 1971-2000. The growth parameters were also tested at various degrees of confidence levels.

### *Determinants of Export*

The following log-linear export turn over function was fitted.

$$\ln EXP_t = a + b_1 \ln T + b_2 \ln NNP_t + b_3 \ln PD_{t-1} + b_4 \ln WT_t + b_5 \ln REP_t + U_t$$

Where,
$EXP_t$ = Export earnings in $t^{th}$ year (Rs in crores)
T = Trend variable
$NNP_t$ = Net National Product at current prices in $t^{th}$ year (Rs in crores)
$PD_{t-1}$ = Lagged Domestic production ('000 Rs)
$WT_t$ = Value of world trade in $t^{th}$ year ('000 Rs)
$REP_t$ = Ratio of export to domestic wholesale price indices in $t^{th}$ year
$b_1$.......$b_5$ are the corresponding elasticities
a is the constant and
$U_t$ is the stochastic error term

### *Export Instability*

Instability indices were measured on export quantity, export earnings and unit export price by examining co-efficient of variation, mean positive and mean negative deviations of bulk tea, packet tea, tea bags and instant tea for the period under study. Three year moving average method was used to examine the percentage deviation and the coefficients of variation were also computed to examine the magnitude of export instability. The analysis was extended further for quantity exported and price of tea. Instability of trend values were was also examined by using C-D-V index of different periods by using the following formulae.

$$\text{C-D-V index} = C.V\ (1-R^2)^{0.5}$$

### *Decomposition Analysis*

The sources of variance of production of tea was examined by decomposing the variance of production into its sources, *viz.*, area variance, productivity variance and area productivity co-variance (Bohrnstedt and Goldberger, 1969).

$$V(Q) = (\text{mean } A)^2 V(Y) + (\text{mean } Y)^2 V(A) + 2 (\text{mean } A) (\text{mean } Y) \text{Cov } (A, Y) - \text{Cov } (AY)^2 + R$$

Where,

V(Q) = Production variance

V(Y) = Productivity variance

V(A) = Area covariance

Cov (A, Y) = Area productivity covariance

Cov $(A, Y)^2$ = Higher order covariance between area and productivity

R = Residual

The data on area, production and productivity of tea were detrended using linear relations of the form :

$$Y_t = a + b \, Te_t$$

Where $Y_t$ = Dependent variable (area, production and productivity)

T = time

$e_t$ = Residual

After detrending, the residuals were centred on the mean of each period, Y, resulting in detrended time series data of the form (mean Y), resulting in detrended time series data of the form:

$$\text{Estimated (mean } Y_t) = e_t + (\text{mean } Y)$$

### *Market Potentiality and Market Opportunity Indices*

Market potentialities in different importing countries were examined by multiplying the number of population in that country with average price and per capita consumption in that particular country. Relative comparison was made to identify the countries with high export market potentiality of Indian tea.

Market Opportunity Index was also developed integrating Category Development Index and Brand Development Index of bulk and value added tea in the selected major importing countries. Market Opportunity Index was examined as the ratio of Category Development Index over Brand Development index; Category Development Index was estimated by deflating percentage of total export of tea over the percentage of total population of that country. Brand Development index was also estimated by deflating the percentage of the total export of the brand by the percentage of the total population. The higher Market Opportunity Index will indicate the export opportunity of Indian tea to that country following appropriate promotional measures.

## Results and Discussion

### *Changes in Area, Production and Productivity of Tea*

Table 2 indicates that change in area during 1971 to 2003 was higher in China followed by Kenya. Area under tea changed marginally in India and it declined in Sri Lanka. Production of tea changed in higher magnitude than area in those countries. It was the highest in Kenya followed by China because of jump in production of tea in the recent years, while it was almost stagnant in India. Productivity change was low in China, but higher in Kenya due to its vigour of productivity in new bushes. In India and China increase in production was due to increase in area while it was the contribution of productivity in Kenya and Sri Lanka.

### *Product Diversification and Value Addition*

India is a prime producer of CTC tea and it produces more than 80.00 per cent of CTC tea and less than 20.00 per cent of orthodox tea. Production of green tea is quite negligible. The demand of CTC tea is high in the domestic market. In the export market there is high demand for green and orthodox tea. India can avoid market glut if it diversifies product more to orthodox and green tea.

Due to change in taste and habit of world consumers there is rising demand for value added tea like tea bags, instant tea and packet tea. Their export growth has been shown in the later part of the paper. Apart from these grades, consumers also prefer herbal tea, masala tea, tea snacks and tea colas. A sizeable quantity can also be converted for chemical extraction for large-scale pharmaceutical uses. Such chemical extracts are catechin, cafein, thiaflavin, thiarubizin, polysaccharide, carotenoids, seponin and thianin.

There is also the growing demand for organic tea in the more advanced segments of the consumers. Currently data are lacking. It is coming up in few gardens of Assam and Arunachal Pradesh. The new planters can go for this venture of high valued produce.

**Table 2. Changes in Area, Production and Productivity of Tea in Major Tea Growing Countries**

| *Countries* | *Area (,000 ha)* | | | *Production (mkg)* | | | *Productivity (kg/ha)* | | |
|---|---|---|---|---|---|---|---|---|---|
| | *1971* | *2003* | *% change* | *1971* | *2003* | *% change* | *1971* | *2003* | *% change* |
| India | 357 | 517 | 1.40 | 435 | 857 | 3.03 | 1218 | 1657 | 1.13 |
| Sri Lanka | 242 | 188 | -0.70 | 218 | 303 | 1.22 | 901 | 1611 | 2.46 |
| China | 303 | 1207 | 9.32 | 180 | 768 | 10.21 | 594 | 636 | 0.22 |
| Kenya | 44 | 131 | 6.18 | 36 | 294 | 22.40 | 818 | 2244 | 5.45 |
| Indonesia | 98 | 155 | 1.82 | 72 | 168 | 4.17 | 735 | 1084 | 1.48 |

## *Export Growth of Indian Tea*

The compound growth rates of export of Indian tea in terms of quantum, export earnings and unit export prices during 1971 to 2000 were estimated (Table 3).

It was observed that quality of tea exported registered a marginal increase in growth (0.58%) in period -1 which increased to 1.07 per cent in period III while a negative non significant growth of (-) 0.59 per cent was observed in period II. The export in quantum reached as high as 241.20 mkgs in 1981 and was as low as 149.30 mkgs in 1994.

**Table 3. Export Growth of India Tea**

| *Periods* | *Quantity* | | *Export earnings* | | *Unit export price* | |
|---|---|---|---|---|---|---|
| | *CGR (%)* | *$R^2$* | *CGR (%)* | *$R^2$* | *CGR (%)* | *$R^2$* |
| 1971-80 | 0.59$^{NS}$ (0.570) | 0.40 | 15.12** (5.714) | 0.80 | 14.26** (6.093) | 0.82 |
| 1981-90 | -0.59$^{NS}$ (0.850) | 0.82 | 9.28*** (4.204) | 0.69 | 9.94** (5.375) | 0.78 |
| 1991-2000 | 1.07$^{NS}$ (0.700) | 0.65 | 9.93*** (4.271) | 0.72 | 8.77*** (4.162) | 0.93 |
| 1971-2000 | -0.61*** (2.702) | 0.21 | 9.07*** (18.795) | 0.93 | 9.69*** (25.645) | 0.96 |

Figures in parentheses indicate 't' values ** Significant at 5 per cent level of probability, *** Significant at 1 per cent level of probability and NS= Non significant.

The CGR for export earnings were significant at 15.12 per cent in period I 9.28 per cent in period II and 9.93 percent in period m. The export unit price varied from Rs 7.61 per kg in 1971 to Rs 109.80 per kg in 1998. The CGR of unit price in all the periods were found to significant in all the sub periods. Export price grew faster in period I. CGR of export earnings, on the other hand, gained significantly after 1971 and it was contributed mainly higher growth in export price of quality tea at higher prices in subsequent years.

*Determinants of Exports*

Relative influence of the factors on tea export from India was examined on the assumption that with the increase in global demand for tea, export of tea will increase after meeting the domestic demand under influence of comparative advantage. The following results of the export turnover for tea will support the hypothesis to a greater extent.

$$\ln EXP_t = (-)\,15.3921 + \underset{(0.464)}{0.0535 \ln T^{NS}} - \underset{(0.540)}{0.1763 \ln SSP_t^{NS}} + \underset{(3.168)}{3.4630 \ln PD_{t-1}^{***}}$$

$$+ \underset{(0.712)}{0.1561 \ln WT_t^{NS}} + \underset{(3.669)}{0.0967 \ln REP_t^{***}} + U_t$$

Figures in parentheses indicate 't' values

*** Indicates significance at 1 per cent level of probability NS = Non significant

The above turnover function indicated that the variables considered in the model explained about 95 per cent of variation in export earnings. The elasticities of export turnover function indicated that an increase in domestic demand on account of an increase in the national income could reduce the quantum of export of tea. The influence of lagged domestic production and relative prices had positive effect on export on tea.

The magnitude of effect of lagged domestic production was higher than relative prices indicating a low competitiveness of Indian tea whose export was predominated by production. Role of prices/or exchange ratio was more visible during this period in tea export. This might be due to withdrawal of simplification of restrictions and regulations for Indian tea export and expanded domestic market with higher elasticities. Relative price ratios were also higher due to high export price for value added tea. Export earnings increased throughout over time and rising world trade had positive impact on tea export. Income and production effects were found to play an important role in export earnings of tea in India. Other factors like growing consumerism, technology options and preferences of buyers from the countries, bilateral relations may affect the export earnings more adversely.

### *Changes in Export and Consumption of Tea*

Domestic production has increased faster annually in China and Kenya and export has declined in India and is quite low in Sri Lanka during 1971 to 2003 (Table 4). Relationship between export and domestic production indicated that Sri Lanka and Kenya export more than 90.00 per cent of domestic production of tea. Domestic consumption of tea as percentage of domestic production increased in India while it declined in other countries in the recent years.

### *Export Instability*

#### *Bulk Tea*

The percentage deviation from three year moving average, mean positive and mean negative deviations and coefficient of variations in export earnings from various types of tea (Table 5) indicates that out of four types of tea, *viz.* bulk tea, tea bags, packet tea and instant tea, export of bulk tea from India was more stable and consistent (26.93%).

#### *Value Added Tea*

It was unstable for value added tea like tea bags and packet tea. The higher coefficient of variation for export earnings of instant tea over quantity indicated that its market was unstable followed by tea bags. The mean positive deviations for quantity was higher for tea bags and it was higher for instant tea in export earnings. This indicated that quantity exported for tea bags was unstable and it was unstable for instant tea in export earnings. Quantity of export and export earnings of value added tea were more unstable influenced by international prices due to rising trend of demand for value added tea in the international market in the recent years.

**Table 4: Changes in Export and Consumption of Tea in Major Producing Countries during 1971-2003**

| *Countries* | *Annual p.c. change of domestic production* | *Annual p.c. change of total export* | *Change in in export as p.c. of domestic production* | *Change in consumption as p.c of domestic production* |
|---|---|---|---|---|
| India | 3.03 | -5.00 | -26.16 | 24.51 |
| Sri Lanka | 1.22 | 1.41 | 4.17 | -0.59 |
| China | 10.21 | 16.70 | 11.07 | NA |
| Kenya | 22.40 | 21.60 | -2.94 | -13.95 |
| Indonesia | 4.17 | 3.75 | -3.18 | -28.10 |

**Table 5. Instability in Export Eearnings from Different Types of Tea (1971-2000)**

| *Instability measures* | *Bulk tea* | *Packet tea* | *Tea bags* | *Instant tea* | *Total value added tea* |
|---|---|---|---|---|---|
| Quantity | | | | | |
| Co-efficient of variation | 26.93 | 71.64 | 88.82 | 63.92 | 93.27 |
| Mean positive deviation | 20.39 | 81.05 | 107.89 | 54.85 | 89.01 |
| Mean negative deviation | 34.35 | 54.03 | 46.24 | 41.95 | 59.34 |
| Export earnings | | | | | |
| Co-efficient of variation | 55.30 | 110.33 | 154.46 | 313.02 | 110.43 |
| Mean positive deviation | 38.27 | 121.52 | 233.93 | 385.45 | 130.45 |
| Mean negative deviation | 43.74 | 81.01 | 71.02 | 77.09 | 75.42 |
| Variation of export earnings | | | | | |
| Attributable to quantity variation | -2.59 | 24.90 | 3.79 | 0.86 | 47.82 |
| Price variation | 38.10 | 250.15 | 26.57 | 31.58 | 302.02 |
| Relative production | -1.03 | -0.87 | -0.25 | -1.93 | -7.62 |
| Interactions | 12.36 | 74.54 | 2.34 | -2.50 | 118.89 |

### *Decomposition Analysis*

It was observed that change in area and yield variance contributed negatively and the change in area yield co-variance shared high (138.19%) in the production variance of in India. Residual change was found to be highest in Tamil Nadu (313.68%) and lowest in Assam (32.98%).

### *Market Potentiality in Different Importing Countries*

It was observed from Table that the high potential markets of Indian tea are located in Japan, USSR/CIS, UK, USA and also in Iran. Japan market is more lucrative for its high unit price of tea while USSR/CIS market is more potential for higher population drinking tea. Similarly, UK market is more lucrative for its high unit price and higher average consumption per capita while US market is more potential for its population as well as high average price. There is scope to increase market potentiality in USA and Japan if the level of average consumption in those countries can be increased through market promotion.

Market potentiality in Pakistan and ARE was moderately higher due to lower per capita consumption of tea and lower unit prices. The low market potential countries of Indian tea were Germany, Irish Republic, Netherlands, Poland, Afganistan, Iraq, Saudi Arabia and Australia. Countries with high average price with low per capita consumption were Germany, Netherlands and Saudi Arabia. The remaining countries have low population with low unit price reducing the market potentiality. Adequate measures for market promotion for increasing the per capita consumption may increase the market potentiality. Diversification of demand oriented products to suit the taste and habits of specific countries may also increase the total export of Indian tea.

### *Market Opportunity*

Indian tea in total export to several countries like USSR, Poland, Irish Republic, Saudi Arabia, Iraq, Germany and UK are moderately high in the recent years. In contrary, the share of Indian tea in total export to the countries like Netherlands, USA, Iran, Japan, ARE and Australia are low. More than 80.00

per cent of bulk tea were exported to Australia, ARE, UK, Netherlands, Iraq, Germany, Irish Republic, Pakistan and Japan. Other countries like Iran, Poland, USSR, USA, Saudi Arabia imported less quantity of bulk tea from India.

Market of packet tea was more lucrative in Saudi Arabia followed by USSR and also in Iran and Japan. Share of tea bags was relatively higher in Poland, Pakistan and also in Iran. Instant tea market was rich in USA, Irish Republic and also in Japan. High Category Development Index for Irish Republic, Saudi Arabia, Poland and Iraq indicated that these countries consumed more tea irrespective of population. Brand Development Index also indicated that bulk tea was more popular and the value added tea was yet to pick up amongst the consumers. In value added tea, status of packet tea was higher in Saudi Arabia, Poland, Afghanistan, Iraq, Iran, Australia and Irish Republic. It was lowest in USSR and USA. It can be inferred from Brand Development Index that there is scope to increase the share of export of value added tea in the countries with high population density.

Market Opportunity Index showed that consumption of bulk tea was higher and near saturated with low market opportunity for bulk tea. In case of packet tea, consumption was more prominent in Irish Republic followed by Netherlands, UK and also in Iraq, Poland, USSR and Australia. European countries, Irish Republic Australia and Saudi Arabia were gaining popularity for tea bags. Tea bags imported from India are becoming popular in Pakistan, Japan, Iran, USA, Poland and Netherlands. Market Opportunity Index was higher for instant tea in Europe and Australia.

### *Sales Growth Through Different Methods of Sale*

It was observed that growth of sale through auction has declined in India during the last decade. It declined faster for London auction. Sales through forward contract gained up to 1990 but declined during the last decade. Ex-garden sale of tea in the country had increased significantly with higher magnitude. Indian auction was almost stable while London auction was more erratic in nature with high CV during the period. The study indicated that sales through forward contract and ex-garden sale gained momentum over auction sale in the country and it needs attention of the planners and tea planters. Tea Board of India should also plan for e-auction program for sale of made tea. During 1971-2000, public auction has declined by -6.64 percent, while ex-garden sale has increased by 5.54 per cent and export under forward contract has increased by 1.10 per cent in India.

### Status of Small Tea Cultivation in India

The tea growers with a holding size upto 10.12 hectares are categorized as small tea growers by the Tea Board of India. The small tea growers are agricultural entrepreneurs rather than simple tea farmers. Most of them grow tea in a close vicinity of the existing tea estates either on ceiling surplus land, grazing land or government land allotted or encroached or on their own land. These small gardens are considered as satellite green production centres for the factories of existing big tea gardens. The underutilized factories of the big tea plantations and the recently installed bought leaf factories are the outlets for the green leaf produced by the small tea growers. The produce is sold to these outlets directly or through agents involved in the trade.

**Table 6. Growth of Small Tea Gardens in India (in Nos.)**

| *Year* | *Assam* | *West Bengal* | *North India* | *Tamil Nadu* | *Kerala* | *South India* | *All India* |
|---|---|---|---|---|---|---|---|
| 1990 | 12<br>(69) | 15<br>(43) | 29<br>(117) | 6565<br>(8143) | 3921<br>(2524) | 10488<br>(10676) | 10517<br>(10793) |
| 1994 | 236<br>(811) | 39<br>(149) | 1962<br>(3138) | 25543<br>(17122) | 5970<br>(4791) | 3534<br>(21996) | 33496<br>(25134) |
| 1998 | 24942<br>(25923) | 809<br>(2996) | 29961<br>(32452) | 50571<br>(31276) | 5970<br>(4796) | 56556<br>(36146) | 86517<br>(68598) |
| 2002 | 42492<br>(41249) | 8398<br>(9500) | 58167<br>(58104) | 61985<br>(43157) | 5999<br>(4810) | 68000<br>(48050) | 126167<br>(106154) |
| Annual increase | 3540<br>(3432) | 698<br>(790) | 4845<br>(4840) | 4618<br>(3596) | 176<br>(190) | 4793<br>(3115) | 9637<br>(7947) |

Figures in parentheses indicate area in hectares.

Table 6 shows that annual change of small tea gardens in number was higher in Tamilnadu and Assam with higher addition of hectares every year. In Assam there is spurt of addition of small growers after 1994 while small tea growers were much higher in Tamilnadu and Kerala since 1990. This sector is growing rapidly in India at the annual rate of 9637 number of small growers with 7947 hectares were put under this plantation crop.

Contribution of small tea growers towards production indicated that its contribution was higher in West Bengal followed by Assam. It was almost stagnant in south India. The small growers primarily depend upon the big growers as they do not have their own factories. This sector is unorganized and not licensed as in the case of Kenya under the umbrella of Kenya Tea Development Authority who is in charge of collection, processing and distribution of made tea. In India, tea quality produced in this sector has been affected due to unscientific practices of cultivation. This is pausing a challenge to this small sector, which is contributing about 20.00 per cent of total tea production in the country.

**Table 7. Production of Tea in Small Tea in Small Tea Sector in Thousand Kilograms**

| *Year* | *Assam* | *West Bengal* | *North India* | *Tamil Nadu* | *Kerala* | *South India* | *All India* |
|---|---|---|---|---|---|---|---|
| 1990 | 200 | 35 | 235 | NA | NA | NA | — |
| 1994 | 567 | 139 | 2034 | 49184 | 983 | 50167 | 52201 |
| 1998 | 24718 | 8057 | 34502 | 59349 | 3435 | 62844 | 97346 |
| 2002 | 62770 | 32245 | 97123 | 78764 | 1969 | 80965 | 178088 |
| Annual increase | 26.07 | 76.69 | 34.36 | 0.05 | 0.08 | 0.05 | 0.20 |

Estimate of annual increase in south India and all India were worked out based on 1994 data for non availability of data in 1990.

## Sustainability of Small Tea Sector

This sector is more promising due higher productivity and reduced cost of production. Employment of family labour is high and escapes the high overhead cost owing to under utilized permanent labour and fixed cost on capital investment. In India, this sector is more unorganized and scattered managed by unskilled labour. The sector needs regulation in quality maintenance, collection and processing in their own factories and marketing in their own brands under one umbrella as in Kenya. The sector can be made more sustainable through integration as a family farm.

## Cost of Production and Competitiveness

The cost of production is the highest in India amongst the major tea growing countries in the world. The average price received per kilogram of made tea is lower than the cost of production per kilogram in India. It has been observed that the average price is comparatively lower in India than its competing countries like Sri Lanka ($1.61/kg) and Kenya($1.53/kg) Table-8). The tea of this country has the strength of producing flavoured tea from Assam, Darjeeling and Nilgiri hills. Main weakness is the high cost of production due to low productivity, high overhead cost for meeting the welfare criteria, seasonality and pest endemic for climatic disadvantage. The threat of Indian tea comes from Kenya which produces more than 80.00 per cent of tea in small tea gardens escaping the overhead costs and produces quality tea through strict regulation and modernization in processing. India's small tea sector has good opportunity if it can be controlled and regulated as in Kenya. Market opportunity in India can be increased if the per head consumption of ta in the domestic market can be increased through market promotion. More than 80.00 per cent of India's total production of tea is consumed in the domestic market.

**Table 8. Comparative Cost of Production and Output Prices (US$ /kg)**

| *Countries* | *Cost of production* | *Average prices* | |
|---|---|---|---|
| | | *2001* | *2003* |
| India | 1.62 | 1.30 | 1.20 |
| Sri Lanka | 1.23 | 1.61 | 1.54 |
| Bangladesh | 0.98 | 1.05 | 1.56 |
| Vietnam | 0.95 | NA | NA |
| Indonesia | 0.58 | 0.97 | 0.96 |
| Kenya | 1.26 | 1.53 | 1.54 |
| Malawi | 0.84 | 0.87 | 0.93 |

### Chemical Residues Affecting Quality of Tea

European union is the major market for Indian tea. Important markets are in UK (17.76 mkg), Ireland (2.9 mkg), Netherland (2.9 mkg), Germany (4.8 mkg), and Poland (5.1 mkg). EU has put Indian tea as high residue group of many banned substances in tea. The ED has decided to bring down the pesticide limit used in India upto the level of 0.01 ppm from 0.22 ppm for ethion which is the most used pesticide in the tea garden. Chlorine content is also high in Indian tea due to high use of dichofol. This has also cut the export of tea to EU union. The tea industry should go for organic cultivation of tea to capture the market segment of highprofile.

### Changes of Import of Tea by Major Importing Countries

The tea market in India was near monopolistic nature in the early years of its growth. Due to emergence of other competing countries like Sri Lanka, Kenya and China import of many countries from India has reduced. The countries like UK, Ireland, Germany, USSR, Pakistan imported 35.09, 39.82, 44.63 and 91.31 per cent of their total import from India during 1971 and declined to 14.83, 17.70, 21.35 and 59.32 per cent respectively in 1999. These counties have shifted their import from China and Kenya and marginally from Sri Lanka and Indonesia and some other countries. UK increased its tea import from 1.61 per cent to 40.43 per cent from China and from 11.25 per cent to 49.88 per cent from Kenya. Germany shifted its import from 1.43 per cent to 23.97 per cent from China and from 8.57 per cent to 21.37 per cent from other countries during the same period. USSR which was the major buyer of Indian tea to the extent of 91.31 per cent in 1971 reduced its import to 59.31 per cent in 1999 and shifted its import from Sri Lanka (4.69-17.68%), China (2.35-9.08%) and some other countries (1.65-13.92%). Similarly import of Japan has shifted from 9.30 to 67.13 per cent during the same period. Iran has also shifted from India to Sri Lanka Kenya and other countries. ARE has also reduced its import from India (45.00-3.43 %) and had shifted its import to Kenya (80.35%). Pakistan has become a major importer of Kenyan CTC tea (25.84-69.95%). Shift to china is mainly for green tea, which has more theuropetical effect while shift to Sri Lanka is mainly for orthodox tea. A swift to Kenya by many countries is for CTC tea at a very cheaper rate and this country is the main competitor of Indian tea as India produces more than 80.00 per cent of CTC tea. However, drought in Kenya during 2006 has put Indian tea price at hike as about 40.00 per cent of production of this country has been affected by drought. Thus India lost the markets of most of the traditional importing countries during the last thirty years.

## Conclusions and Policy Imperatives

Indian tea plantation sector needs massive replanting and replacement planting program as about half of tea bushes are in the low productive age groups affecting cost and profit of the gardens. Export of value added tea needs priority to match the rising demand of value added tea in the world. Export turn over function indicated that increase in the national income could reduce the quantum of export of tea. Rising domestic demand of tea in India is a positive sign of market security. Lagged domestic production and relative prices had positive effect on export of tea. Other factors like growing consumerism, technology options and preferences of buyers from other countries, bilateral relationships might affect the export earnings more adversely. Kenya and China are emerging as market giants in the competition of global tea. India consumes more than 80.00 per cent of total domestic production of tea while Kenya and Sri Lanka export more than 90.00 per cent of their total production of tea. Sales through auction were found to decline while sales through forward contract were found to increase. Ex-garden sale in the country was found to increase significantly. High potential markets of Indian Tea were located in Japan, USSR, UK, USA and Iran and also in Pakistan and ARE due to demand of quality tea and higher per capita consumption of tea as in UK. There is scope to popularise Indian tea in Germany, Irish Republic, Netherlands, Poland, Afghanistan, Iraq, Saudi Arabia and Australia. Market opportunity of value added tea was increasing over bulk tea in many Asian and European countries and needs strong market strategies with promising brands. Value added tea has high potentiality to be promoted in the thickly populated countries. Production from the small tae sector is about 20.00 per cent in the country. This sector is more promising if it can be controlled and organized as in case of Kenya which is producing quality tea at the lowest cost in the world. The main challenges of Indian tea in the global competition are high overhead cost, social obligations to labour and high cost of management, pest endemic for which incidence of chemical residues is high, prone to adverse climate with dormancy in production process adding to overhead cost, low productivity and higher cost of production due to old bushes and lack of modernization in the processing industries.

## REFERENCES

Sahewalla, I.P. (2005), Dynamics of tea production pattern, market structure changes and growth in export of Indian tea with special reference to Assam, Ph.D. thesis submitted to Assam Agricultural University, Jorhat.

Talukdar, K.C. and I. P. (1997), Sahewalla, Growth and instability of exports of Indian tea to major European countries. Paper presented in 24th World Marketing Congress on Indo-European Co-operation for Marketing and Management: Vision for 21st century organized by Institute of Marketing and Management, New Delhi, 1997, Jan. 5-8.

Talukdar, K.C., Commodity Report on Tea, Agricultural Marketing Intelligence Unit, Department of Agricultural Economics, Assam Agricultural University, Jorhat.

# 13

# AN AGRO ECONOMIC SCENARIO—THE MENACE OF SEZs

*B. Datta Ray*

In 2005, the United Nations Department of Economic and Social Affairs revealed that 80 percent of worlds gross Domestic Products belong to 1 billion people of the first world's developed countries and the rest 20 percent is shared by 5 billion people of second and third world countries. For them better living standard, health care social justice for all people, remain a distant dream and people are restless. The distributive equality has proved to be ineffective and poverty becomes deep in all third world countries. The social wealth is concentrated in the hand of the few who have the lots. Thus is sharp class difference, between have not and have then lots in all countries. The growing restlessness in the countryside is bound to engulf the whole of the country. It is not a law and order problem in the conventional sense. There is now increasing competition for the scarce natural resources including land in the tribal world. Since 1990, with the new strategy of the first world countries and under the WTO regime, the problem has been accentuated, globalization could not solve the basic problem of poverty we face now the prospect of jobless growth. We have failed to define the social goals in terms of employment and poverty elimination. The GDP growth is not a social objective in itself. The Grows Domestic Product growth of 9.2 percent in the current fiscal shows buoyancy of the economy except in agriculture, the industrial growth fell short of target of 10 percent in the 10th plan given the recent performance, the 11th plan target of 10 percent annual industrial growth appears achievable. The momentum of GDP growth in the service sector has been maintained with growth of 11 percent in 2006-207.

Buoyed by 9.2 percent GDP growth in 2006-2007, the Planning Commission said that the country must aim for higher economic performance but agriculture remains an area of concern. The agricultural sector is estimated by the Central Statistical Organization, to grow in 2006-07 at 2.7 percent. This is lower than the projected desired growth rate of 4 percent. At the base year of 2005-06, the growth rate was at a height of 6 percent. The lack lusture performance of agriculture has forced the Central Government to lower the estimates of wheat and rice production to 72.5 million tones against the initial estimates of 74 million tones.

2007-2008 is the first year of Eleventh Five Year Plan (2007-12). The Approach Paper on the 11th Plan has acknowledged that the regional disparities have been accentuated during the past and the emphasis should be laid on inclusive development. The Approach Paper has laid great importance on doubling the food grain production in the next 10 years from 210 million tones per year at present to 420 million tones. For this utmost importance will have to be laid on agriculture. With two- thirds of the rural population still directly dependent on agriculture, it is imperative that we inject fresh dynamism into agriculture.

Despite tall claims, both the agricultural production and productivity have declined in North East India. For example, the growth rate of agriculture which was 1.3 during the Eight Plan period dropped to 0.39 percent during Ninth Plan period and it may be negative growth on (-) 0.27 percent at the end of Tenth Five Year Plan period. The poverty ratio is 36.09 percent. All India average is 26 percent.

We are witnessing a situation where growth rate of GDP is higher than the inflation at 6.58 percent. It is the rising price of agricultural commodities that has pushed inflation 6.58 percent. There are, no doubt many known short term and medium term measures to tackle inflation with monetary measures. But a close look at the nature of inflation would show that it is prices of essential agricultural commodities that are rising fast as the demand for agricultural products far outstrips their supply. The only way this problem can be tackled by increasing the supply of agricultural products. Imports of agro- products cannot be a permanent solution. The solution lies in increasing the agricultural products and productivity by a second agricultural green revolution. The first green revolution has bypassed North East India. We now need massive public sector investment in agriculture. The outlays on agriculture sector was 16.7 percent till 5 year Plan period put it came down to 11.3 percent in the 10th Plan growth in the rural specific infrastructure - road connectivity, ware houses, cold storages, food processing of perishable food and packaging of the food products should be exclusively in the public sector.

Special Economic Zones (SEZs) are conclaves where obligations of corporations as entities operating in a democratic country are diluted so also tax holidays, break from the normal labour laws and subsidies on basic utilities like water and electricity that the rest of the people pay for. How such iniquitous structure can bring about social and distributive justice for the country?

Today SEZs have become the main instrument of dispossession of peasants and farmers. The SEZs are now the combination of private interest and State collusion. SEZs are costly ways of promoting elitist export aimed industrialization. They get wholly undeserved tax cuts to promoters and inflict loss upon the government of at least Rs. 1, 50,000 crores every year. The government has approved 237 SEZs with 34,500 hectors of land and notified 63 of them. Another 165 SEZs have been approved in principle for which another 148,665 hectors are to be acquired from the peasant farmers. Applications for another 300 SEZs are pending with the government.

SEZs have not proved success in most countries including China. In fact Shenzhem, China's best know Special Economic Zone has turned out a nightmare. Above all SEZs are a gigantic land scam. Most are meant to grab land close to the big cities and get monopoly profits.

SEZs have brought displacement without any proper rehabilitation and skill development with disastrous consequences—social, economic and cultural. SEZs have created conflicts due to dispossessions and displacements. SEZs are economically irrational, socially divisive and fully inequitable and socially unjustified. Fortunately for North East India, there is hardly any attempt to create SEZs. The type of land to be acquired, if any, should be non-agricultural land in the hills. States of Rajasthan and Gujarat are different when they have larger tracts of non-agricultural land available. Multi-crop land anywhere should not be touched.

It is uncomfortable to know that the 63 notified SEZs would involve Rs. 35,561 crores of rupees and the 234 approved SEZs would invest Rs. 3000,000.

SEZs are not the answer to our economic problem. Our growth since 1991 is lopsided. We have targeted a high growth of eight or nine percent. But we have failed to define in clear terms the social objectives of growth in terms of employment and poverty eradication. The 63 notified Zones would create only 18,457 jobs and the 234 approved SEZs may create 40 lakh jobs. The 63 notified SEZs would eat up 67 sq. km of land and 234 SEZs will take 350 sq. km of land and 162 cleared in principle SEZs would take 1,400 sq, km of land. Country exported in 2005-2006 Rs. 22, 840 crores rupees worth of goods from SEZs which might rise up Rs. 67,3000 crores of rupees in 2007-2008. The benefits go to the have-lot, the peasant-farmers lose the land. The GDP growth cannot be viewed as an objective in itself. Though the Gross Domestic Product growth has not declined, but agriculture and small-scale industry have shown decline. Investment in agriculture has virtually come down by 50 percent in the past few years. The public sector must now invest more money to achieve a growth rate of 3.5 percent in agricultural sector from the current average growth of 2.0 percent.

In the Human Development Index in 2003, India came 127$^{th}$, out of 170 countries. Maternal mortality is 540 per 100,000 live births while infant morality is 67 per 1000. There are 927 females born in 2001 down from 945 in 1991. In 2001, the percapita consumption of cereal dropped to an all time low of 140.56 kg per head; pulses at 10.26 kg were also at their lowest. These are significant factors in cutting the purchasing power of the poor. Farmers' suicide will be there while there would be significant jobless growth of Gross Domestic Products in the absence of distributive justice in the countryside.

In the modified SEZs regulation, the states will be barred from acquiring land for this purpose, the developers would do so on their own and developers are required to devote 50 percent to processing. The Central Government has ordered 5000 hectare ceiling of SEZs. It is no answer to agricultural problems of the country.

Asian Development Bank (ADB) has also criticized India for offering unnecessary tax incentives to SEZs developers and units. Special Economic Zone tax inducements are expensive and come at a time when the Government is struggling to provide adequate infrastructure in the wider economy as stated in the Annual Report, Asian Development Outlook 2007 released recently. Tax breaks are unnecessary. ADB's report said special tax incentives would also open loopholes for tax evasions and undermine investments in firms located outside the SEZs. Better infrastructure and greater bureaucratic efficiency in SEZs would be sufficient to stimulate investment there from potential investors. The outlook said that some of the loudest political oppositions to SEZs projects come from the landless, who may not receive compensation for the land conversion and who lack the capital to become self-employed. Centre for the time being has put on hold its Special Economic Zone policy ending the formulation of national rehabilitation policy. It should be buried for all time.

# PART II

# STATE SPECIFIC & CROP SPECIFIC ISSUES ON AGRICULTURAL DEVELOPMENT

# 14

# AGRICULTURAL GROWTH AND THE CROPPING PATTERN CHANGES IN ASSAM DURING 1951-52 TO 1999-2000*

*Pravat Kumar Kuri & Bidyut Kumar Ghosh*

## Introduction

Assam is essentially an agricultural State; around 70 percent of the total working force of the State depends upon agriculture and allied activities directly or otherwise. According to the provisional estimates of land utilization statistics, Assam had an estimated 39.44 lakh hectares of gross cropped area in 1998-1999, of which net sown area is about 27.02 lakh hectares which is about 34.4 percent of the total geographical area of the State. The importance of agriculture in the total economic activities of the State can be gauged from the percentage contribution of agriculture in the net state domestic product (NSDP). A close examination of the trends of NSDP shows that despite declining contribution of agriculture and allied sector to the State income, agrarian sector continues to be dominant in the economy of Assam. The share of agricultural sector in the State domestic product of Assam at constant prices (1993-94) was 36.35 percent in 1999-2000 (37.83 percent at current prices). With a declining share of agriculture in the total State income together with a constant share of agricultural employment in the total population, per capita agricultural income tended to show a decline relative to per capita income in the non-agricultural sector (Economic Survey,

2000-2001). Although agriculture continues to be a major sector in Assam, the growth performance of agricultural production and productivity is not satisfactory in the State. This paper seeks to analyze the growth performance of agriculture and the nature of changes in cropping pattern in the State of Assam over a period of 49 years commencing from 1951-52. For this study the estimates of area, production and yield of principal crops in Assam were collected from the Directorate of Economics and Statistics, Government of Assam. The time series data of area, production and yield of major crops of the state have been compiled for the period 1951-52 to 1999-2000. It is to be noted in this context that in 1951-52 there were only 8 districts in Assam. With the creation of new districts from time to time, there are altogether 23 districts in the State at present. Thus, for the purpose of time series analysis it has become imperative to group the date according to the old districts. For convenience, the paper is divided into four sections. Section I deals with the growth performance of major crops of Assam. This section also deals with the nature of deceleration/acceleration of agricultural growth rates along with the variability and instability of agricultural production and productivity. Section II analyzes the nature of changes in cropping pattern in terms of area under different crops and the corresponding relative share of each crop in gross cropped area. The decomposition of output change in terms of area effect and yield effect is also considered in Section II. Section III explains the changes in cropping pattern in the State in terms of substitution and expansion effect. Conclusion appears in Section IV.

### Growth of Area, Yield and Production of Foodgrains

Agricultural production including both food and non-food crops constitutes the single largest economic enterprise in Assam. However, excluding tea, Assam is basically a food grain producing State. Foodgrain crops including both cereals and pulses constituted 84.40 per cent of the total cropped area in the State in 1951-52. This percentage has increased to 84.60 percent in the year 1999-2000. Rice occupies the most important place among the crops of the state both in terms of area under cultivation and the level of production. The production of rice dominates among the cereals as well as among total food grain crops. In 1951-52, 79.990 per cent of the total cropped area was under rice, which still remains as high as 78.00 per cent in 1999-2000. Moreover, 96 percent of the area under cereals constitutes the area under rice. In fact, three seasonal rice varieties are grown in the state viz., winter rice *(Sali)*, autumn rice *(Ahu)* and Summer rice *(Boro)*. Among these three varieties, winter rice dominates in the agriculture of Assam. The production of total rice in the state was 13.98 lakh tones in 1951-52, out of which 11.84 lakh tones i.e. 84.70 per cent was winter rice. Similarly, in 1999-2000 total rice production was 38.61 lakh tones with 69.70 percent contribution from winter rice. Thus, the trends of production of food grains in general and rice in particular carry a considerable weight on the overall growth pattern of agriculture in Assam. The exponential growth rates of area, production and yield of major crops in Assam over a period of 49 years since 1950-51 is shown in Table 1 below:

**Table 1: Exponential Rates of Growth of area, Production and Yield of some important crops in Assam during 1951-52 to 1999-2000**

| *Crops* | *Production* | *Area* | *Yield* |
|---|---|---|---|
| Winter Rice | 1.71(21.57) | 0.95(16.92) | 0.76(7.22) |
| Autumn Rice | 1.65(12.14) | 1.29(13.76) | 0.36(2.43) |
| Summer Rice | 8.59(19.41) | 7.53(23.72) | 1.06(4.91) |
| Total Rice | 1.90(25.28) | 1.18(30.46) | 0.72(7.98) |
| Wheat | 11.60(13.86) | 9.94(14.13) | 1.66(8.03) |
| Total Cereals | 2.03(19.48) | 1.28(31.77) | 0.75(9.32) |
| Gram | 0.64(0.94) | 0.72(1.13) | -0.08(-0.86) |
| Tur | 3.24(11.65) | 2.84(10.84) | 0.40(2.61) |
| Total Pulses | 1.88(16.94) | 1.26(12.82) | 0.62(6.85) |
| Total Food grain | 2.02(30.21) | 1.25(26.98) | 0.77(8.73) |
| Jute | 0.057(0.16) | -0.87(-8.29) | 0.93(2.63) |
| Rape and Mustard | 3.18(14.98) | 2.64(17.41) | 0.54(4.69) |
| Tobacco | -4.72(-13.74) | -3.90(-13.88) | -0.08(-0.86) |
| Potato | 4.84(15.92) | 3.27(25.28) | 1.57(6.65) |

*Source:* Directorate of Economics and Statistics, Government of Assam, Guwahati.

*Note:* (i) exponential growth rates are estimated using SPSS Computer package. (ii) Figures within parenthesis are the 't' values.

Total food grain production in the State has grown at an average rate of 2.02 per cent per annum. Interestingly, about 62 per cent of this growth is due to the growth of area. Yield rate grew only at the rate of 0.77 percent per annum. Thus, the expansion of are under food grain crops is mainly responsible for the increase in the production of food grain during the period under study. Inter-districts analysis also reveals the same pattern (see appendix Table A1). United M and N.C Hill experienced the highest production growth rate of 5.84 percent per annum, followed by Nagaon 2.63 percent per annum and Lakhimpur 2.32 per cent per annum. The lowest growth rate was observed in the Darrang district of Assam (1.28 percent per annum). The variation in the growth performance across districts is mainly because of the widening inter-district disparities in the use of HYV seeds, the consumption of fertilizer and the availability of irrigation facilities (Phukan, 1990, Kuri, 2000). Total food grain is the sum total of total cereal and total pulse. The exponential growth rate of cereal production during the period under study was 2.03 per cent per annum while it was 1.88 per cent per annum for total pulse production. Like total food grain crops, the expansion of area contributed significantly in the growth of production of total cereals and total pulses. Rice is the major cereal crop in Assam. The production of total rice grew at the rate of 1.90 percent per annum while area registered a growth rate of about 1.18 per cent per annum accounting for as much as 62.10 per cent of the growth of output. The yield of rice grew only at the rate of 0.72 per cent per annum. Like the total foodgrain, there is substantial inter-district variation in the growth performance of total rice in Assam (See appendix Table-A2). The highest production growth rate of total rice was observed in United M. and N.C. Hills (5.79 per cent per annum), followed by Nagaon (2.53 per cent per annum) and Lakhimpur (2.24 per cent annum), while the lowest production growth rate was observed in the Darrang district of Assam (0.91 percent per annum). The production of winter rice is found to grow at the exponential rate of 1.71 per cent per annum with 55.6 percent contribution from the expansion of area. The production of summer rice experienced with a significant growth over the period under

## Table 2
## Discontinuous Kinked Exponential Estimates of Rates of Growth of Major Foodgrain Crops in Assam, 1951-52 to 1974-75 and 1975-75 to 1999-2000

| *Crops* | | *Total* | *Discontinuous* | | *Kinked* | |
|---|---|---|---|---|---|---|
| | | *1951-52 to 1999-2000* | *1951-52 to 1974-75* | *1975-76 to 1999-2000* | *1951-52 to 1974-75* | *1975-76 to 1999-2000* |
| Foodgrain | P | 2.02(30.21) | 1.59(9.09) | 2.33(13.05) | 1.66(11.95) | 2.39(17.18) |
| | A | 1.25(26.98) | 1.53(26.29) | 0.73(5.96) | 1.67(20.66) | 0.85(10.50) |
| | Y | 0.77(8.73) | 0.06(0.40) | 1.60(7.35) | -0.01(0.05) | 1.54(10.14) |
| Total Rice | P | 1.90(25.28) | 1.43(9.00) | 2.39(11.02) | 1.73(9.45) | 2.39(15.82) |
| | A | 1.18(30.46) | 1.38(29.93) | 0.76(7.06) | 1.50(21.54) | 0.86(12.33) |
| | Y | 0.72(7.98) | 0.05(0.29) | 1.63(7.23) | -0.08(-0.5) | 1.53(9.76) |
| Winter Rice | P | 1.71(21.57) | 1.17(6.78) | 1.99(8.37) | 1.31(7.87) | 2.11(12.65) |
| | A | 0.95(16.92) | 0.95(19.09) | 0.47(2.52) | 1.22(10.14) | 0.69(5.78) |
| | Y | 0.76(7.22) | 0.22(1.31) | 1.52(4.64) | 0.09(0.45) | 1.42(6.74) |
| Autumn Rice | P | 1.65(12.14) | 2.06(4.89) | 1.49(4.11) | 1.97(5.62) | -0.59(-.96) |
| | A | 1.29(13.76) | 2.42(31.45) | 0.09(0.66) | 2.49(14.73) | -2.38(-8.0) |
| | Y | 0.36(2.43) | -0.36(-.82) | 1.40(5.06) | -0.52(-1.7) | 1.79(3.34) |
| Summer Rice | P | 8.59(19.41) | 11.30(9.84) | 12.20(13.71) | 7.86(7.92) | 9.35(9.43) |
| | A | 7.53(23.72) | 10.10(11.44) | 8.98(13.71) | 7.92(11.07) | 7.10(9.92) |
| | Y | 1.06(4.91) | 1.12(1.93) | 3.25(7.44) | -0.06(-.14) | 2.25(5.02) |

*Source:* Directorate of Economics and Statistics, Government of Assam, Guwahati.

*Note:* (i) exponential growth rates are estimated using SPSS Computer package. (ii) Figures within parenthesis are the 't' values.

study (8.59 per cent per annum) with 87.70 percent contribution from the area effect. Among cereal crops, wheat occupies the second position both in terms of area under cultivation and the level of production. In 1951-52, wheat was produced only in 0.1 per cent of the total cropped area of the state, which increased to 2.30 per cent of the total cropped area in 1999-2000. During 1951-52 to 1999-2000 wheat production registered a robust growth rate of 11.60 per cent per annum with 85.70 percent contribution from the expansion of area. The growth rate of wheat was more than six times higher than the growth rate of total rice. It is to be noted here that this high growth rate of wheat is mainly because of low base period production and partly because of the impact of green revolution in Assam agriculture. Rapeseed and Mustard are two oilseed crops in Assam. The production of Rape and Mustard together has grown at the exponential rate of 3.18 percent per annum. The production of jute, a traditional crop, declined during the period under study. The area under jute registered a negative growth rate of -0.87 percent per annum.

The estimates of decadal growth rates of major foodgrain crops as presented in appendix Table A7 reveal that the State of Assam achieved highest production growth rate in the period 1961-62 to 1969-70 (at the rate of 1.94 per cent per annum). This is true for all most all the foodgrain crops. Thus, the period of green revolution could not give any boost to foodgrain production in the State. The growth rate of area is found to be highest during the period 1960-61 to 1979-80. The discontinuous kinked[1] exponential estimates of rates of growth of the major foodgrain crops reveals (Table 2) that the State achieved higher growth rates during the sub-period 1975-76 to 1999-2000 compared to the sub-period 1951-52 to 1974-75.

To test the acceleration or deceleration in growth trends, we have used log-quadratic form of equation[2] where a significant positive coefficient of time square indicates an accelerating growth rate while a significantly negative value indicates deceleration. The log-quadratic estimates of production and yield of principal crops in Assam during 1951-52 to 1999-2000 is presented in appendix Table A3. Total Rice, total cereals, and total foodgrain experienced significant acceleration in the growth rate of production, while summer rice, and wheat experienced significant deceleration in growth rate during period under review. An analysis of the fluctuations[3] in area, production and yield of major foodgrain crops in Assam reveals that the influence of area fluctuations carries much greater weight in the variability of production (see appendix Table A4 & A5). The production fluctuation is largely associated with area fluctuations rather than yield fluctuations. To understand the trends of instability[4] in foodgrain production, the deviation of the observed production form the predicted levels of production over time has been used as an indicator of instability. The instability coefficients as presented in appendix Table A6 indicate that except wheat (in which there is significant decline in instability) 'C' coefficients are not statistically significant from zero. In the case of total foodgrain production, the coefficient appears to be negative implying decline in instability but 't'

**Table 3: The Cropping Pattern in Assam During 1951-52 to 1999-2000**

*(Area in 000 hectares)*

| Crops | 1951-52 | 1960-61 | 1970-71 | 1980-81 | 1990-91 | 1999-2000 |
|---|---|---|---|---|---|---|
| Winter Rice | 1220.92 | 1304.26 | 1421.31 | 1621.97 | 1801.22 | 1793.99 |
| Autumn Rice | 323.45 | 394.64 | 526.67 | 571.72 | 608.15 | 557.22 |
| Summer Rice | 3.00 | 17.26 | 26.27 | 35.24 | 117.37 | 294.65 |
| Total Rice | 1547.38 | 1716.16 | 1974.24 | 2228.93 | 2526.7 | 2645.87 |
| Wheat | 1.87 | 3.69 | 20.08 | 91.68 | 83.97 | 76.31 |
| Maize | 1.56 | 6.51 | 12.23 | 22.46 | 20.57 | 19.71 |
| Other Cereals | 2.17 | 3.64 | 4.36 | 20.16 | 10.54 | 10.68 |
| Total Cereals | 1552.99 | 1729.99 | 2011.01 | 2363.21 | 2641.8 | 2752.57 |
| Gram | 4.16 | 1.46 | 1.60 | 4.02 | 3.19 | 2.34 |
| Tur | 3.46 | 2.85 | 4.11 | 6.45 | 7.18 | 7.27 |
| Other Pulses | 73.54 | 69.94 | 97.07 | 100.79 | 103.00 | 106.58 |
| Total Pulses | 81.17 | 77.46 | 84.80 | 111.26 | 113.04 | 116.18 |
| Total Food grain | 1634.16 | 1790.14 | 2085.68 | 2474.47 | 2755.20 | 2862.75 |
| Mustard | 117.9 | 114.32 | 137.49 | 187.02 | 294.89 | 286.24 |
| Linseed | 1.57 | 0.88 | 0.77 | 5.34 | 8.19 | 10.25 |
| Castar | 1.11 | 2.32 | 1.84 | 2.19 | 1.90 | 1.73 |
| Seasamum | 4.45 | 7.73 | 9.75 | 12.24 | 14.70 | 15.14 |
| Cotton | 1.20 | 5.68 | 4.07 | 4.24 | 2.15 | 1.64 |
| Jute | 123.38 | 125.65 | 128.01 | 105.50 | 95.52 | 73.55 |
| Mesta | 0.21 | 5.02 | 6.73 | 13.32 | 7.43 | 5.18 |
| Tobacco | 7.63 | 9.43 | 8.35 | 5.70 | 2.60 | 1.85 |
| Chilies | 3.55 | 5.69 | 8.14 | 10.91 | 12.90 | 14.0 |
| Sugarcane | 23.66 | 26.94 | 32.06 | 48.03 | 35.80 | 36.4 |
| Potato | 18.09 | 21.97 | 24.25 | 37.35 | 59.06 | 76.75 |
| Total Cropped Area | 1936.91 | 2115.77 | 2447.14 | 2906.20 | 3290.34 | 3391.47 |
| Gross Cropped Area | 2204.5 | 2465.9 | 2882.5 | 3120.51 | 3204.75 | 3944.14 |

*Source:* Directorate of Economics and Statistics, Government of Assam, Guwahati.

*Note:* Total cropped area is less than the gross cropped area because of the non-inclusion of some minor crops in the analysis.

value shows that this trend is not statistically significant. From this simple measure of instability, it appears that though there is a substantial fluctuation in the production of foodgrain crops but it has not been experienced with greater instability during the period under study.

## Changes in Cropping Pattern

Rice is the major foodgrain crop in Assam. In most places three seasonal paddy are grown in a year—winter paddy occupies the first position both in terms of area under cultivation and the level of production, followed by Autumn paddy and Summer paddy. Next to paddy, the important cereal crops are wheat and Maize. However, other than cereals and pulses, spices, fruits, vegetables and cash crops viz mustard, seasumum, Jute, Sugarcane etc. are also grown in the State (Government of Assam, 1999-2000). During the period under study the cropping pattern was actually moved towards the intensification of food grain production in the State of Assam.

During the 49 years under review, the gross cropped area in the State of Assam increased by 1721.90 thousand hectares from 2204.5 thousand hectares in 1951-52 to 3944.14 thousand hectares in 1999-2000. The increase in rice area was 1098.5 thousand hectares, that is, its share in gross cropped area increase was 63.80 percent. The share of mustard acreage in gross cropped area increase was 9.80 per cent, of wheat 4.30 Per cent and of potato 3.40 Percent. The share of total food grain acreage in gross cropped area increase was 71.70 per cent. One of the striking features of the cropping pattern in Assam is that irrespective of yield growth rates the cropping pattern in the State witnessed a built in bias towards the production of food grain crops in general and rice in particular. The area expansion effect exerts greater impact on the expansion of production rather than the growth of yield (Table 4).

The shift of cropping pattern, as it is observed, in Assam during 1951-52 to 1999-2000 can vividly be analyzed with the help of cropped area gross cropped area elasticity (E)[6] (Venkataramanan and Prahaladachar, M., 1978). It is assumed that E will take the value unity for all individual crops in a situation of unchanged cropping pattern where the rate of growth of area under individual crops is equal to the rate of growth of gross cropped area during the same period of time. If the rate of growth of area under individual crops were found to be different from the rate of growth of gross cropped area, this would provide evidence on the changes in cropping pattern. Obviously, E will take the value greater than unity if the rate of growth of area under a crop exceeds the growth rate of gross cropped area and so, the cropping pattern would shift in favour of this crop. Similarly, E will lie between zero and unity if the rate of growth of area under crop is less than the rate of growth of gross-cropped area, while E will take negative value if the rate of growth of area under crop is found to be negative. The three-fold classification of crops according as $E< 1$, $E>1$, $0<E<1$ and $E<0$ is shown in Table below:

**Table 4**

**The Decomposition of the Total Increase in Output in Terms of Area Effect, Yield Effect, and the Crop Area-Gross Cropped Area Elasticity Under Crops in Assam During 1951-52 to 1999-2000**

| *Crops* | *Area Effect (000 tones)* | *Yield Effect (000 tones)* | *Total increase in output (000 tones)* | *Increase in area under crops (000 ha)* | *'E'* |
|---|---|---|---|---|---|
| Winter Rice[2] | 838.70 | 669.70 | 1508.40 | 573.10 | $0<E<1$ |
| Autumn Rice[2] | 233.90 | 69.10 | 303.00 | 233.80 | $0<E<1$ |
| Summer Rice[1] | 571.00 | 80.10 | 651.10 | 291.70 | $E>1$ |
| Total Rice[2] | 1529.30 | 933.30 | 2462.60 | 1098.50 | $0<E<1$ |
| Wheat[1] | 63.90 | 10.70 | 74.60 | 74.40 | $E>1$ |
| Total Cereals[2] | 1625.30 | 950.50 | 2575.80 | 1199.60 | $0<E<1$ |
| Gram[3] | -1.10 | -0.10 | -1.20 | -1.80 | $E<0$ |
| Tur[1] | 3.50 | 0.50 | 4.00 | 4.40 | $E>1$ |
| Total Pulses[2] | 24.40 | 12.00 | 36.40 | 35.00 | $0<E<1$ |
| Total Food grain[2] | 1616.90 | 995.20 | 2612.10 | 1234.60 | $0<E<1$ |
| Mustard[1] | 65.20 | 13.40 | 78.60 | 168.30 | $E>1$ |
| Jute[3] | -41.90 | -43.80 | -85.70 | -49.80 | $E<0$ |
| Tobacco[3] | -13.10 | -2.70 | -15.80 | -2.50 | $E<0$ |
| Potato[1] | 413.30 | 198.10 | 611.40 | 58.70 | $E>1$ |

*Source:* Directorate of Economics and Statistics, Government of Assam, Guwahati

*Note:* (i) 'E' stands for cropped area-gross cropped area elasticity

(ii) Total increase in output is decomposed into its area and yields constituents with the help of exponential growth rates of area, production and yield as presented in Table 1 above.

(iii) 1,2,3 denote three different categories of crops according as E>1, 0<E<1 and E<0.

It is found that summer rice, wheat, tur, Mustard and potato constitutes category 1 where E>1. It is interesting to note that in all these crops, yield growth rate is found to be much lower than the area growth rate. The category 2 includes winter rice, autumn rice, total rice, total cereals, total pulses and total food grain, while third category encompasses gram jute, and tobacco. The expansion of area under the crops of category 1 and 2 is found to be much higher than the decline in area under the third category of crops. This clearly indicates that large-scale crop-substitution did not take place in Assam during 1951-52 to 1999-2000, only marginal crops were substituted for major crops and the changes in area allocation under crops were observed mainly because of the expansion effect of gross cropped area.[5]

The changes in cropping pattern imply the changes in the allocation of area under different crops. Total change in cropping pattern is the resultant of both substitution effect and expansion effect. For a given gross cropped area, the substitution effect is defined as the relative decline in area under some crops and the compensating increase in area of the substitutable crops. The expansion effect, on the other hand, is defined as the expansion of the cropped area due to the expansion of the gross cropped area. The contributions of substitution and expansion effects in the overall change in cropping pattern in the state during 1951-52 to 1999-2000 are shown in Table below.

### Table 5
### Cropping Pattern Change in Assam: Substitution Effect and Expansion Effect (1951-52 to 1999-2000)

| *Crop* | *Substitution Effect (000 hectares)* | *Crop* | *Substitution and Expansion Effects (000 hectares)* |
|---|---|---|---|
| Gram | -1.80 | Total Food grain | 1234.60 |
| Jute | -49.80 | Mustard | 168.30 |
| Tobacco | -2.50 | Potato | 58.70 |
| Total | -54 .10 | Total | 1461.60 |

*Source:* Directorate of Economics and Statistics, Government of Assam, Guwahati.

*Notes:* (1) total change in crop area under major crops = 1461.60 thousand hectares.

(2) Substitution effect= -54.10 thousand hectares.

(3) Expansion effect = (1461.60-54.10) = 1407.50 thousand hectares.

(4) Total change in Gross cropped area during the period= 1721.60 thousand hectares.

(5) Minor crops have been excluded from the analysis.

The above Table shows that total decline in the area under gram, jute and tobacco was 54.10 thousand hectares and so, this would give a measure of substitution effect on the assumption that this decline was taken place with no change in gross cropped area. On the other hand, the area under total food grain, mustard and potato was increased by 1461.60 thousand hectares. This increase in area was either due to compensating increase in area by substitution effect or by the expansion effect of the gross cropped area. Thus the expansion effect alone led to an increase of area under these crops by (1461.60-54.10) or 1407.50 thousand hectares. Interestingly, during 1951-52 to 1999-2000, the gross cropped area of the State was increased by 1721.60 thousand hectares. The balance of 1721.60 and 1407.50 i.e., 314.10 thousand hectares might be because of the non-inclusion of some minor crops in the analysis and partly because of the error of aggregation.

So, the empirical evidence and the subsequent analysis reveal that it was expansion effect which exclusively driven the changes in cropping pattern in the State. The expansion effect along could explain 97 percent of the change. The contribution of substitution effect on the overall change in cropping pattern in the state was found to be negligible (only 3 percent).

## Conclusion

The overall performance of agriculture in the state does not show any significant trend of prosperity. The increase in yield rates, which is very important under the current circumstances, has been very low and even negative for many crops. Therefore, to raise productivity a conscious effort is necessary involving agricultural scientists, extension and development agencies to unveil the untapped potential of agriculture in the State to obtain the targeted growth in yield of crops. In fact, in the absence of any appreciable change in relative yield growth rates between crops, the farmers in the State did not make any significant changes in their cropping pattern. Instead, there seems to be a gradual shift of cropping pattern from diversification towards unification favouring food grain crops. The main changes in cropping pattern that have occurred during the period are the relative growth of area under food grain crops and the relative decline in area under some minor crops viz gram, jute and tobacco. The expansion of gross cropped area was led mostly to the expansion of area under food grain crops. In effect, the expansion effect of gross cropped area could explain 97 per cent of the total changes in area allocation under crops during the 49 years since 1951-52.

*Notes*

1. The kinked exponential model has been specified by re-normalizing time such that t=0 at the break point (year 1974-75). The sub period growth rates have been estimated with a joint intercept: $\text{Log } Y_t = a + b_1 D_1 t + b_2 D_2 t + u_t$. Where D is a dummy variable, which takes the value 1 in the jth sub period and 0 otherwise. The break point has been chosen the year 1974-75 because in the State of Assam the use of HYV seeds, fertilizers etc. started popularized from this period only.

2. The log-quadratic function $Y_t = a + bt + ct^2 + U_t$ is fitted to examine the acceleration or deceleration in growth trends (Boyce, 1987). If c is found to be significantly different from zero, it can be used as a measure of the rate of acceleration or deceleration in growth trends.
3. To measure the variability in area, yield and production of major crops of Assam, the statistical index the coefficient of variation (C.V.) has been used. The higher value of C.V. implies greater fluctuation and vice-versa.
4. Instability is measured in terms of trend in fluctuation, that is, by examining changes over time in the deviation of actual production from the estimated levels of production (Saha and Swaminathan, 1994). Symbolically, $z_t = y_t - \acute{y} \mid / \ y = a + c_t$; the residual, $z_t$ has been regressed on time, where ý refers to the estimated level of production from the OLS trend equation.
5. It is to be noted that during this period there were significant increase in the area under linseed (552.90 percent), seasamum (240.20 per cent), mesta (2366.70 percent) and chilies (234.10 per cent). Since these crops occupied only a small share of gross cropped area, their shares in the increase in gross cropped area were also small. There was a significant decline in area under tobacco (75.80 percent), gram (43.80 percent) and jute (40.40 percent).
6. By definition, 'E' is the ratio of the percentage change of the area under crop and the percentage change in gross cropped area.

## APPENDIX

### Table A1: Exponential growth rate of Total Food grain in the Districts of Assam During 1951-52 to 1999-2000

| *Districts* | *Growth rate percentage of* | | |
|---|---|---|---|
| | *Production* | *Area* | *Yield* |
| Goalpara | 1.84(21.36) | 4.53(16.26) | 0.31(2.51) |
| Kamrup | 1.76(16.91) | 0.82(20.91) | 0.94(0.01) |
| Darrang | 1.28(12.21) | 1.07(21.74) | 0.21(2.04) |
| Nagaon | 2.63(15.69) | 1.58(29.05) | 1.05(7.47) |
| Sibsagar | 1.92(18.75) | 0.81(14.82) | 1.11(12.66) |
| Lakhimpur | 2.32(16.84) | 1.93(23.26) | 0.39(3.32) |
| Cachar | 1.39(6.52) | 0.29(2.06) | 1.10(7.89) |
| United M & N.C. Hills | 5.84(18.29) | 4.75(18.18) | 1.09(8.94) |

*Source:* Directorate of Economics and Statistics, Government of Assam.

*Note:* Figures in the parentheses are the 't' values.

### Table A2:
### Exponential growth rate of Total Rice in the Districts of Assam During 1951-52 to 1999-2000

| *Districts* | *Growth rate percentage of* | | |
|---|---|---|---|
| | *Production* | *Area* | *Yield* |
| Goalpara | 1.57(17.27) | 1.31(15.13) | 0.26(2.07) |
| Kamrup | 1.64(15.14) | 0.71(20.21) | 0.93(8.75) |
| Darrang | 0.92(8.13) | 1.19(16.97) | -0.28(-2.31) |
| Nagaon | 2.53(13.87) | 1.42(21.42) | 1.11(6.90) |
| Sibsagar | 1.88(17.72) | 0.79(9.88) | 1.09(10.38) |
| Lakhimpur | 2.25(22.67) | 1.85(25.11) | 0.04(5.95) |
| Cachar | 1.44(9.65) | 0.47(9.56) | 0.97(6.83) |
| United M & N.C. Hills | 5.79(17.61) | 4.73(17.44) | 1.06(8.74) |

*Source:* Directorate of Economics and Statistics, Government of Assam.

*Note:* Figures in the parentheses are the 't' values.

## Table A3: Log-quadratic estimates of trend of production and yield of major crops in Assam during 1951-52 to 1999-2000

| *Crops* | *Production* | | *Yield* | |
|---|---|---|---|---|
| | *b* | *c* | *b* | *c* |
| Total Rice | 0.008819<br>(3.28) | 0.000205*<br>(3.93) | -0.009240<br>(-3.32) | 0.0003296*<br>(6.12) |
| Winter Rice | 0.009082<br>(2.99) | 0.000160<br>(2.73) | -0.006340<br>(-1.69) | 0.0002779*<br>(3.81) |
| Autumn Rice | 0.020730<br>(3.68) | -0.000081<br>(-0.743) | -0.014800<br>(-2.93) | 0.0003713*<br>(3.795) |
| Summer rice | 0.229000<br>(9.14) | -0.00209*<br>(-4.30) | -0.014800<br>(-1.26) | 0.0004167<br>(1.83) |
| Wheat | 0.254000<br>(9.304) | -0.00277*<br>(-5.22) | 0.226000<br>(10.46) | -0.002540*<br>(-6.05) |
| Total Cereals | 0.012190<br>(4.77) | 0.000162<br>(3.27) | -0.008690<br>(-3.93) | 0.000323<br>(7.54) |
| Total Pulses | 0.018420<br>(4.01) | 0.000008<br>(0.095) | -0.006040<br>(-1.851) | 0.000245*<br>(3.88) |
| Total Foodgrains | 0.012300<br>(4.95) | 0.000158*<br>(3.29) | -0.007650<br>(-2.76) | 0.000306*<br>(5.699) |
| Rape & Mustard | 0.017940<br>(2.11) | 0.000277<br>(1.68) | -0.001880<br>(-0.409) | 0.0001454<br>(1.63) |
| Jute | 0.003047<br>(0.20) | -0.000049<br>(-0.171) | -0.003890<br>(-0.268) | 0.0002645<br>(0.94) |

*Source:* Directorate of Economics and Statistics, Government of Assam

*Note:* (1) Figures in the parentheses are the 't' values

(2) * Implies significant at 5 per cent level.

## Table: A4

## The co-efficient of variation of area, yield and production of major food grain crops in Assam during 1951-52 to 1999-2000

| *Districts* | *Coefficient of Variation (%)* | | |
|---|---|---|---|
| | *Area* | *Yield* | *Production* |
| Total Rice | 16.90 | 14.40 | 29.40 |
| Winter Rice | 14.90 | 15.20 | 26.40 |
| Autumn Rice | 18.90 | 14.50 | 26.40 |
| Summer Rice | 112.30 | 26.10 | 144.60 |

| | | | |
|---|---|---|---|
| Wheat | 76.90 | 27.80 | 86.00 |
| Total Pulses | 20.40 | 13.10 | 29.00 |
| Total Cereals | 18.10 | 13.90 | 30.50 |
| Total Food grains | 18.00 | 14.70 | 30.40 |

*Source:* Directorate of Economics and Statistics, Government of Assam.

## Table A5: Contribution of Area and Yield Variations in total Production Variation in Assam during 1951-52 to 1999-2000

| *Crops* | *Variance component* | | | |
|---|---|---|---|---|
| | *Area(%)* | *Yield(%)* | *Interaction(%)* | *Total Production (%)* |
| Total Rice | 37.50 | 23.30 | 39.20 | 100 |
| Winter Rice | 32.90 | 33.60 | 33.50 | 100 |
| Autumn Rice | 56.80 | 29.10 | 14.10 | 100 |
| Summer Rice | 73.50 | 4.10 | 22.40 | 100 |
| Wheat | 73.00 | 2.80 | 24.20 | 100 |
| Total Pulses | 48.80 | 18.80 | 32.40 | 100 |
| Total Cereals | 39.70 | 19.80 | 40.50 | 100 |
| Total Food grains | 39.00 | 22.10 | 38.90 | 100 |

*Source:* Directorate of Economics and Statistics, Government of Assam

## Table A6: Instability of Foodgrain Production in Assam during 1951-52 to 1999-2000

| *Crops* | *Instability coefficient* |
|---|---|
| Total Rice | -0.000761(-1.302) |
| Wheat | -0.0219(-6.88) |
| Total Cereals | -0.00131(-2.204) |
| Total Pulses | -0.000518(-0.910) |
| Total Foodgrains | -0.00135(-2.311) |

*Source:* Directorate of Economics and Statistics, Government of Assam.

*Note:* Figures in the parentheses are the 't' values.

**Table A7: Decadal Growth Rates for Major Food grain crops in Assam during 1951-52 to 1999-2000**

| *Crops* | | *1951-52 to 1959-60* | *1960-61 to 1969-70* | *1970-71 to 1979-80* | *1980-81 to 1989-90* | *1990-91 to 1999-2000* |
|---|---|---|---|---|---|---|
| Total Foodgrain | P | 1.25(2.58) | 1.94(2.29) | 0.93(1.48) | 1.21(1.39) | 1.23(2.82) |
| | A | 1.13(5.29) | 1.61(8.59) | 1.74(7.77) | 0.53(0.65) | 0.07(0.34) |
| | Y | 0.12(0.22) | 0.33(0.37) | -0.81(-1.76) | 0.68(0.44) | 1.16(4.08) |
| Total Rice | P | 1.24(2.48) | 1.67(2.37) | 0.72(1.21) | 0.54(0.71) | 1.05(2.19) |
| | A | 1.07(4.95) | 1.48(8.94) | 1.53(7.08) | 1.41(2.44) | -0.002(-0.001) |
| | Y | 0.17(0.31) | 1.18(0.25) | -0.81(-1.79) | -0.87(-0.97) | 1.05(3.54) |
| Winter Rice | P | 0.97(1.16) | 1.02(1.49) | 0.84(1.88) | 0.48(0.46) | -0.05(-0.09) |
| | A | 0.55(2.36) | 0.93(6.51) | 1.53(11.11) | 1.58(1.65) | -0.72(-2.05) |
| | Y | 0.42(0.54) | 0.08(0.12) | -0.69(-1.75) | -1.10(-0.63) | 0.67(2.18) |
| Summer Rice | P | 20.20(4.45) | 16.40(3.93) | -0.88(-0.26) | 11.30(5.14) | 11.40(5.56) |
| | A | 17.40(5.47) | 12.50(3.13) | 3.47(2.92) | 8.20(7.97) | 8.51(5.89) |
| | Y | 2.86(1.02) | 3.90(12.12) | -4.35(-1.46) | 3.07(2.16) | 2.89(4.03) |
| Autumn Rice | P | 1.54(0.61) | 3.42(2.62) | 0.44(0.34) | -0.93(0.064) | -0.28(-0.27) |
| | A | 2.54(6.69) | 2.63(7.88) | 1.44(2.82) | 0.06(0.09) | -0.99(-2.58) |
| | Y | -0.99(-0.37) | 0.79(0.59) | -0.99(-1.01) | -0.99(-1.03) | 0.71(0.81) |
| Total Cereals | P | 1.25(2.51) | 1.69(2.43) | 0.93(1.45) | 1.08(1.54) | 1.02(2.22) |
| | A | 1.12(5.06) | 1.53(9.21) | 1.75(7.65) | 0.67(2.56) | 0.03(0.14) |
| | Y | 0.13(0.24) | 0.16(0.22) | -0.82(-1.80) | 0.42(0.75) | 0.99(3.37) |
| Total Pulses | P | 1.28(0.96) | 3.12(3.26) | 0.95(0.65) | 1.17(0.98) | 3.75(6.46) |
| | A | 1.48(1.67) | 2.12(8.07) | 1.38(3.22) | 0.61(0.58) | 0.87(1.61) |
| | Y | -0.20(-0.16) | 0.86(0.93) | 0.42(0.50) | 0.68(1.39) | 2.88(5.41) |

*Source:* Directorate of Economics and Statistics, Government of Assam.
*Note:* Figures in the parentheses are the 't' values.

* An earlier version of this paper was presented at the 66th Annual Conference of Indian Society of Agricultural Economics, at Shillong, Meghalaya in November, 2006.

## REFERENCES

Boyce, J.K. (1987): Agrarian Impasse in Bengal: Agricultural growth in Bangladesh and West Bengal 1949-1980, Oxford University Press.

Govt. of Assam (2000): Economic Survey, Directorate of Economics and Statistics, Guwahati.

-----------------(1986-86): Agricultural Census, Directorate of Economics and Statistics, Guwahati.

Kuri, P.K.(2000): 'Changes in Cropping Pattern in Assam during 1950-51 to 1994-94', *Arunachal University Research Journal*, Vol. 3 (1).

Phukan, U. (1990): *Agricultural Development in Assam*, Mittal Publication, New Delhi.

Saha, A. and Swaminathan, M. (1994): 'Agricultural growth in West Bengal in the 1980s: A Disaggregation by Districts and crops', *Economic and Political Weekly*, Vol. XXIX No. 13.

Venkataramanan, L.S. and Prahaladachar M. (1978): 'Study of Cropping Pattern Changes in Andhra Pradesh during 1950-75', *Indian Economic Review*, Vol. XIII, New Series, No. 2, Oct.

# 15

# FOOD GRAINS PRODUCTION IN ASSAM: AN ANALYSIS OF GROWTH AND INSTABILITY

*Juthika Barman & K.K. Barman*

## Introduction

Assam is the easternmost state of the Indian union and is located between 24°08′N and 27°58′ N latitudes and 89°42′E and 96°01′E longitudes. The economy of the state is predominantly agricultural with a total cultivable area of 39 lakh hectares (1998-99), which constitutes 50.2 per cent of the total geographical area.

Agriculture in the state continues to face various challenges. Most of the farmers are small and marginal and being poverty stricken; they find it difficult to accept any risks and hence are less responsive to any change in technology. The fertilizer consumption figures for Assam and India for 1999-2000 are 27.94 and 93.81 kg\hectares respectively (CMIE, Dec., 2002). The state is far behind in terms of irrigated area as it has covered only 33 per cent of irrigable area against the national average of about 80 per cent (Govt. of Assam, 2000-2001). Most of the crops in the state are grown under rainfed conditions.

Assam produces both food crops and cash crops. But the cropping pattern of the state is dominated by food grains particularly paddy crops. Rice is a staple food of the people of Assam and it is grown everywhere by all sections of the society in the hills and the plains. In spite of the importance of rice in the economy of Assam, the average productivity of food grains in general and rice in particular is one of the lowest in Assam amongst Indian states.

Further, fluctuations in crop output particularly in the food grains production happens to be a regular feature in Assam's agriculture. The main reason for this is the dependence on nature i.e. rainfall for the cultivation of crops in the state of Assam. Natural hazards like flood and drought are also responsible for such fluctuations. Inadequate irrigation facilities and a lack of permanent measures to control flood and erosion have compounded the problem in the state.

It is in the light of the problem discussed above an analysis of the growth performance and instability of food grains production in Assam is made for a period of 50 years from 1951-52 to 1999-2000. The analysis is made for the entire period and also for two periods, namely period I- Pre Green Revoluytion Period (1951–52 to 1966–67) and period II- Post green Revolution Period (1967–68 to 1999–2000). Period II is further divided into three parts – the First Part (1967–68 to 1977–78), the Second Part (1978–79 to 1988–89) and the Third Part (1989–90 to 1999–2000) in order to examine the decade wise growth and instability.

## Methodology

Compound growth rates of area, production and productivity have been estimated for four crop groups namely 'Rice', 'Cereals', 'pulses' and 'Food Grains' for the state of Assam by fitting a semi logarithmic trend function of the following form:

$Y = e^{a+bt}$

Where, y = Area/Production/Yield

a = Constant

b = Growth rate

t = time

For measuring fluctuations, co-efficient of variations have been calculated for each crop group for different periods.

To examine the source of instability, variance of production is decomposed into its constituent sources viz. area variance, yield variance, area-yield co-variance and higher order interaction between area and yield.

Let Q denote production, A the area sown and Y, yield. Then for each crop and district

$Q = AY$

The variance of production V (Q) can be expressed as

$V(Q) = \ddot{A}^2 V(Y) + Y^2 V(A) + 2 AY COV (A_2Y) - COV (A_2Y)^2 + R$

Where $\ddot{A}$ = Mean area

Y = Mean yield

R = Residual term (which is expected to be small)

The above equation is not only a function of variances of yields and areas sown, but also the mean area and yield and of co-variance between areas and yields.

## Growth rates in Area, Production and Productivity of Food grains

Table 1 and 2 present the annual compound growth rates in area, production and productivity of four crop groups in Assam for different periods and sub periods. Table 1 shows that during the Pre Green Revolution Period, the area under all the four crops registered significant growth rate . As compared to period I, the area growth rate is found to have declined for all the crop groups during period II. Area growth rate is also found to be significant in the state during the whole period under study.

**Table 1**

**Compound Growth Rates in Area, Production and Productivity of Different Food Grain Crops in Assam**

*(In percent)*

| *Crops* | *Period I* | | | *Period II* | | | *Period III* | | |
|---|---|---|---|---|---|---|---|---|---|
| | *1951-52 to 1966-67* | | | *1967-68 to 1999-00* | | | *1951-52 to 1999-00* | | |
| | *A* | *P* | *Y* | *A* | *P* | *Y* | *A* | *P* | *Y* |
| Rice | 0.53* | 0.37** | -0.16 | 0.41* | 0.96* | 0.55* | 0.51* | 0.84* | 0.33* |
| Cereals | 0.55* | 0.38** | -0.16 | 0.43* | 0.97* | 0.54* | 0.56* | 0.88* | 0.32* |
| Pulses | 0.46** | 0.80** | 0.34 | 0.43* | 0.90* | 0.47* | 0.55* | 0.83* | 0.28* |
| Total Food Grains | 0.54* | 0.39** | -0.15 | 0.43* | 0.97* | 0.53* | 0.56* | 0.88* | 0.32* |

*Source:* Estimated from data collected from Directorate of Agriculture (Statistical wing), Khanapara, Assam.

*significant at 10% level, **significant at 5% level, ***significant at 1% level, A-Area, P-production, Y-productivity, no star-non significant.

As far as the yield is concerned, it remained stagnant for all the crop groups during Period I. Due to stagnancy in productivity, the production growth rate of food grains during Period I was only 0.39 per cent in Assam, which however increased to 0.97 per cent during the Post Green Revolution period. The significant growth in productivity throughout the state during Period II might be due to the use of a package of practices like fertilizer, pesticides, HYV rice and extension of irrigation facilities in the state. For the entire period, growth rate of area under food grain (0.56 per cent) is found to be higher than the growth rate of productivity (0.32 per cent). It is important to note that the growth rates in area, production and productivity for rice , cereals and total food grains have shown almost similar pattern. This is due to the fact that rice is the most important cereal crop which occupies more than 90 per cent of the cropped area under food grain.

Table 2 exhibits compound growth rates of rice, cereals, pulses and total food grains of Assam for four sub-periods in terms of area, production and productivity. The sub Period I covers the period 1951–52 to 1966–67 i.e., the Pre Green Revolution Period.

It can be observed from Table 2 that area continued to be the major component to increase production of food grains in Sub Period II. As compared to sub-Period I, the state exhibits higher growth in area with 0.88 per cent for total food grains during sub Period II. The high growth rate in area during this period might be due to the fact that more and more land in the state had been brought under cultivation through reclamation of land. For instance, during the period from 1967–68 to 1977–78, net area sown in the state increased from 2192 thousand hectares to 2651 hectares. As far as the productivity is concerned it remained stagnant for all the four crop groups in this sub period. The low productivity during this period was mainly due to persistence of traditional form of agricultural technology and many bottlenecks in the input supply system. The infrastructure for rapid agricultural development in the form of transport and communications, storage facilities, electrification, irrigation was not adequate. The recurrent floods in the plain districts of the state also affected agricultural productivity adversely. Despite stagnancy in productivity, production growth rate improved in the state to the level of 1.03 per cent for total food grains in sub-period II as against 0.39 per cent during sub Period I. This indicates that area played the major role in raising production of food grain in Assam till 1977 – 78. In sub Period III, area continued to grow in the state for all the crop group. It is important to note that productivity which remained stagnant during the earlier years, grew significantly for all the crop groups except pulses during sub Period III. Under the impact of area and yield, production of food grains increased by 0.90 per cent per annum in Assam in this sub period.

**Table 2: Decadewise Compound Growth Rate in Area, Production and Productivity of different Food Grain Crops in Assam**

*(in percent)*

| *Crops* | *Sub-Period I 1951-52 to 1967-68* | | | *Sub-Period II 1978-79 to 1977-78* | | | *Sub-Period III 1989-90 to 1988-89* | | | *Sub-PeriodIV 1966-67to 1999-2000* | | |
|---|---|---|---|---|---|---|---|---|---|---|---|---|
| | A | P | Y | A | P | Y | A | P | Y | A | P | Y |
| Rice | 0.53* | 0.37** | -0.16 | 0.73* | 0.86** | 0.13 | 0.28*** | 0.88*** | 0.60*** | 0.05 | 0.71*** | 0.64** |
| Cereals | 0.55* | 0.38** | -0.16 | 0.90* | 1.04** | 0.13 | 0.29*** | 0.89*** | 0.59*** | 0.05 | 0.69*** | 0.64** |
| Pulses | 0.46** | 0.80** | 0.34 | 0.54** | 0.58 | 0.04 | 1.28** | 1.62** | 0.33 | 0.30 | 1.48* | 1.18* |
| Total Food Grains | 0.54* | 0.39** | -0.15 | 0.88* | 1.03** | 0.14 | 0.34** | 0.90*** | 0.56*** | 0:06 | 0.70** | 0.61** |

*Source:* Estimated from data collected from Directorate of Agriculture (Statistical wing), Khanapara, Assam.

*significant at 10% level, **significant at 5% level, ***significant at 1% level, A-Area, P-production, Y-productivity, no star-non significant.

Area in the state which increased significantly in the earlier sub-periods remained stagnant during sub period IV. The growth rate of production during this sub Period was mainly due to significant growth in productivity.

The growth rate in productivity in the state during 1977-78 to 1999-2000 might be due to growing use of HYV seeds, chemical fertilizer, modern implements etc. From 1975-76 to 2000-2001 the area under HYV paddy in Assam gradually increased from 2.87 lakh hectares to 14.01 lakh hectares , which is around 34.2 per cent of the total cropped area of the state.

With the adoption of new agricultural strategy, the consumption of fertilizer by the agricultural sector of the state started to increase gradually. During the period from 1976-77 to 1999-2000 consumption of fertilizer per hectare of land in Assam has increased from 1.3 kg to 27.94 kg.

The above analysis indicates that till the end of seventies, growth in production of food grains was possible only by bringing more land under cultivation as the new agricultural strategy had little impact on agricultural production of Assam. It is only after 1977-78 the state has started to experience to a limited extent the new agricultural strategy with the growing use of HYV seeds, chemical fertilizer, modern implements etc. So we can conclude that Assam has been experiencing some impact of Green Revolution in the production of food grains during the eighties and nineties.

### Measurement of Fluctuation

For measuring fluctuations, co-efficient of variations of Area, production and productivity have been calculated for the four crop groups for the period of Pre-Green Revolution, Post Green Revolution, the whole period under study. Table 3 depicts the co-efficient of variation (C.V.) of area, production and productivity of four crop groups.

**Table 3: Coefficient of Variations of Area, Production and Productivity of Food Grain Crops in Assam**

| *Crops* | *Period I 1951-52 to 1966-67* | | | *Period II 1967-68 to 1999-2000* | | | *Period III 1951-52 to 1999-2000* | | |
|---|---|---|---|---|---|---|---|---|---|
| | *A* | *P* | *Y* | *A* | *P* | *Y* | *A* | *P* | *Y* |
| Rice | 5.98 | 6.16 | 5.44 | 9.79 | 23.45 | 14.29 | 16.49 | 29.53 | 14.08 |
| Cereals | 6.15 | 6.21 | 5.43 | 10.45 | 23.50 | 14.05 | 18.08 | 30.53 | 13.84 |
| Pulses | 7.33 | 12.86 | 8.80 | 14.85 | 23.05 | 14.13 | 20.47 | 29.54 | 13.91 |
| Total Food Grains | 6.08 | 6.24 | 5.34 | 10.46 | 23.41 | 13.95 | 18.07 | 30.45 | 13.75 |

*Source:* Estimated from data collected from Directorate of Agriculture (Statistical wing), Khanapara, Assam.

A-area, P-Production, Y-Productivity.

It has also been observed from the Table 3 that for the state of Assam, co-efficient of variation of production was much higher as compared to area and productivity variation during the period 1951-52 to 1999-2000. During the Pre-Green Revolution period, such degree of fluctuation for production was slightly higher than area and yield fluctuation. But during the Post Green Revolution period the production recorded very high fluctuation around 23 per cent (Table 3) for food grain. In this period the area fluctuation for food grains was around 10 per cent, while the fluctuation in productivity was around 14 per cent. It has also been observed that in period I, the area variations were slightly higher than yield variation, but during period II, yield has shown much higher fluctuation than the area. This may be due to the fact that with the adoption of new agricultural strategy, the farmers of Assam have also started to adapt multiple cropping system in the agricultural operation. The success of this system depends on assured and controlled water supply. As the irrigation facilities in the state is not properly developed, farmers have to depend on rainfall, which is very irregular and uneven, causes ups and downs in the agricultural production. Moreover, occurrence of flood, which is an assured feature in the state, also causes fluctuation in agricultural production.

### *Sources of Variances*

From the above discussion, it has been found that there exists a considerable variation in the production of food grains in the state of Assam. To find out the possible sources that causes variance, the variance of production of food grains has been decomposed into area variance, productivity variance and area productivity co-variance for different time period separately using Hazell's (1982) decomposition model. To apply Hazell's variance decomposition method, the total time span covered by the data has been divided into two periods. The first period covers 25 years starting from 1951–52 to 1975 –76 and the second period covers 24 years beginning from 1976–77 to 1999–2000. Data on yield and area of or total rice, total cereals, total pulses and total food grains for each district were detrended separately for each period using linear trends. For each variables, the residual of the trend equation were added to the mean of the data series to generate detrended data centered on the mean. The detrended production was obtained by multiplying detrended yield and detrended area. The final analyses were conducted using these detrended data. The result has been presented in Table 4.

**Table 4: Sources of Variance in Production**

*(in per cent)*

| | Period | Area Variance | Yield Variance | Area yield covariance | Higher order interaction | Residual | Total |
|---|---|---|---|---|---|---|---|
| | | V(A) | V(Y) | Cov(A,Y) | Cov (A,Y) | R | 100 |
| | I | 80 | 12 | 4 | 0 | 4 | 100 |
| Rice | I I | 48 | 9 | 32 | 0 | 11 | 100 |
| | I I I | 101 | 12 | -15 | 0 | 2 | 100 |
| | I | 64 | 17 | 15 | 0 | 4 | 100 |
| Cereals | I I | 52 | 7 | 30 | 0 | 11 | 100 |
| | I I I | 111 | 19 | -32 | 0 | 2 | 100 |
| | I | 72 | 14 | 14 | 0 | 0 | 100 |
| Pulses | I I | 46 | 71 | -17 | 0 | 0 | 100 |
| | I I I | 67 | 74 | -35 | 0 | -6 | 100 |
| | I | 64 | 16 | 15 | 0 | 5 | 100 |
| Food grain | I I | 1 | 65 | 12 | 0 | 22 | 100 |
| | I I I | 114 | 20 | -36 | 0 | 2 | 100 |

*Source:* Estimated from data collected from Basic Agricultural Directorate of Agriculture, Statistics (statistical wing), Khanapara , Assam .

Period I 1951-52--1975-76 Period II 1976-77--1999-2000

Period III 1951-52--1999-2000

The decomposition of production variance into sources of variance for the state of Assam reveals that area variance has contributed the major share to the variance in production in the state of Assam during the period from 1951–52 to 1975–76 (Period I) for all the four crop groups. Area variance has continued to contribute the major share of production variance for the crop group of rice and cereals during Period II (1976–77 to 1999–2000), but for total pulses and food grains productivity variance is found to have contributed the major share. In the entire period from 1951–52 to 1999–2000, area variance is the main source of production variance for total food grains, total cereals and total rice in the state of Assam. While for pulses, both area variance (67 per cent) and productivity variance (74 per cent) contributed in a simmilar manner to the production variance. The negative contribution of area productivity covariance for all the crop groups during period III indicate that it has a stabilizing effect on the instability of food grains production in the state.

Thus, the above analysis has brought out the conclusion that area variance is the dominant source of production variance for the crop group of rice, cereals and food grains in all the districts of Assam during the period from 1951–52 to 1999–2000. This has been proved by the study undertaken by Deka (2000) that during the period 1971–72 to 1996–97, area variance shared the major portion (79 per cent) of production variance of rice in Assam. However, Hazell (1982) concluded that productivity variance was the important source of production variance of cereal crops in India.

### *Components of Change in Variance of Production*

In order to examine the sources of instability more precisely the change in variance in production for each crop group was decomposed into several components. The effect of different components in the changes in the variance of production for the four crop groups—total rice, total cereals, total pulses and total food grains for the state of Assam is presented in table 5. It is observed from the table that changes in the variance of yields account for the largest shares of changes in variance of production for all crop groups except pulses. They account for 22 per cent of the increase in the variance of total rice and total cereal production and 23 per cent for the increase in the variance of total food grain production. For pulses, the change in the variance of yields had a negative effect (-7 per cent) to the changes in production variance.

Changes in mean yields account for small shares of the changes in the variance of production for all the crop groups; they account for only 1 per cent of the increase in the variance of total rice, total cereals and total food grains and 4 per cent of the increase in the variance of total pulses.

**Table 5: Components of Change in the Variance of Production of Food Grains in Assam (1951-52-1975-76 to 1976-77-1999-2000)**

*(in per cent)*

| | *Rice* | *Cereals* | *Pulses* | *Food Grains* |
|---|---|---|---|---|
| Change in mean yield | 1 | 1 | 4 | 1 |
| Change in mean area | 6 | 9 | 47 | 9 |
| Change in yield variance | 22 | 22 | -7 | 23 |
| Changes in area variance | 5 | 4 | 78 | 3 |
| Interaction between changes in mean yield and mean area | 0 | 0 | 0 | 0 |
| Changes in area-yield covariance | 22 | 19 | -26 | 17 |
| Interaction between changes in mean area and yield variance | 17 | 20 | -7 | 21 |
| Interaction between changes in mean yield and area variance | 2 | 2 | 24 | 1 |
| Interaction between changes in mean area and yield and changes in area-yield covariance | 13 | 12 | -16 | 11 |
| Change in residual | 12 | 12 | 1 | 14 |
| Total | 100 | 100 | 100 | 100 |

*Source :* Estimated from data collected from Basic Agricultural Statistics (statistical wing). Directorate of Agriculture Khanapara, Assam.

Changes in the variance of area account for the largest share of the changes in variance of production for pulses. They account for 78 per cent of the increase in the variance of total pulses production. For total rice, total cereals and total food grains, the share of this component to the increase in variance of production is 5 per cent, 4 per cent and 3 per cent respectively.

For total pulses, changes in mean area account for the second largest component (47 per cent) of the changes in variance of production. This component also had destabilizing effects on total rice (6 per cent), total cereals (9 per cent) and total food grains (9 per cent).

Changes in the co-variances between areas and yields account for large part of the changes in the variance of production for total rice, total cereals and total food grains. They account for 22 per cent for total rice, 19 per cent for total cereals and 17 per cent for total food grains. But it had a stabilizing effect (-26 per cent) to the changes in variance of production of pulses.

It can be observed from the table that interaction terms are also an important contributing factor to the changes in variance of production of total rice, total cereals and total food grains. The interaction terms in all accounted for 32 to 33 per cent of the increase in variance of all the crop groups except pulses. Interaction terms had little effect on the change in variance of total pulses production in the state of Assam.

Finally, changes in the residual term also contribute to the changes in the production variance of total rice (12 per cent), total cereals (12 per cent), total food grains (14 per cent). But it accounts for only 1 per cent of the changes in variance of pulses production.

## Conclusion

In the context of the foregoing discussion, it is observed area expansion was the major component of the growth of food grains production in Assam prior to the period 80's; but during 90,s the area growth rate was insignificant. This indicates that there is little scope to increase agricultural production in the state by extending cultivation to new areas. But there is ample scope to raise the intensity of cropping, which is only 149 percent in 1999-2000. The study has also shown that Assam has been experiencing some improvement in raising the productivity of food grain crops during 80's and 90's , but this is not sufficient to meet its domestic demand for food as the population of the state is increasing at an alarming rate. Further, it is observed there has always been instability in the growth of food grains production.

Such fluctuation in production is due to both area and yield fluctuations. But during Post-Green Revolution period yield has shown much higher fluctuation than the area.

Thus we can conclude that in the context of the rapid growth of population and little scope for extension of cultivation to new areas, the future strategy of raising food grains production in Assam should be concentrated on increasing cropping intensity and greater emphasis should be laid on increase per hectare yield of crops. For this purpose farmers of Assam should be provided with all necessary facilities such as quality seeds, adequate fertilizer and water management, credit facilities, plant protection measures, marketing and storage facilities etc. Steps should also be taken by the Government to distribute land in such a way that each farmer gets sufficient land for remunerative production. Further, appropriate measures should be taken to control flood which causes much damage to crops every year in all localities of the state. As the intensity and complexity of flood have been increasing year after year, control of flood should be treated as national problem and as such it should be taken over by the Government of India.

## REFERENCES

Bandyopadhya, Alok (1989), 'Growth and Instability in the production of main cereal crop of West Bengal and Punjab-Haryana, 1950-51 to 1984-85—A Note', *Indian Journal of Agricultural Economics (IJAE)*, Vol. XXXXIV, No.2, April–June.

Das, Purendra Sekhar (1978), 'Growth and Instability in Crop Output in Eastern India', *Economic and Political Weekly (EPW)*, Vol. XIII. No.41, October.

Deka, N. (2000), Determinants of Instability and Efficiency in Rice Production under Rainfed Agriculture in Assam, Ph. D. Thesis, Assam Agricultural University, Jorhat.

Hazell, Peter, B. R. (1982), 'Instability in Indian Foodgrain Production', *Research Report 30*, International Food Policy Research Institute, May.

Kalita, D. R. and Baruah, B. K. (1992), 'Growth in Rice Production in Assam', *A S I*, Vol. XLVII, No. 4, July.

Khan, S. A. (1964), Growth of Food Grain Production in Bihar, *IJAE*, Vol. XXIV, No. 2, April-June.

Mahendradev, S. (1987) 'Growth and Instability in Foodgrains—An Interstate Analysis', *EPW*, Vol. XXII, No. 39, September.

# 16

# GROWTH OF CROPS, EXTENT OF DIVERSIFICATION AND EFFECT OF TECHNOLOGY WITH SPECIAL REFERENCE TO RICE IN THE FLOOD PLAINS OF ASSAM

*K.C. Talukdar*

## Introduction

Diversification of farms enterprises has received impetus in the flood affected areas to overcome risks in generating income and employment for sustained living. Adjustment of cropping pattern is more crucial as rearing of livestock is more risky in the flood prone areas. Generation of income can be increased through diversification with high valued crops in the farm families with skillful management of seasonal and perennial crops having market demand. Value based crop diversification will also encourage land-saving cropping pattern increasing per capita farm income. In broad terms, diversification of crops encompasses the situations like shifting of crops or enterprises to more profitable crops or enterprises, use of resources in diverse but complementary activities.

The marginal and small group of farms in a subsistence economy needs farm diversification for various reasons. The prime reason of such diversification is food security while income generating crop activities are often neglected by the farmers of low economic status in marginal holdings. Their marketed surplus is low and negative indicating a non linear relationship with size of holding particularly in cereals in India (Rudra 1982, Talukdar 1984). In flood affected areas farmers are more concerned for crop production before and after flood situation and the cropping pattern depends on various factors. The important ones are land type and its topography, irrigation, food requirements, economic status of the farmers, accessibility to markets, availability of farm labour and other inputs, commercial aptitude of farmers and general farm environment. It is used as a risk and resource management tool. It provides eco system check and balance mechanism for controlling pests (Jha, 1996). It is also important to manage soil, water, plant nutrients and effective labour distribution for growth and sustainability (Joshi, 2001). Flood affected areas are fragile and there is uncertainty of use before and after flood. There is much scope to diversify crops in the flood affected areas as per land type.

Intensity of flood damage is higher in Nagaon district and crop area was highly affected during 1994 to 2000 in the state. During 1984-89 higher area affected was accompanied by higher affect of total cropped area. Cropping intensity is also higher in Nagon district. On an average, about two lakh hectares are affected annually by flood in the state and it was found to increase after 1989.It is evident that flood recurs in Assam every year and the farmers are to adjust their cropping pattern incorporating high valued crops so as to sustain their food and income.

Rice is the staple food of every Assamese and its production is declining in the state. During 2003-04, total production of rice was 38.81 lakh mt while it had declined to 34.71 lakh mt during 2004-05. Productivity per hectare had also declined from 1794 kg to 1476 kg with the decline in rice area from 25.30 lakh hectares to 23.84 lakh hectares during the same period. Lack of a system approach for input and output distribution and procurement, timely and need based extension services, lopsided attention to develop horticulture, lower profitability, flood havoc and inadequate budget for rural extension were some of the reasons for declining trend of rice in the state. Technology used and needs at farm level are also to be assessed so that potentiality to increase rice production can be examined for plan strategy development.

There are number of studies on crop and farm diversification in different parts of the country. However, studies on crop diversification in the flood prone areas are quite limited in this country for formulating strategic farm planning. Assam is a flood prone state and the study has more practical relevance to formulate farm planning for marginal and small farmers. The study, therefore, is an attempt to examine the pattern of crop diversification and potentialities of its growth, technological assessment in rice crop ,the impact of diversification on farm economy and high value alternatives in flood affected areas of Assam.

## Data and Methodology

The micro level data related to Nagaon district of Assam during 2002-03 and the macro level data were collected for the state pertaining from 1951 to 1998. The state level data were restricted to 1998 as published data were not available when the study was in progress. Nagaon district was particularly selected as the district is agriculturally advanced and cropping intensity is also higher. The results derived from such research efforts may be a good yardstick for farm planners to develop strategic planning for the other flood affected districts of the Brahmaputra plains. Selection of subdivision, blocks and villages was made purposively in consultation with District Agricultural Officer so that the results are based on a real ground situation for consistent development of farm plans. In the district, sub-division with the highest area affected by flood was selected in the first stage, two blocks from the sub-division having the highest area affected was selected at the second stage and five villages from each block were selected randomly at third stage. Altogether 200 marginal and 200 small farmers were selected randomly for primary data analysis. Farm level data were collected through structured pre tested questionnaires designed for the farmers of flood affected areas. A tabular analysis was preferred over complex mathematical decomposition due to partial and slow growth of structural changes in the flood prone areas with subsistence farming of marginal and small farmers having substantial effect on socio-economic conditions of the farmers and the results were discussed through simple expressions for wide coverage of ground level situations. Temporal changes in parameters in the district were examined by using exponential growth models and were tested at different levels of significance. Extent of diversification was also examined by using Simpson Index (Joshi, 2001). The value of the index ranged between zero and one and a higher value indicated higher extent of diversification.

## Results and Discussions

### *Changes of Cropping Pattern and Extent of Crop Diversification*

### *Temporal Changes of Cropping Pattern*

In Nagaon district, cropping pattern for cereals increased significantly upto 1990 after which it declined. Out of total cereals, summer rice grew significantly and its growth was slowed down after 1991. Cropping pattern for pulses grew positively during 1966-90 and then declined. In contrary, cropping pattern for oilseed crops had grown significantly from 1951-90 and was found to decline faster in the later years. Sesamum crop in oilseeds was found to gain in area in the district. Relative to potato, sugarcane as cash crop was found to gain area in the district out of gross cropped area. Area under spices gained faster while jute area as proportion to gross cropped area declined through out. Decline in area in crops might have been distributed to other crops like cereals, spices and cash crops and added to current fallows due to flood. Growth of crop diversification declined during 1951-65, gained there after and then marginally declined. Crop diversification was moderate in the district (APPENDIX-I &II).

### *Shift in Cropped Area*

During 1951-65 and 1966-90, most of the crops experienced marginal growth in area. The area under gram and jute declined significantly. However, growth rates of area of several non-remunerative crops after liberalization were found to decline while sesamum, jute and potato areas had gained. Summer rice in the state exhibited a high growth rate after 1991 while wheat area had declined in the state. The promising crops after 1991 were sesamum, jute, potato and summer rice (Table 1). Extension of this analysis in different flood prone

Table 1- NATURE OF CROPS AND THEIR GROWTH RATES FOR AREA IN ASSAM

| Years | Negative growth | Below marginal growth (0-1%) | Marginal growth (1.10-3)% | Moderate growth (3.10-6)% | High growth (Above 6)% |
|---|---|---|---|---|---|
| 1951-65 | Gram* | Winter rice**, Mustard** Jute** | Autumn rice**, Total rice**, Onion**Sugarcane**,Potato**, | Sesamum** | Summer rice** Wheat** |
| 1966-90 | Jute* | Autumn rice* | Winter rice**, Total rice**, Gram*, Sesamum**,Onion*, Sugarcane** | Summer rice **, Potato**, Mustard** | Wheat** |
| 1991-1998 | Autumn rice**, Winter rice**, Total Rice**,Wheat NS, Gram NS, MustardNS, Sugarcane** | ---- | Sesamum**, Jute NS | Potato* | Summer rice** |

** Indicates significance at one per cent level of probability, * indicates significance at ten per cent level of probability, NS indicates non significance

districts of Assam and its classification into different growth ranges indicated that summer rice in Kamrup and Sibasagar districts, onion in Nagaon district and sesamum in Lakhimpur district exhibited a more rapid growth. It was also observed that more than 80.00 per cent of the field crops exhibited stagnant and decline growth in the districts. The most promising crops in the districts were summer rice, onion, sesamum and potato (Table 2).

### *Shift in Technology and Productivity of Rice*

Under rainfed condition use of technology was higher in boro and Sali rice cultivation mainly due to use of chemical fertilizer in marginal farms. The technology used by the best farmers was higher in boro rice cultivation followed by early ahu rice with higher shift in technology. The average productivity of rice under rainfed condition varised from 15.40 quintals in boro rice to 29.70 quintals in Sali rice. The best farmers' productivity varied from 31.18 qtls in ahu rice to 36.24 quintals in Sali rice. The shift was higher in boro rice and was lower in ahu rice. With the use of higher level of technology there is potentiality to increase the average productivity in Sali and boro rice under rainfed condition. Under unirrigated condition technology use was higher in

**TABLE 2 - CLASSIFICATION OF CROPS IN TERMS OF AREA GROWTH IN THE FLOOD PRONE DISTRICTS OF ASSAM DURING 1951-98**

| Districts | More rapid growth | Slow growth | Stagnant and decline growth | More rapid decline growth |
|---|---|---|---|---|
| Kamrup | Summer rice | --- | Winter rice, Autumn rice, total rice, Gram, Mustard,Sesamum, Sugarcane, Jute, Potato, Onion | Wheat |
| Angoon | Onion | Sesamum | Winter rice, Summer rice, Total rice, Mustard, Jute | Wheat, Gram, Autumn rice, Sugarcane, Potato |
| Sibsagar | Summer rice | --- | Autumn rice, Sesamum, Jute | Winter rice, Total rice Mustard, Sugarcane, Potato |
| Lakhimpur | Sesamum | Summer rice, Potato | Winter rice, Total rice, Sugarcane | Autumn rice, Mustard, Jute |
| Assam | ----- | Summer rice, | Winter rice, Autumn rice, Total rice, Gram, Sugarcane, Sesamum, Mustard, Potato, Jute, Onion | Wheat |

transplanted early ahu followed by boro rice. It was lower in Sali rice due to high risk of rainfall. Technology shift was higher in Sali rice over the average farmers. Productivity of average farmers was higher in boro rice. If cultivation of boro rice can be economically sustained the state can gain surplus production. The small farmers on the other hand, used higher technology in boro and ahu rice cultivation uner rainfed condition. It was observed that technology shift was higher in Sali and ahu rice cultivation with lower shift in productivity. The return to cost ratio was lower than 2.00 in rice cultivation. Under irrigated condition cost of irrigation and chemical fertilizer was higher in boro rice. The average productivity varied from 28.05 qtls in Sali rice to 48.14 qtls in boro rice per hectare. The best farmer's productivity was higher in boro rice and the yield gap was higher in Sali and ahu rice (APPENDIX-III, IV).

### *Size of Farms, Technology and Extension Gaps in Summer Rice*

Attempt was made to examine the technology and extension gaps in summer rice as this crop is growing faster in the state. Technology gap was examined by estimating the difference of potential yield and the on farm yield while the extension gap is the difference of on-farm yield and the farmers' yield. It was observed that the technology and extension gaps were quite less in summer rice (Table-3). It was 1.20 and 1.26 t/ha for marginal farms and 1.12 and 1.15 t/ha for small farms respectively. It was 1.36 and 1.04 t/ha for all farms in the district. It was also observed that the average productivity of local boro rice under rain fed condition was3.45 and 3.36 t/ha for marginal and small farms. In case of HYV boro rice it was 3.66 and 5.65 t/ha respectively under irrigated condition. The average productivity of rain fed boro and HYV boro rice was 3.50 and 5.89 t/ha respectively in all farms and was found to be higher than the average productivity of rice in the state. Under normal practice, the productivity of local rainfed winter rice was found to be 24.48 quintals in the district and the productivity of HYV rain fed and HYV irrigated winter rice was higher than its normal productivity by 47.35 and 143.22 per cent respectively. The productivity of local rain fed *ahu* rice was 20.05 quintals and the productivity of HYV rain fed and HYV irrigated *ahu* rice shifted by 64.40 and 139.90 per cent respectively over the rainfed local ahu rice. The productivity of local *boro* rice was 34.90 quintals per hectare in the traditional areas and its productivity was higher by 65.50 per cent for irrigated HYV boro rice. It was also found that if boro rice is cultivated instead of ahu and Sali rice escaping the flood, there would be a productivity gain of 188.07 and 135.95 per cent without using additional technologies.

**Table 3**

**Use of Technologies and Levels of Productivity of HYV Irrigated Summer Rice in Nagaon (Morigaon) District**

| *Observations* | *Marginal (irrigated)* | *Small (Irrigated)* |
|---|---|---|
| A. FYM | | |
| a. Quantity used (kg/ha) | 2750.00 | 2450.00 |
| b. Farmers' using FYM (%) | 56.52 | 53.57 |
| c. Area covered (%) | 56.47 | 58.70 |
| d. Use gap (%) | 72.50 | 75.50 |
| B. Chemical fertilizer | | |
| a. NPK(kg/ha) | 81.20 | 88.10 |
| b. Farmers' using NPK (%) | 86.96 | 92.86 |
| c. Area covered by NPK(%) | 92.00 | 97.85 |
| d. Use gap (%) | 67.00 | 64.20 |
| C. Irrigation expenditure (Rs/ha) | 1749.00 | 1705.00 |
| a. Farmers using irrigation(%) | 100.00 | 100.00 |
| b. Area covered (%) | 100.00 | 100.00 |
| D. Productivity | | |
| a. Potential (kg/ha) | 678.00 | 678.00 |
| b. Actual (kg/ha) | 432.00 | 452.00 |
| c. Gap (%) | 36.28 | 33.33 |
| E. Output-labour ratio (Rs) | 248.40 | 254.48 |
| F. Output-capital ratio (Rs) | 3.59 | 3.70 |

**Table 4 Estimation of Technology and Extension Gaps of Summer Rice in Nagoon (Morigaon) District**

| *Farms* | *Potential yield (t/ha)* | *On farm Yield (t/ha)* | *Average farmer's yield (t/ha)* | *Technology gap (t/ha)* | *Extension gap (t/ha)* |
|---|---|---|---|---|---|
| Marginal | 6.78 | 5.56 | 4.32 | 1.20 | 1.26 |
| Small | 6.78 | 5.56 | 4.52 | 1.12 | 1.15 |
| Mean | 6.78 | 5.56 | 4.38 | 1.05 | 1.12 |

### *Shift in Cropping Pattern*

The percentage of cropped area to gross cropped area and their growth rates from 1951-98 in the flood from prone districts of Assam indicated that growth of cropping pattern in the state for most of the crops exhibited a stagnant and decline growth . Wheat, jute, sugarcane and mustard area declined more rapidly relative to gross cropped area. Assam being predominant in rice, area under winter and autumn rice declined proportionately. The proportionate area under summer rice in Kamrup and Sibasagar districts and in the state, potato in Kamrup and onion in Nagaon districts showed a rapid growth after 1951. The analysis indicated that cropping pattern for more than 50.00 per cent of the crops either remained stagnant or declined in the state (Table 5).

### *Shift in Cropping Intensity and Diversification Indices*

Cropping intensity and diversification indices in the state was found to decline after 1991 for different groups of crops (Table-6). The cropping intensities of cereals, pulses, oilseeds and cash crops were found to gain significantly up to 1990 and then declined. The cropping intensity of jute declined faster up to 1990 and it gradually gained after 1991. Simpson Index of diversification indicated that during 1951-65 extent of crop diversification was found to decline showing a mono crop situation in the state. During 1966-90 extent of diversification had grown significantly due to initiatives of the government through special rabi programs. After 1991 it has again slowed down due to special emphasis put for cereal cultivation in the state. The negative cropping intensity and diversification indices are a causing concern in cultivation of different crops in the state of Assam.

**TABLE 5- CLASSIFICATION OF CROPS IN TERMS OF GROWTH OF CROPPING PATTERN IN THE FLOOD PRONE DISTRICTS OF ASSAM DURING 1951-98**

| Districts | More rapid growth | Slow growth | Stagnant and decline growth | More rapid decline growth |
|---|---|---|---|---|
| Kamrup | Summer rice, Potato | Autumn rice | Winter rice, Autumn rice, Mustard, Sesamum, Sugarcane, Other crops | Wheat, Jute |
| Angoon | Onion | Autumn rice, Gram, Potato | Winter rice, Summer rice, Mustard, Sesamum, Jute, Other crops | Wheat, Sugarcane |
| Sibsagar | Summer rice | Autumn rice, Sesamum, Other crops | Jute, Winter rice, , Sugarcane | Mustard, Jute |
| Lakhimpur | --- | Sesamum, | Autumn rice,Winter rice, Sugarcane, Summer rice, Potato,, Other crops | Mustard, Jute |
| Assam | Summer rice | Sesamum, Gram, Other crops | Winter rice, Autumn rice, Sugarcane, Potato, Onion ,Jute, | Wheat, Mustard |

**Table 6: Growth of Cropping Intensity and Diversification Indices in Assam**

| *Crops* | *CGR(%) of cropping intensity* | | |
|---|---|---|---|
| | *1951-65* | *1966-90* | *1991-98* |
| Cereals | 1.0130** | 1.994** | -0.632NS |
| Pulses | -15.285** | 2.153* | -1.194NS |
| Oil seeds | 1.732** | 2.163** | -8.712NS |
| Cash crops | 0.893* | 6.130** | -0.659NS |
| Fibre crops | -0.859* | -5.113** | -0.728NS |
| Simpson Index (CGR) | -0.043 NS | 0.632** | -0.246NS |

### *Cropping Intensity and Diversification Indices According to Land Types*

Cropping intensity varied according to land topography and it was higher under medium and up land situations as low and deep water land were less suitable to cultivate more number of crops except rice and jute (Table 7). Up land was used for seed bed preparation and in cultivating vegetables, pulses, spices and condiments and oil seed crops. Medium land was used partly for vegetables, spices and condiments and oil seeds, rice and jute. Under irrigated condition summer rice was extensively cultivated as a rescue crop in medium land in the flood affected areas. Low land was dominated mostly by rice and jute and cropping intensity was quite low. Under irrigated condition cropping intensity was higher in up land and medium land situations due to cultivation of vegetables while oil seeds like rapeseed and mustard as well as sesamum and pulses were also cultivated under unirrigated conditions. There is much scope to increase the intensity of use of low land before and after flood in the district under irrigated and unirrigated conditions.

**Table 7: Cropping Intensity and Diversification Indices in Nagaon District**

| *Size of farms* | *Irrigated* | | | | *Unirrigated* | | | |
|---|---|---|---|---|---|---|---|---|
| | *Up land* | *Medium land* | *Low land* | *Deep water* | *Up land* | *Medium land* | *Low land* | *Deep water* |
| Marginal Cropping intensity | 175.00 | 197.38 | 64.62 | 0.00 | 161.36 | 172.81 | 37.17 | 0.00 |
| Diversification Indices | 0.78 | 0.08 | 0.13 | 0.00 | 0.94 | 0.92 | 0.57 | 0.00 |
| Small Cropping intensity | 155.02 | 147.80 | 114.29 | 0.00 | 143.62 | 159.59 | 113.76 | 0.00 |
| Diversification Indices | 0.89 | 0.80 | 0.47 | 0.00 | 0.89 | 0.74 | 0.70 | 0.00 |

### *Profit Classes of Crops, Income and Employment*

The crops cultivated by the marginal and small farms were classified according to profit classes and return per hectare. The prevailing prices of the crops when compared with average variable and average total costs and net return per quintal it was observed that the crops fell into five profit and return classes. It was seen that the marginal farmers could obtain higher prices over average total cost with high net return for most of the crops, which covered 79.23 per cent of gross cropped area sharing 88.10 per cent of gross return. The one reason of higher prices obtained by them is retailing of small surplus in the market instead of farm gate sale. Man days generated by them was also higher. It was followed by the food crops, particularly cereals whose prices were lower than average variable cost per unit with negative return per hectare. Employment generated was also lower. Crops under this group could not be observed in the small farms. Area and return shared by the crops under super normal profit low return and super normal profit medium returns were marginal to low generating lower level of employment. The small farmers in contrary could earn 55.19 per cent of gross return from the crops which shared 38.44 per cent of gross cropped area under super normal profit high return. They could earn equally large amount of gross return with higher employment level under super normal profit low return and super normal profit medium return as they had to depend their sales largely on middlemen. Crops sowing normal profit high return were quite low. The marginal farms were better off than the small farms in terms of profit and return of crops in the district (Table 8).

**Table 8: Relative Profitablity, Return and Employment Generated by Different Classes of Crops in Marginal and Small Farms**

| *Profitability and return classes* | *Marginal* | | | *Small* | | |
|---|---|---|---|---|---|---|
| | *P.C. share of area* | *P.C. share of gross return* | *P.C. share of mandays* | *P.C. share of area* | *P.C. share of gross return* | *P.C. share of mandays* |
| Below normal profit negative return | 13.74 | 7.77 | 16.85 | --- | --- | --- |
| Normal profit high return | --- | --- | --- | 1.78 | 1.41 | 3.02 |
| Super normal profit low return | 1.72 | 0.87 | 1.41 | 31.78 | 22.54 | 27.98 |
| Super normal profit medium return | 5.31 | 3.26 | 3.95 | 28.00 | 20.86 | 55.33 |
| Super normal profit high return | 79.23 | 88.10 | 77.79 | 38.44 | 55.19 | 13.67 |

### *Economic Rationality in Land Use*

More number of marginal farmers participated in crop cultivation and crop diversification was more in up land and medium land showing high magnitude of indices. Lowland was diversified more in small farms. Medium land shared higher proportion of Gross cropped area in both the farms. The average marginal farmers did not cultivate deep-water rice while the small farmers cultivated it in marginal area. Up land and medium land become free from excess soil moisture before on set of winter and the land can be ploughed back for rabi crops. Lowland has its own limitation and deep water land is more specific for cultivating perennial rice and seasonal tall crop like jute in some patches. It is practiced mostly by the immigrants in the chaar areas (riverine tract of Brahmaputra) where jute is harvested with the help of banana raft and boat during flood.

Table 9 shows that cropping intensity is higher in

## Table 9: Economic Rationality and Efficiency in Use of L.ınd in Nagaon District

| *Particulars* | *Marginal farmers* | | | | *Small farmers* | | | |
|---|---|---|---|---|---|---|---|---|
| | *Up land* | *Medium land* | *Low land* | *Deep water* | *Up land* | *Medium land* | *Low land* | *Deep water* |
| No of farmers utilizing the land | 41.00 | 92.00 | 61.00 | --- | 34.00 | 70.00 | 31.00 | 2.00 |
| P.C of total farmers | 41.00 | 92.00 | 61.00 | --- | 34.00 | 70.00 | 31.00 | 2.00 |
| No of crops cultivated | 22.00 | 30.00 | 3.00 | -- | 22.00 | 29.00 | 8.00 | 1.00 |
| Area under the crops(ha) | 1.43 | 92.41 | 27.87 | -- | 33.50 | 110.47 | 30.07 | 0.80 |
| P.C of GCA | 10.04 | 69.11 | 20.84 | --- | 19.36 | 63.18 | 17.20 | 0.45 |
| P.C share of income from food crops | 24.54 | 20.50 | 19.39 | --- | 20.70 | 17.00 | 16.67 | 7.45 |
| P.C share of income from non food crops | 10.67 | 12.42 | 12.48 | --- | 12.81 | 12.57 | 12.81 | --- |
| Man days generated per cropped hectare | 96.69 | 82.01 | 88.03 | --- | 89.52 | 70.34 | 77.80 | 36.50 |
| Annual per capita farm income (Rs) | 736.57 | 4195.95 | 1168.76 | --- | 1325.78 | 4053.33 | 701.75 | 13.38 |
| Output-labour ratio | 300.63 | 293.41 | 252.48 | --- | 250.15 | 270.71 | 223.85 | 230.55 |
| Output-Capital ratio | 2.89 | 3.04 | 2.48 | -- | 2.64 | 2.39 | 2.16 | 2.65 |
| Cropping intensity(%) | 166.68 | 146.67 | 99.08 | --- | 147.79 | 154.71 | 113.89 | 100.00 |
| Diversification Index | 0.920 | 0.93 | 0.474 | --- | 0.888 | 0.819 | 0.713 | 0.00 |
| Annual per capita food production (Kg) | 725.21 | 4005.36 | 1054.93 | -- | 1491.89 | 4215.82 | 1147.55 | 13.38 |
| Annual per capita non food production (Kg) | 11.36 | 190.58 | 113.83 | --- | 107.36 | 95.98 | 282.08 | -- |
| Productivity per gross cropped ha of food crops (quintal) | 61.59 | 68.10 | 40.90 | --- | 68.59 | 41.10 | 33.21 | 15.30 |
| Returns per gross cropped hectare of non food crops (quintal) | 29650.73 | 24773.52 | 23423.20 | --- | 23393.34 | 19195.77 | 18833.42 | 8415.00 |
| Productivity per gross cropped hectare of non food crops (quintal) | 17.34 | 23.57 | 23.79 | --- | 22.88 | 20.10 | 22.88 | --- |
| Returns per gross cropped hectare of non food crops (quintal) | 12891.65 | 15008.98 | 15082.96 | -- | 14478.00 | 14199.72 | 14478.00 | --- |
| Total return per gross cropped hectare(Rs) | 29067.97 | 24062.46 | 22226.18 | --- | 22400.68 | 19042.00 | 17413.96 | 8415.00 |

marginal farms particularly in upland to accommodate more of horticultural crops for income security. The small farmers cultivated more of agronomical field crops and foodgrain crops in low and deep water land. Marginal farms were also more productive to labour and capital with high rate of diversification. Annual per capita food production was higher in small farms due to higher size of operational holding while non-food production was higher also in small farms due to more area put under jute. Up land and medium land was relatively with higher returns in the district.

### *Constraints of Crop Diversification*

There are various constraints, which affect crop diversification in a region. They are managerial, social, economic and physical constraints. The case study in Nagaon district of Assam indicated that under managerial constraints lack of extension facilities affected the marginal (45.59%) and small (39.59%) farms followed by shortage of family members for agricultural works. The important social constraints were lack of contact in marginal (42.52%) and small (38.77%) farms followed by scarcity of labour, lack of experience and community functions in improved farming. Economic constraints outweighed all other constraints for crop diversification. The important economic constraints in marginal and small farms were high cost of cultivation, high cost of inputs, low output price, non availability of credit and its inadequacy, low level of own capital, high rate of interest, risk in agriculture, low rate of profit in agriculture and which were equally perceived by the marginal and small farmers. The important physical constraints were lack of preferred market, lack of storage facility, lack of quality seeds, lack of agro service centres, lack of irrigation, soil quality, poor communication, irregular rainfall, untimely flood, problems of pest and disease, problems of weeds, water logging, poor draught power and problems created by birds and animals perceived equally by marginal and small farmers.

## Conclusions and Policy Issues

Several districts of Assam are in the flood prone zones affecting two lakh hectares annually and the farmers are to sustain their livelihood adjusting their cropping pattern before and after flood. Cropping pattern in the state is pre-dominated by cereals whose proportionate area is declining in the recent years and the state has witnessed a moderate growth of diversification. In cereals summer rice has emerged as a rescue crop in the flood affected areas. Area growth of some commercial cash crops like sesamum, jute, onion, potato have emerged in the flood affected areas against flood risk. The state has witnessed a declining growth of cropping intensity and crop diversification in all crop groups in the recent years. Case studies indicated that up land and medium land were more intensively used with higher magnitude of diversification under irrigated and unirrigated conditions. Special emphasis may be put for land based crop planning to increase cropping intensity and diversification indices. Marginal and small farmers could generate higher return and employment increasing cropping intensity and realizing higher prices by cultivating high valued crops. Medium land was highly productive with a good base of food security while up land was good for income source through cultivation of high valued crops. Return and profitability were higher for boro rice with full technology, pulses and oilseed crops due to higher market prices, and vegetables adjusted in early seasons could fetch higher prices generating higher return and higher profit. Most of the spices and condiments were found to be highly remunerative. Farm strategies may be formulated for commercial cultivation of vegetables, pulses, oilseed crops, spices and condiments for farm income generation and to bridge the demand supply gap of pulses, oilseeds and spices. This will also strengthen the state exchequer. There exists wide gap between average productivity and best farmer's productivity of rice in the state due to low level of technology used. There is scope to increase productivity of rice with proper use of technology and crop planning. The important constraints to diversification like strengthening of input and output markets, irrigation, roads and communication, mechanization, credit and storage facilities and agroservices should be minimized for increasing cropping intensity and crop diversification in the state. Extension for commercialization of crop cultivation will be highly imperative for generating income and employment in the farm sector.

*Notes*

Rice means total irrigated HYV and rainfed traditional varieties cultivated by the farmers. Below normal profit: prevailing market price of the produce is less than AVC of production per unit, Normal profit: prevailing market price of the produce is more than AVC of production per unit, but less than its ATC of production per unit, Super normal profit: prevailing market price of the produce is more than ATC of production per unit, Low return: A crop whose net return is lower than 10.00 per cent over cost C2 per hectare, Medium return: A crop whose net return is 10.01-20.00 per cent over cost C2 per hectare, High return: A crop whose net return is higher than 20.00 per cent over cost C2 per hectare. Crops under other classes were not observed.

Analytical frame work

The extent of diversification on crop was studied using, Simpson indices

Simpson Index of Diversification

$I_i = 1 - (\Sigma S_i^2) / (\Sigma S_i)^2$

When Si is the share of crop i in gross cropped area. A high Simpson index indicates greater diversity.

Cost of cultivation procedures followed by CACP

Cost $A_1$= All actual expenses in cash and kind incurred in production by owner

Cost $A_1$ = Cost $A_1$ + rent paid for leased in land

Cost $A_2$ +FL = Cost $A_2$ + imputed value of family labour

Cost $B_1$ = Cost $A_1$ + interest on value on own capital assets excluding land

Cost $B_2$ = Cost $B_1$ + rental value of land (net of land revenue)

Cost $C_1$ = Cost $B_1$ + imputed value of family labour

Cost $C_2$ = Cost $B_2$ + imputed value of family labour

Cost C = Cost $C_2$ + 10% of Cost $C_2$ on account of managerial functions performed by farmers

Total Cost= Total Variable Cost + Total Fixed Cost + cost of marketing and post harvest measures

Gross return/ha= Total quantity produced times farm harvest price+ value of by-products if any

Net return/ha = Gross return/ha- Total cost/ha

## REFERENCES

Arora, V.P.S. and Srivastava, S.K (1996). Diversification of Cropping Pattern and Food Grain Mix in India: Pace, Magnitude and Implications, *Indian Journal of Agricultural Economics*, 51(4).

Chand, Ramesh (1996). Agricultural Diversification and Small Farm Development in Western Himalayan Region, in *Small Farm Diversification: Problems and Prospects*, T. Haque (Ed). National Centre for Agricultural Economics and Policy Research, New Delhi.

Foster, John, H. (1996). Constraints to Dry land Double Cropping in Northern Madhyapradesh, *Indian Journal of Agricultural Economics*, 43(2):148-155

Jha, Brajesh K.(1966). Farm Level Diversification: Some Disconcerting Evidences from the North West India, *Agricultural Economics Research Review*, 9(1):49-56.

Joshi, P.K. (2001). Crop Diversification in Rainfed Agriculture, *Agricultural Economics Research Review*, Conference Issue.

Pope, R.D. and Richard Presscot (1980). Diversification in Relation to Farm and other Socio-economic Characteristics, *American Journal of Agricultural Economics*, 62(3).

Rahim, M.B. Kazi and N. Ali (1992). Technological gaps as the Determinants of Yield Gaps in Rice Varieties in Birbhum District of West Bengal, *Indian Journal of Agricultural Economics*, 47(30):545.

Rudra, Asok (1983). On so called Marketing Efficiency, *Indian Economic Review*, 15(4).

Siddique, F.A. (1999). Rice: Not a Distant Dream, *The Hindu Survey of Indian Agriculture*, Section II:39-46.

Talukdar, K.C. (1994). A Study of Efficiency and Equity in Marketing of Major Agricultural Commodities in Assam State, Ph. D. thesis submitted to Indian Agricultural Research Institute, New Delhi.

Talukdar, K.C. (1998). Study of Cropping Pattern and the Factors Affecting it in Assam, ICAR ad-hoc project, Department of Agricultural Economics, Assam Agricultural University, Jorhat.

Talukdar, K.C. and B.C. Deka, (2005).Cultivation of Summer Rice in the Flood Plains of Assam—An Assessment of Economic Potential on Marginal and Small Farms, *Agricultural Economics Research Review*, 18(1):21-37.

Vyas, V.S. (1996). Diversification in Agriculture: Concepts, Rationale and Approaches, *Indian Journal of Agricultural Economics*, 51(4).

## APPENDIX-I

### CROPPING PATTERN AND EXTENT OF CROP DIVERSIFICATION IN ASSAM

| Proportionate area(%) under | 1951 | 1955 | 1960 | 1965 | CGR of proportionate area (%) | 1966 | 1970 | 1975 | 1980 | 1985 | 1990 | CGR of proportionate area (%) | 1991 | | 1998 | CGR of proportionate area (%) |
|---|---|---|---|---|---|---|---|---|---|---|---|---|---|---|---|---|
| A.Cereals | 63.91 | 67.46 | 70.22 | 67.60 | 1.013** | 69.61 | 72.62 | 77.78 | 72.50 | 72.24 | 82.04 | 1.994** | 77.96 | 78.43 | 83.60 | -0.632NS |
| i. Autumn rice | 12.20 | 11.98 | 12.01 | 11.52 | -0.122 | 12.31 | 16.22 | 18.27 | 14.58 | 18.97 | 19.19 | 1.533** | 17.65 | 18.51 | 19.15 | 0.911 NS |
| ii.Winter rice | 51.50 | 54.82 | 54.45 | 55.33 | 0.121 NS | 55.78 | 55.16 | 53.72 | 51.52 | 46.83 | 45.97 | -0.972** | 42.19 | 42.00 | 45.02 | 0.0002 NS |
| iii.Summer rice | 0.12 | 0.44 | 2.74 | 0.75 | 8.444* | 1.50 | 0.72 | 3.11 | 3.13 | 3.28 | 14.26 | 7.487** | 13.83 | 14.45 | 19.39 | 2.969 NS |
| iv. Wheat | 0.09 | 0.22 | 0.02 | 0.004 | -20.038** | 0.02 | 0.52 | 2.68 | 3.27 | 3.16 | 2.62 | 19.177** | 4.29 | 3.47 | 0.04 | -16.094 NS |
| B. Pulses | 0.88 | 0.59 | 0.15 | 0.06 | -15.285** | 0.06 | 0.07 | 0.06 | 0.12 | 0.15 | 0.07 | 2.153* | 0.09 | 0.06 | 0.07 | -1.194 NS |
| i. Gram | 0.88 | 0.59 | 0.15 | 0.06 | -15.285** | 0.06 | 0.07 | 0.06 | 0.12 | 0.15 | 0.07 | 2.153* | 0.09 | 0.06 | 0.07 | -1.194 NS |
| C. Oil seeds | 7.82 | 7.93 | 7.15 | 8.51 | 1.732** | 8.50 | 8.24 | 7.05 | 9.70 | 9.03 | 6.45 | 2.163* | 8.22 | 7.40 | 6.46 | -8.712 NS |
| i. Mustard | 7.63 | 7.53 | 6.74 | 8.13 | 0.750* | 8.10 | 7.79 | 6.51 | 9.27 | 8.60 | 6.22 | 0.559 NS | 7.90 | 7.00 | 6.23 | -9.908 NS |
| ii. Sesamum | 0.19 | 0.40 | 0.41 | 0.39 | 2.980* | 0.40 | 0.45 | 0.54 | 0.43 | 0.43 | 0.23 | -1.348* | 0.32 | 0.41 | 0.23 | 1.665 NS |
| D. Cash crops | 1.65 | 1.81 | 1.71 | 1.68 | 0.893 | 1.84 | 1.47 | 3.09 | 3.33 | 4.20 | 3.37 | 6.130** | 3.33 | 4.05 | 3.77 | -0.659 NS |
| i. Potato | 0.53 | 0.68 | 0.52 | 0.59 | 0.463 NS | 0.74 | 0.36 | 0.58 | 0.86 | 1.21 | 1.12 | 4.610** | 1.11 | 1.66 | 1.45 | -0.605 NS |
| ii. Sugarcane | 1.12 | 1.13 | 1.19 | 1.09 | -0.240 NS | 1.10 | 1.11 | 2.51 | 2.47 | 3.00 | 2.25 | 4.345** | 2.22 | 2.39 | 2.32 | 0.498 NS |
| E. Spices | — | — | -- | — | -- | — | — | -- | 0.19 | 0.05 | 0.17 | 0.289 NS | 0.19 | 0.34 | 0.27 | 8.171* |
| i. Onion | -- | — | -- | — | -- | — | -- | -- | 0.19 | 0.05 | 0.17 | 0.289 NS | 0.19 | 0.34 | 0.27 | 8.171* |
| F. Fibre crops | 17.08 | 16.38 | 14.71 | 15.09 | -0.859* | 13.86 | 12.42 | 6.54 | 7.12 | 7.12 | 3.09 | -5.113* | 4.88 | 4.67 | 2.68 | -0.728 NS |
| i. Jute | 17.08 | 16.38 | 14.71 | 15.09 | -0.859* | 13.86 | 12.42 | 6.54 | 7.12 | 7.12 | 3.09 | -5.113* | 4.88 | 4.67 | 2.68 | -0.728 NS |
| G. Other crops | 8.62 | 5.83 | 6.10 | 7.03 | 0.786 NS | 6.11 | 5.23 | 5.54 | 6.92 | 7.18 | 4.78 | 0.458 NS | 5.32 | 5.05 | 2.76 | -3.318 |
| H. Simpson Index | 0.677 | 0.649 | 0.647 | 0.646 | -0.043 NS | 0.643 | 0.644 | 0.633 | 0.692 | 0.723 | 0.723 | 0.632** | 0.757 | 0.756 | 0.716 | -0.246 NS |

APPENDIX-II

CHANGES OF CROPPING PATTERN ND EXTENT OF CROP DIVERSIFICATION IN NAGAON DISTRICT OF ASSAM DURING 1951-98

| Proportionate Area under | 1951-1965 | | | | CGR of proportionate area (per cent) | 1966-1990 | | | | | | CGR of proportionate area (per cent) | 1991-1998 | | | CGR of proportionate area (per cent) |
|---|---|---|---|---|---|---|---|---|---|---|---|---|---|---|---|---|
| | 1951 | 1955 | 1960 | 1965 | | 1966 | 1970 | 1975 | 1980 | 1985 | 1990 | | 1991 | 1995 | 1998 | |
| A. Cereals | 63.91 | 67.46 | 70.22 | 67.60 | 1.013** | 69.61 | 72.62 | 77.78 | 72.50 | 72.24 | 82.04 | 1.994** | 77.96 | 78.43 | 83.60 | -0.632$^{NS}$ |
| i. Autumn rice | 12.20 | 11.98 | 12.01 | 11.52 | -0.122$^{NS}$ | 12.31 | 16.22 | 18.27 | 14.58 | 18.97 | 19.19 | 1.553** | 17.65 | 18.51 | 19.15 | 0.911$^{NS}$ |
| ii. Winter rice | 51.50 | 54.82 | 54.45 | 55.33 | 0.121$^{NS}$ | 55.78 | 55.16 | 53.72 | 51.52 | 46.83 | 45.97 | -0.972** | 42.19 | 42.00 | 45.02 | 0.002$^{NS}$ |
| iii. Summer rice | 0.12 | 0.44 | 2.74 | 0.75 | 8.448* | 1.50 | 0.72 | 3.11 | 3.13 | 3.28 | 14.26 | 7.487** | 13.83 | 14.45 | 19.39 | 2.969$^{NS}$ |
| iv. Wheat | 0.09 | 0.22 | 0.02 | 0.004 | -20.038** | 0.02 | 0.52 | 2.68 | 3.27 | 3.16 | 2.62 | 19.177** | 4.29 | 3.47 | 0.04 | -16.094$^{NS}$ |
| B. Pulses | 0.88 | 0.59 | 0.15 | 0.06 | -15.285** | 0.06 | 0.07 | 0.06 | 0.12 | 0.15 | 0.07 | 2.153* | 0.09 | 0.06 | 0.07 | -1.194$^{NS}$ |
| i. Gram | 0.88 | 0.59 | 0.15 | 0.06 | -15.285** | 0.06 | 0.07 | 0.06 | 0.12 | 0.15 | 0.07 | 2.153* | 0.09 | 0.06 | 0.07 | -1.194$^{NS}$ |
| C. Oilseeds | 7.82 | 7.93 | 7.15 | 8.51 | 1.732** | 8.50 | 8.24 | 7.05 | 9.70 | 9.03 | 6.45 | 2.163** | 8.22 | 7.40 | 6.46 | -8.712$^{NS}$ |
| i. Rapeseed & Mustard | 7.63 | 7.53 | 6.74 | 8.13 | 0.750* | 8.10 | 7.79 | 6.51 | 9.27 | 8.60 | 6.22 | 0.559$^{NS}$ | 7.90 | 6.99 | 6.23 | -9.908$^{NS}$ |
| ii. Sesamum | 0.19 | 0.40 | 0.41 | 0.39 | 2.989* | 0.40 | 0.45 | 0.54 | 0.43 | 0.43 | 0.23 | -1.348* | 0.32 | 0.41 | 0.23 | 1.665$^{NS}$ |
| D. Cash crops | 1.65 | 1.81 | 1.71 | 1.68 | 0.893* | 1.84 | 1.47 | 3.09 | 3.33 | 4.20 | 3.37 | 6.130** | 3.33 | 4.05 | 3.77 | -0.659$^{NS}$ |
| i. Potato | 0.53 | 0.68 | 0.52 | 0.59 | 0.463$^{NS}$ | 0.74 | 0.36 | 0.58 | 0.86 | 1.21 | 1.12 | 4.610** | 1.11 | 1.66 | 1.45 | -0.605$^{NS}$ |
| ii. Sugarcane | 1.12 | 1.13 | 1.19 | 1.09 | -0.240$^{NS}$ | 1.10 | 1.11 | 2.51 | 2.47 | 2.99 | 2.25 | 4.345** | 2.22 | 2.39 | 2.32 | 0.498$^{NS}$ |
| E. Spices | - | - | - | - | - | - | - | - | 0.19 | 0.05 | 0.17 | 0.289$^{NS}$ | 0.19 | 0.34 | 0.27 | 8.171* |
| i. Onion | - | - | - | - | - | - | - | - | 0.19 | 0.05 | 0.17 | 0.289$^{NS}$ | 0.19 | 0.34 | 0.27 | 8.171* |
| F. Fibre crops | 17.08 | 16.38 | 14.71 | 15.09 | -0.859* | 13.86 | 12.42 | 6.54 | 7.12 | 7.12 | 3.09 | -5.113** | 4.88 | 4.67 | 2.68 | -0.728$^{NS}$ |
| i. Jute | 17.08 | 16.38 | 14.71 | 15.09 | -0.859* | 13.86 | 12.42 | 6.54 | 7.12 | 7.12 | 3.09 | -5.113** | 4.88 | 4.67 | 2.68 | -0.728$^{NS}$ |
| G. Others | 8.62 | 5.83 | 6.10 | 7.03 | 0.786$^{NS}$ | 6.11 | 5.23 | 5.54 | 6.92 | 7.18 | 4.78 | 0.458$^{NS}$ | 5.32 | 5.05 | 2.76 | -3.318$^{NS}$ |
| Simson Index of diversification | 0.677 | 0.649 | 0.647 | 0.646 | *-0.043*$^{NS}$ | 0.643 | 0.644 | 0.633 | 0.692 | 0.723 | 0.723 | *0.632*** | 0.757 | 0.756 | 0.716 | *-0.246*$^{NS}$ |

Figures in italics indicate CGR of Simpson Index of diversification

## APPENDIX-III

### LEVEL OF TECHNOLOGY USED AND PRODUCTIVIY OF MAJOR CROPS IN MARGINAL FARMS IN NAGAON DISTRICT OF ASSAM

| Crops | Levels of technology used | | | | | Average productivity (q/ha) | Best farmers productivity (q/ha) | Higher cost on technology (Rs./ha) | Percentage shift in technology used | Percentage shift in productivity | Cost-return ratio at $C_2$ |
|---|---|---|---|---|---|---|---|---|---|---|---|
| | Organic manure (Rs./ha) | Chemical fertilizer (Rs./ha) | Cost of irrigation (Rs./ha) | Other chemicals (Rs./ha) | Total cost of technology (Rs./ha) | | | | | | |
| Cereals | | | | | | | | | | | |
| Ahu Rice | | | | | | | | | | | |
| Rainfed | 0.00 | 673.60 | 0.00 | 0.00 | 673.60 | 28.82 | 31.18 | 1055.26 | 54.66 | 8.19 | 1.51 |
| HYV Irrigated | 351.10 | 1612.00 | 1755.00 | 0.00 | 3718.10 | 35.11 | 44.52 | 3356.16 | No shift | 26.81 | 1.37 |
| Sali Rice | | | | | | | | | | | |
| Rainfed | 28.95 | 771.10 | 0.00 | 26.32 | 592.00 | 29.74 | 36.24 | 844.88 | 2.24 | 21.86 | 1.75 |
| HYV Irrigated | 495.00 | 990.00 | 0.00 | 0.00 | 2685.00 | 36.00 | 43.83 | 3400.07 | 26.93 | 21.76 | 1.75 |
| Boro Rice | | | | | | | | | | | |
| Rainfed | 0.00 | 787.70 | 0.00 | 0.00 | 8.72 | 787.70 | 35.39 | 1230.00 | 56.15 | 56.51 | 1.87 |
| HYV Irrigated | 19.19 | 563.70 | 2442.00 | 2442.00 | [illegible] | 3033.61 | 60.00 | 2972.94 | No shift | 1.92 | 2.13 |

APPENDIX-IV

LEVEL OF TECHNOLOGY USED AND PRODUCTIVIY OF MAJOR CROPS IN SMALL FARMS IN NAGAON DISTRICT OF ASSAM

| Crops | Levels of technology used | | | | | Average productivity (q/ha) | Best farmers productivity (q/ha) | Highest cost on technology (Rs./ha) | Percentage shift in technology used | Percentage shift in productivity | Cost-return ratio at $C_2$ |
|---|---|---|---|---|---|---|---|---|---|---|---|
| | Organic manure (Rs./ha) | Chemical fertilizer (Rs./ha) | Cost of irrigation (Rs./ha) | Other chemicals (Rs./ha) | Total cost of technology (Rs./ha) | | | | | | |
| **Cereals** | | | | | | | | | | | |
| Ahu rice | | | | | | | | | | | |
| Rainfed | 382.30 | 757.30 | 0.00 | 0.00 | 1139.60 | 27.07 | 28.00 | 1380.00 | 21.09 | 3.43 | 1.61 |
| HYV irrigated | 477.60 | 1101.00 | 1579.00 | 0.00 | 3157.60 | 33.55 | 41.45 | 3765.00 | 19.24 | 23.55 | 1.54 |
| Sali rice | | | | | | | | | | | |
| Rainfed | 120.50 | 684.60 | 0.00 | 0.00 | 805.10 | 24.08 | 26.00 | 1125.00 | 39.73 | 7.97 | 1.55 |
| HYV irrigated | 0.00 | 682.60 | 1472.00 | 0.00 | 2145.60 | 28.05 | 35.60 | 2167.50 | 1.02 | 26.92 | 1.50 |
| Boro rice | | | | | | | | | | | |
| Rainfed | 608.90 | 1096.40 | 0.00 | 0.00 | 1705.30 | 37.61 | 43.00 | 1750.00 | 2.62 | 14.33 | 1.90 |
| HYV irrigated | 373.60 | 1096.40 | 1726.00 | 0.00 | 3196.00 | 48.14 | 58.62 | 3080.00 | No shift | 21.77 | 1.82 |
| Bao rice | | | | | | | | | | | |
| Rainfed | 0.00 | 0.00 | 0.00 | 0.00 | 0.00 | 15.30 | 17.00 | 0.00 | 0.00 | 11.11 | 1.90 |
| HYV irrigated | 0.00 | 0.00 | 0.00 | 0.00 | 0.00 | 0.00 | 0.00 | 0.00 | 0.00 | 0.00 | 0.00 |

# 17

# POST-HARVEST LOSS AND FOOD SECURITY—A STUDY ON FRUITS AND VEGETABLES OF ASSAM

*C. Hazarika*

## Introduction

The world food hunger is widespread and growing day by day. Billions of people suffer from severe malnutrition in Asian, African and Latin American countries. Apart from their caloric needs, there is severe shortage of food materials such as vegetables, tubers, root crops and fruit nuts, which are the most important plant foods that supply man with many of his nutritive requirement including minerals, vitamins, proteins, starches, fats and sugars (Salunkhe and Desai, 1984). Increasing food supplies can obviously solve the problem of population-food imbalance.

Fruits and vegetables are highly perishable food products in nature and in the process of supply from the farm level to the table, there are many opportunities for these products to be lost through water loss and post harvest decay. The seasonal factors also considerably influence both the vegetable production and their post harvest losses. The seasonal gluts and lack of infrastructures and marketing facilities in the developing countries have significant effect on the extent of post harvest losses of vegetables. It is being increasingly realized that the production of more and better food alone is not enough; the food must also be delivered to the ultimate consumers through the post harvest system without loss.

Horticultural crops have got a unique role in developing countries both in economic and social spheres in respect of improving income and nutritional status particularly of rural masses. It not only generates employment for rural masses and conserve natural resources but also have several other advantages like value addition and foreign exchange earnings.

India being a vast country with varied soil and agro-ecological conditions has grown a wide variety of fruits and vegetables. India is the second largest producer of vegetables in the world next only to China with an estimated production of about 50.09 million tones from an area of 4.50 million hectares. India shares about 12.00 per cent of the world output of vegetables from about two per cent of the cropped area in the country and contributes about 24.00 per cent of gross agricultural output (Verma, 2004). Lack of efficient methods of harvesting and non-availability of adequate facilities like transportation, packaging, storage, processing and market accessibility are the other problems in vegetable cultivation.

The post harvest loss of fruits and vegetables are a matter of grave concern for India's agricultural sector. The fruits and vegetables being a highly perishable are lost after harvest due to inefficient methods of harvest, decay, over-ripening, mechanical injury, weight loss, trimming, sprouting etc. Therefore, these aspects are to be looked into critically if any improvement in the present state of fruits and vegetable industry, especially in the context of processing and marketing are to be accomplished. Post harvest loss generally led to a social and economic loss to the farmers, wholesalers, retailers and also to the consumers. It is understood that nearly 35-40 per cent of agricultural produce are wasted, which amount to a staggering loss of RS. 50,000 crores annually. The amount of fruits and vegetables wasted in India is sufficient to meet the requirements of United Kingdom (Yadav, 2003). Hence, careful handling of fruits and vegetables by selecting appropriate post harvest means will be of paramount importance to reduce the handling losses to a considerable extent.

Assam with wide diversity in topography, rainfall, soil type and climatic conditions has immense potential for growing a wide range of fruits (tropical and sub-tropical) and vegetables. However, cultivation of fruits and vegetables in the state is carried out mostly on small farms with out any organized back up for packaging, storage, transport and marketing. Assam produces a large number of fruits, vegetables and spices having high commercial value. There is a wide scope for vertical and horizontal expansion of these crops in the state. With the intervention of improved production technology and faster dissemination of technological know how, there has been a manifold increase in the magnitude of production of fruits and vegetables in last couple of years. It is noteworthy that mere emphasizing quantitative increase will not carry meaning unless appropriate post harvest management practices are devised and practiced (Deka, 2005).

In Assam an efficient marketing system is of crucial importance for commercial production of horticultural crops. Development of new marketing system supported with cost effective, properly designed storage facilities with optimum storage conditions are the need of the time to lower down the post harvest losses. By reducing the post harvest losses, one can increase the per capita availability of food with the existing level of production technology (Hazarika, 2003). For this, it is necessary to know the quantum of post harvest losses at various levels and during different operations. This will help the policy makers to formulate proper measures to reduce the post harvest losses. This study is an attempt to estimate the post harvests losses of major perishable horticultural crops in Assam. This study would certainly be useful for the planners, policy makers, researchers, academicians in general and the farmers in particular.

## Methodology

The major high valued perishable fruits and vegetables undertaken for the study were Pineapple, Orange, Ginger and Tomato. These crops were mainly selected on the basis of their area and production of these crops in the state of Assam. Four major growing districts for each crops was selected based on secondary data collected from the Directorate of Agriculture, Government of Assam. To select the sample farmers for each crop five highest growing villages were identified in the respective districts in consultation with the State Department of Agriculture. A list of farmers with area and production has been collected from the District and Sub-Divisional Agricultural Offices of respective districts. The sample farmers were selected by applying multistage stratified random sampling technique. A total of 200 sample farmers for individual crops were selected covering 50 farmers from the respective four districts. As a whole, 800 farmers were selected for the study in the state. The primary data were collected by personal interview method with the help of specially designed comprehensive pre-tested schedule- cum-questionnaires. Tabular and percentage analysis were mainly undertaken in this study to arrive at the results. For conducting an intensive and in-depth study of various aspects of post harvest losses at middlemen's level, 20 middlemen for each crop from each district was selected and altogether a sample of 320 middlemen were interviewed for the study.

## Results and Discussion

The utilization patterns of all the selected crops are presented in Table 1. The quantity of pineapple and oranges were measured in numbers whereas tomato and ginger were measured in quintals. It was evident from the table that all these crops were extensively grown as cash crops as the farm family consumption was found to be very small and the crops were mainly grown for marketing purpose. The home consumption was found to be 4.00 per cent, 1.71 per cent, 0.87 per cent and 0.09 per cent of the total production respectively for pineapple, orange, tomato and ginger. Similarly, the kind payments were also found to be very low which ranges from as low as 0.07 per cent of the total production for ginger to 3.03 per cent for pineapple. It was also evident from the table that only the ginger growers kept an amount of 2.29 per cent of the total production for seed purposes to be used in the next year. For all the crops, the marketed surplus was found to be very high. It was 88.75 per cent, 90.04 per cent, 85.25 per cent and 92.88 per cent of the total production respectively for pineapple, orange, tomato and ginger. Being perishable in nature and the high volume of marketed surplus post harvest management of these crops assumes a high level of importance. The post harvest loss at growers level were found to be 4.22 per cent, 6.59 per cent, 13.07 per cent and 4.67 per cent in case of pineapple, orange, tomato and ginger respectively.

The post harvest losses at growers level were analyzed according to various post harvest operations performed by the grower and the results were presented in Table 2. Out of the total loss, the post harvest loss was maximum due to delayed harvesting (18.09 per cent) followed by loading and unloading (11.96 per cent) in pineapple. In case of orange, loss was found to be maximum due to road blockage or mis-happening of vehicles (12.38 per cent) followed by quantity discarded for low quality (11.50 per cent). In tomato also, low quality (19.20 per cent) and spoilage and wastage due to delayed harvesting (18.88 per cent) were found to be the major items of post harvest loss at growers level. In ginger, quantity discarded for low quality (17.08 per cent) and loss due to lack of market demand (15.75 per cent) were found to be the major items of post harvest losses. It can be inferred from the table that the low quality of produce is a major problem of the growers and hence proper training may be imparted to the grower to maintain the quality for increasing the market demand and to reduce the post harvest loss at the growers level. This will also lead to increase the market access of these commodities.

The post harvest loss at the market middlemen level in various operations for different types of middlemen involved in the marketing channels were also estimated. Table 3 presents the post harvest losses at market middlemen level for pineapple in Assam. In pineapple marketing, middlemen involved were Wholesaler (WS), Retailer (R), Wholesaler cum Pre Harvest Contractor (WS-C-PHC), Wholesaler via Commission Agent (WS via CA) and Pre Harvest Contractor (PHC). From the table, it was evident that the post harvest loss was found to be highest at the level of Pre Harvest Contractor (10.46 per cent) and lowest in the Wholesaler via Commission Agent level (2.60 per cent). The loss was maximum during transportation in case of Wholesaler cum

Pre Harvest Contractor level (23.63 per cent) and Wholesaler via Commission Agent level (20.51 per cent). In the retailer level, the loss was found to be maximum during the storage operations (27.52 per cent). In case of wholesaler, Pre Harvest Contractor and for the whole marketing channel, the loss was found to be maximum during loading/unloading and the loss being 41.17 per cent, 29.41 per cent and 22.39 per cent respectively. It can be inferred from the table that the post harvest loss in pineapple market is mainly occurred during transportation and loading/unloading of commodities. The overall post harvest loss of pineapple at the market level was found to be 5.03 per cent of the total quantity transacted in the market.

Similarly, Table 4 presents the post harvest loss at market middlemen level for tomato in Assam. In tomato marketing, middlemen involved were Wholesaler (local), Wholesaler, Wholesaler via Commission Agent (WS via CA) and Primary Wholesaler (PWS). From the table, it was evident that the post harvest loss was highest at the Wholesaler level (12.00 per cent). Except the retailer, the loss was maximum for all other middlemen in loading/unloading and loss incurred for low quality of products. But in the retailer level, the loss was found to be maximum due to delay in sale (43.46 per cent), produce rejected by agent (18.85 per cent) and space problem in the market (10.91 per cent). The total post harvest loss of tomato at the market level was found to be 9.55 per cent of the total quantity transacted in the market.

Table 5 presents the post harvest loss at market middlemen level for oranges in Assam. In orange marketing, middlemen involved were Wholesaler (WS), Retailer (R), Wholesaler cum Pre Harvest Contractor (WS-C-PHC), Wholesaler via Commission Agent (WS via CA) and Pre Harvest Contractor (PHC). From the table, it was evident that the post harvest loss was highest at the Retailer level (9.53 per cent) and lowest in the Wholesaler via Commission Agent level (6.90 per cent). For all the middlemen, the loss was found to be maximum through spoilage in storage and loading/unloading. In the aggregate level, the loss was 27.16 per cent in spoilage in storage and 15.63 per cent in loading/unloading. The total post harvest loss of orange at the market level was found to be 7.25 per cent of the total quantity transacted in the market.

The post harvest loss at market middlemen level for ginger in Assam is presented in Table 6. In ginger marketing, middlemen involved were Wholesale (local), Retailer (R) and Wholesaler. From the table, it was evident that the post harvest loss was highest at the retailer level (12.50 per cent). In ginger, maximum loss occurred during storage through spoilage and weight loss. Weight loss occurred both during storage and in the market. The total post harvest loss of ginger at the market level was found to be 9.98 per cent of the total quantity transacted in the market.

Table 7 represents the extent of post harvest losses for all the four selected commodities in various stages. The loss was expressed in terms of percentage to the total production of each commodity. It was evident from the table that the maximum post harvest loss was found to be 22.62 per cent for tomato followed by ginger (14.66 per cent), orange (13.85 per cent) and pineapple (9.25). Out of the total post harvest loss, loss in the market was found to be higher than the loss at the growers level in all the commodities except tomato. Only in case of tomato, the loss at growers level was higher (13.07 per cent) than the loss at the market (9.55 per cent). The table also revealed that for all the commodities the post harvest loss was very high during transportation and storage. This might be attributed to the high perishability of the crop and the very low level of marketing and infrastructure development in the state of Assam.

## Conclusion and Policy Implications

It may be concluded from the study that in Assam, the post harvest loss of pineapple, tomato, orange and ginger is very high and proper post harvest management is the need of the hour to reduce these losses. This will help to increase the per capita availability of these crops at the existing level of production technology. Suitable policies may be framed for efficient marketing as well as to develop a sound inter sectoral marketing linkages which will in tern improve the rural livelihood of the state and the food security in near future.

Appropriate strategies like adoption of market driven production system, planned truck farming, appropriate post harvest handling through promotion of pack house concept including storage, processing and value addition, creation of adequate transportation infrastructure and organized marketing system are some of the measures necessary to reduce the post harvest losses. Few policy options are presented below.

By reducing the number of intermediaries involved in the marketing channels of these crops, total post harvest loss could be reduced significantly.

Improvement in transportation, infrastructure, storage etc can also help in reducing the loss.

Cold chain facilities should also be extended to sub-and –main market system. By doing this, the bargaining power of the producer would be increased sufficiently. Besides, it could also reduce the post harvest loss.

Agro-processing units must be started in the production zones to arrest the excess market arrivals during the glut period. This would not only raise the income of the growers but also help in minimising the severe fluctuations of prices and post harvest losses.

Introduction of futures trading to some extent will facilitate the reduction of post harvest loss, since the traders will collect the products immediately after harvest based on the agreement executed in advance.

Institutional support should also be made available to start production-center's–cum-business–incubators at the block level to minimise wastage and enhance value addition of farm produce.

## REFERENCES

Deka, B.C. & S. Saikia (2005), "Marketing Strategies for Some Horticultural Produces of Assam", Souvenir of the 1st District Level Horticultural Show and Competition – 2005, Jorhat, Assam, pp. 39-42.

Hazarika, C. (2003), Report on Economics of Post Harvest Management of Perishable Agricultural Commodities in Assam", Department of Agricultural Economics, Assam Agricultural University, Jorhat, Assam.

Salunkhe, D. K. and B. B. Desai, (1984), Post harvest biotechnology of vegetables, CRC press, INC. Florida, Vol. 1, pp. 1-14.

Verma, M.M. and R.K. Agarwal, Rural and Agricultural Marketing, Forward Publishing Company, New Delhi, 1997, p. 11.

Yadav, Technological Options can check Post Harvest Loss, NIAM, Online edition of the *Hindu*, August 07, 2003.

### Table 1: Utilization Pattern of Selected Fruits and Vegetables by the Sample Growers in Assam

| *Crops* | *Total production* | *Home consumption* | *Kind payment* | *Seed* | *Total post harvest loss at producers level* | *Marketed surplus* |
|---|---|---|---|---|---|---|
| Pineapple (nos) | 475000 | 18990 | 14405 | - | 20067 | 421538 |
| | (100) | (4.00) | (3.03) | | (4.22) | (88.75) |
| Orange (nos) | 7111843 | 121550 | 118300 | - | 468317 | 6403676 |
| | (100) | (1.71) | (1.66) | | (6.59) | (90.04) |
| Tomato (qts) | 21298.00 | 184.73 | 172.66 | - | 2782.92 | 18157.69 |
| | (100.00) | (0.87) | (0.81) | | (13.07) | (85.25) |
| Ginger (qts) | 5475.43 | 5.09 | 3.74 | 125.28 | 256.05 | 5085.27 |
| | (100) | (0.09) | (0.07) | (2.29) | (4.67) | (92.88) |

Figures within parentheses indicate percentage to total production.

### Table 2: Extent of Post-Harvest Loss at Growers Level During Various Activities in Assam

| *Items* | *Pineapple (nos)* | *Orange (nos)* | *Tomato (qts)* | *Ginger (qts)* |
|---|---|---|---|---|
| 1. <u>During Assembling</u> | | | | |
| a. Non-availability of labour | 977 | 30250 | 23.04 | 3.804 |
| | (4.87) | (6.46) | (0.83) | (1.48) |
| b. Insect pest infestation | 1487 | 7097 | 89.11 | 5.896 |
| | (7.41) | (1.52) | (3.20) | (2.30) |
| c. Climatic condition | 1465 | 35333 | 95.16 | 9.101 |
| | (7.30) | (7.54) | (3.42) | (3.55) |

| | | | | |
|---|---|---|---|---|
| d. Spoilage/wastage due to delayed harvesting | 3631 (18.09) | 37599 (8.03) | 525.46 (18.88) | 27.903 (10.89) |
| e. Others | 663 (3.30) | 3888 (0.83) | 20.08 (0.72) | - |
| 2. <u>During Grading and packing</u> | | | | |
| a. Quantity discarded for low quality | 1555 (7.75) | 53841 (11.50) | 536.00 (19.26) | 43.734 (17.08) |
| b.Non availability of packing material | 220 (1.10) | 16950 (3.62) | 61.04 (2.19) | - |
| 3. <u>During Storage</u> | | | | |
| a. Spoilage | 815 (4.06) | - | - | 9.219 (3.60) |
| b.Weight loss | - | - | - | 32.784 (12.80) |
| 4. <u>During Transportation</u> | | | | |
| a. Loading/unloading | 2401 (11.96) | 42323 (9.04) | 127.15 (4.57) | 32.362 (12.64) |
| b.Breakage of boxes/baskets | - | 52222 (11.15) | 69.57 (2.50) | 11.309 (4.42) |
| c.Road blockage /mishappening of vehicles | 648 (3.23) | 57968 (12.38) | 43.86 (1.58) | - |
| d.Others | 430 (2.14) | 4631 (0.99) | 64.27 (2.31) | - |
| 5. <u>Losses in the Market</u> | | | | |
| a. Produced rejected by agents | 1345 (6.70) | 16656 (3.56) | 257.09 (9.24) | 7.543 (2.94) |
| b.Space problem for unloading | 871 (4.34) | 24863 (5.31) | 82.96 (2.98) | 3.068 (1.19) |
| c.Delay in auction/ sale | 1605 (8.00) | 45379 (9.69) | 278.47 (10.01) | 9.068 (3.54) |
| d.Loss due to lack of market demand | 1360 (6.78) | 35333 (7.54) | 495.38 (17.80) | 40.323 (15.75) |
| e.others | 594 (2.96) | 3984 (0.85) | 14.28 (0.51) | 19.942 (7.79) |
| Total Loss | 20067 (100.00) (4.22)* | 468317 (100.00) (6.59)* | 2782.92 (100.00) (13.07)* | 256.05 (100.00) (4.67)* |
| Total Production | 475000 (100.00) | 7111843 (100.00) | 21298.00 (100.00) | 5475.43 (100.00) |

Figures within ( ) indicate percentage to total loss

Figures within ( )* indicate percentage loss to total production

**Table- 3**
**Post Harvest Loss of Pineapple at Middlemen Level During Various Stages of Marketing**

| Type of middle Men | Qty. Transacted | During Assembling | | During Grading for | During Storage | During Transportation | | Loss in the market | | | | Total Loss | PH loss in total quantity transacted (%) |
|---|---|---|---|---|---|---|---|---|---|---|---|---|---|
| | | Transportation | Spoilage | Low quality | | Loading /Unloading | Road blockage | Produce Rejected | Space Problem | Delay in sale | Others | | |
| Whole Seller | 448093 | 1523 (8.23) | 570 (3.08) | 1825 (9.86) | 1582 (8.55) | 7619 (41.17) | 596 (3.22) | 1830 (9.89) | 1212 (6.55) | 1603 (8.66) | 146 (0.79) | 18506 (100) | 4.13 |
| Retailer | 478992 | 1387 (4.42) | 1362 (4.34) | 4402 (14.03) | 8634 (27.52) | 3828 (12.20) | 524 (1.67) | 4897 (15.61) | 1142 (3.64) | 4292 (13.68) | 906 (2.89) | 31374 (100) | 6.55 |
| WS-C-PHC | 87409 | 746 (23.63) | 273 (8.64) | - | 364 (11.53) | 591 (18.73) | - | 364 (11.53) | 291 (9.22) | 418 (13.26) | 109 (3.46) | 31560 (100) | 3.61 |
| WS (Via CA) | 136636 | 729 (20.51) | 37 (1.03) | 346 (9.75) | 474 (13.33) | 455 (12.82) | 182 (5.13) | 510 (14.36) | - | 674 (18.97) | 146 (4.10) | 3553 (100) | 2.60 |
| PHC | 23732 | 274 (11.03) | 73 (2.94) | 292 (11.76) | 256 (10.30) | 730 (29.41) | - | 292 (11.76) | - | 475 (19.12) | 91 (3.68) | 2483 (100) | 10.46 |
| Total | 1174862 | 4655 (7.88) | 2316 (3.92) | 6870 (11.63) | 11306 (19.14) | 13223 (22.39) | 1299 (2.20) | 7892 (13.36) | 2646 (4.48) | 7461 (12.63) | 1401 (2.37) | 59072 (100) | 5.03 |

**Figures within ( ) indicate percentage to total loss**

Table-4
Post Harvest loss of Tomato at Middlemen Level During Various Stages of Marketing

| Type of middlemen | Qty. Transacted | During assembling | | | During grading and packaging | | During storage | | During Transportation | | | | Loss in the market | | | Total loss | Loss (%) |
|---|---|---|---|---|---|---|---|---|---|---|---|---|---|---|---|---|---|
| | | Insect/pest infestation | Transportation | Spoilage | Low quality | Lack of packing material | Spoilage | Weight loss | Loading /unloading | Basket/ boxes problem | Road blockage | Weight loss | Produce rejected by agent | Space problem for unloading | Delay in auction/ sale | | |
| Retailer | 2374 | 3.29 (1.98) | 6.59 (3.97) | 8.24 (4.96) | - | - | 6.59 (3.97) | .29 (1.98) | 16.48 (9.92) | - | - | - | 31.32 (18.85) | 18.13 (10.91) | 72.25 (43.46) | 166.18 (100) | 7.00 |
| WS (L) | 84276 | 151.69 (2.00) | 257.13 (3.39) | 243.47 (3.21) | 1667.91 (21.99) | 270.78 (3.57) | 37.92 (0.50) | 37.92 (0.50) | 1611.78 (21.25) | 1001.96 (13.21) | 840.40 (11.08) | 227.54 (3.00) | 460.40 (6.07) | 386.07 (15.09) | 389.87 (5.14) | 7584.84 (100) | 9.00 |
| WS (CA) | 11870 | 27.16 (2.08) | 40.74 (3.12) | 45.31 (3.47) | 307.10 (23.52) | 39.17 (3.00) | 9.79 (0.75) | 7.18 (0.55) | 244.82 (18.75) | 183.45 (14.05) | 155.64 (11.92) | 31.73 (2.43) | 78.34 (6.00) | 64.76 (4.96) | 70.51 (5.40) | 1305.7 (100) | 11.00 |
| WS | 16618 | 38.69 (1.94) | 159.33 (7.99) | 48.46 (2.43) | 408.40 (20.48) | 51.85 (2.60) | 19.94 (1.00) | 19.97 (0.50) | 509.51 (25.55) | 297.73 (14.93) | 83.16 (4.17) | 79.77 (4.00) | 138.59 (6.95) | 107.28 (5.38) | 41.47 (2.08) | 1994.16 (100) | 12.00 |
| PWS | 3561 | 5.92 (2.08) | 9.88 (3.47) | 7.89 (2.77) | 64.29 (22.57) | 11.91 (4.18) | 3.13 (1.10) | 1.14 (0.40) | 62.02 (21.77) | 39.57 (13.89) | 27.69 (9.72) | 9.88 (3.47) | 17.8 (6.25) | 13.84 (4.86) | 9.92 (3.47) | 284.88 (100) | 8.00 |
| TOTAL | 118699 | 226.75 (2.00) | 473.67 (4.18) | 353.37 (3.12) | 2447.7 (21.59) | 373.71 (3.29) | 77.37 (0.68) | 59.5 (0.52) | 2444.61 (21.56) | 1522.71 (13.43) | 1106.89 (9.76) | 348.92 (3.08) | 726.45 (6.41) | 590.08 (5.20) | 584.02 (5.18) | 11335.76 (100) | 9.55 |

**Table-5**

**Post Harvest Loss of Orange at Middlemen Level During Various Stages of Marketing**

| Type of middleman | Quantity transacted | During assembling | | During grading and packaging | | During storage | During Transportation | | | Loss in the market | | | Total loss | Loss (%) |
|---|---|---|---|---|---|---|---|---|---|---|---|---|---|---|
| | | Trans Porta-tion | Spoilage | Low quality | Lack of packing material | Spoilage | Loading/ Unloading | Basket/ boxes problem | Road Blockage | Produce rejected by agent | Space problem for un-loading | Delay in auction /sale | | |
| Wholesaler | 11179361 | 72180 (9.25) | 38625 (4.95) | 68278 (8.75) | 40615 (5.20) | 215520 (27.62) | 125000 (16.02) | 41569 (5.33) | 32385 (4.15) | 47830 (6.13) | 44088 (5.65) | 54230 (6.95) | 780320 (100) | 6.98 |
| Retailer | 947403 | 7989 (8.85) | 4720 (5.23) | 7205 (7.98) | 5595 (6.20) | 25275 (28.00) | 12640 (14.00) | 6320 (7.00) | 4740 (5.25) | 3746 (4.15) | 5645 (6.25) | 6395 (7.08) | 90270 (100) | 9.53 |
| PHC –cum-wholesaler | 3031692 | 14080 (5.98) | 22720 (9.65) | 18840 (8.00) | 12455 (5.29) | 61238 (26.01) | 35889 (15.24) | 16430 (6.98) | 15895 (6.75) | 13780 (5.85) | 10832 (4.60) | 13321 (5.66) | 235480 (100) | 7.77 |
| Wholesaler (via CA) | 2463249 | 12320 (7.25) | 11218 (6.60) | 17540 (10.32) | 7700 (4.53) | 46570 (27.40) | 25190 (14.82) | 7055 (4.15) | 7940 (4.67) | 11645 (6.85) | 6580 (3.87) | 16207 (9.54) | 169965 (100) | 6.90 |
| PHC | 757923 | 4480 (7.88) | 2350 (4.13) | 6610 (11.63) | 2170 (3.82) | 13440 (23.64) | 9650 (16.98) | 3540 (6.23) | 3380 (5.95) | 2800 (4.93) | 2770 (4.87) | 5654 (9.95) | 56844 (100) | 7.50 |
| Total | 18379628 | 111049 (8.33) | 79633 (5.97) | 118473 (8.89) | 68535 (5.14) | 362043 (27.16) | 208369 (15.63) | 74914 (5.62) | 64340 (4.83) | 79801 (5.99) | 69915 (5.25) | 95807 (7.19) | 133287 9 (100) | 7.25 |

Figures within ( ) indicate percentage to total loss

**Table-6**

**Post Harvest Loss of Ginger at Middlemen Level During Various Stages of Marketing**

| Type of middlemen | Quantity Transacted | During assembling | | During grading and packaging | During storage | | During Transportation | | Loss in the market | | | Total loss | Loss (%) |
|---|---|---|---|---|---|---|---|---|---|---|---|---|---|
| | | Transportation | Spoilage | Low quality | Spoilage | Weight loss | Loading/ unloading | Basket/ boxes problem | Produce rejected by agent | Delay in auction /sale | Weight loss | | |
| **Retailer** | 120 | 1.20 (8.00) | 0.90 (6.0) | 0.60 (4.00) | 6.00 (40.00) | 3.00 (20.00) | 0.75 (5.00) | | 0.65 (4.33) | 0.90 (6.00) | 1.00 (6.67) | 15.00 (100.00) | 12.50 |
| **Whole Saler (L)** | 16800 | 75.60 (4.50) | 92.40 (5.50) | 84.00 (5.00) | 117.60 (7.00) | 680.40 (40.50) | 109.20 (6.50) | 58.80 (3.50) | 75.80 (4.51) | 50.20 (2.99) | 336.00 (20.00) | 1680.00( 100) | 10.00 |
| Whole Saler | 2982 | 43.61 (15.00) | 20.35 (7.00) | 14.54 (5.00) | 87.22 (30.00) | 45.43 (15.62) | 30.00 (10.32) | 13.08 (4.50) | 11.63 (4.00) | 12.89 (4.43) | 12.00 (4.13) | 290.75 (100) | 9.75 |
| **T0TAL** | 19902 | 120.41 (6.06) | 113.65 (5.72) | 99.14 (5.00) | 210.82 (10.62) | 728.83 (36.70) | 139.95 (7.05) | 71.88 (3.62) | 88.08 (4.43) | 63.99 (3.23) | 349.00 (17.57) | 1985.75 (100) | 9.98 |

Figures within ( ) indicate percentage to total loss

**Table 7: Extent of Post-Harvest Loss of Selected Commodities in Assam**

*(Loss expressed in % of total production)*

| *Sl. No.* | *Stages of Loss* | *Pineapple* | *Tomato* | *Orange* | *Ginger* |
|---|---|---|---|---|---|
| 1 | At growers' level | | | | |
| | i. During assembling | 1.73 | 3.54 | 1.61 | 0.85 |
| | ii. During grading | 0.37 | 2.81 | 1.00 | 0.80 |
| | iii. During storage | 0.17 | 0.00 | 0.00 | 0.77 |
| | iv. During transportation | 0.73 | 1.43 | 2.21 | 0.80 |
| | v. Loss in the market | 1.22 | 5.29 | 1.77 | 1.46 |
| 2 | Total loss at growers' level | 4.22 | 13.07 | 6.59 | 4.68 |
| 3 | At market middlemen level | | | | |
| | i. During assembling | 0.59 | 0.89 | 1.04 | 1.18 |
| | ii. During grading | 0.59 | 2.37 | 1.02 | 0.50 |
| | iii. During storage | 0.96 | 0.12 | 1.97 | 4.72 |
| | iv. During transportation | 1.24 | 4.57 | 1.89 | 1.06 |
| | v. Loss in the market | 1.65 | 1.60 | 1.34 | 2.52 |
| 4 | Loss at market middlemen level | 5.03 | 9.55 | 7.26 | 9.985 |
| 5 | Total post-harvest loss | 9.25 | 22.62 | 13.85 | 14.66 |

# 18

# STATUS AND CONSTRAINTS OF POTATO PRODUCTION IN ASSAM: AN EMPIRICAL STUDY

*Arun Pandit, N.K. Pandey, Rajesh K. Rana, N.R. Kumar & K.P. Chandran*

## Introduction

The economy of Assam is primarily agriculture based. Major crops of this state are rice, jute, sugarcane, pulses, tea, oilseeds and vegetables. In recent years, potato is grown extensively in the state with a share of 6% in area and 2.8% in production (Average of 1990-'91 to 2004-'05). The popularity of potato in the state has been increased tremendously over the years. Bhattacharyya et al. (2001) reported that in 1997-98 the potato acreage has been increased by 22.52% over 1988-89 and 222.65% over 1962-63. In 2004-05 potato was grown in about 73 thousand hectares of land and the total production was about 589 thousand tons in the state, which possesses about 59 percent area and about 64 percent production of potato in the northeastern states of India (DES, 2005). The major potato producing districts are Barpeta, Kamrup and Darrang (Bist and Sharma, 1997). Vadivelu et al. (2003) found that Potato was suitable in the post-rainy season in Jorhat district also. However, Barpeta is the most important potato growing district not only in the state but also in the region. The popular potato varieties are Kufri Jyoti and Kufri Sindhuri.

Though potato is a major winter vegetable grown in Assam and cultivated in a number of districts of the state, yield is very low (only 8 ton/ha against national average of 19 t/ha). Hence there is a huge gap between demand and supply of potato in Assam necessitating import from neighbouring states like West Bengal, Meghalaya and even U.P and Punjab every year. The present study was conducted with the objectives of economic analysis, estimating the factors affecting yield and the technical efficiency of potato cultivation in the Barpeta district of Assam.

## Methodology

### *Data Collection*

Barpeta, being the highest potato-producing district of potato in Assam, was purposively selected for the present study. Multi-stage random sampling technique was followed to select three blocks, namely Bhawanipur, Mandia and Sarukshetri. Thirty six farmers were selected for interviews from the selected blocks. The study was conducted during the crop year of 2004-05.

### *Analytical Framework*

The economics of potato production was worked out with the help of simple tools of mathematics, i.e. averages; percentages etc. An attempt has also been made to estimate the contribution of different factors in yield variation. The following model was found suitable for the data.

Yi = a X1i b1 X2i b2 ...Ui

where, Yi is the yield of $i^{th}$ farm and Xj's are the factors considered and Ui is the random term. a and bi's are the coefficients to be estimated.

Technical efficiency is the ability of a farm/firm to achieve maximum possible output with the available resources. In this study, Data envelopment analysis (DEA), a nonparametric technique of technical efficiency estimation is employed. It uses a mathematical program to estimate the efficiency frontier. It does not need the pre-specification of the production function coefficients. Unlike parametric approaches, DEA makes no assumption of the distribution of the underlying data, and all deviations are assumed to be due to inefficiency

(Banker et al., 1989). DEA analyzes farms separately while measuring its efficiency relative to all the observations in the sample.

In this study, simple case of single output was considered. Let X be the input matrix of order k×n and Y the output vector. Here k is the number of inputs. Thus, for ith farm, Xi and Yi represent the respective inputs and output. Now the problem reduces to obtaining a ratio measure $\mu'Yi / v'Xi$ where $\mu$ and $v$ are the output and input weights, respectively. Optimal weights are obtained by solving the following mathematical program:

$$\text{Max}_{\mu,v} \; (\mu_i' Y_i / v_i' X_i)$$

Subject to

$$\mu'Yj/v, Xj \leq 1 \quad j = 1, \ldots, n$$

$$\mu, v \geq 0$$

In order to avoid infinite number of solutions, imposing a constraint $v'Xj = 1$, we get

$$\min_{\theta,\lambda} \theta$$

subject to

$$-yi + Y\lambda \geq 0$$

$$\theta xi - X\lambda \geq 0$$

$$\lambda \geq 0$$

Here, $\theta$ is a scalar and $\lambda$ is an n×1 vector of optimal weights. $\theta$ represents the technical efficiency (TE) corresponded to constant return to scale (CRS). Imposing an additional constraint $1'\lambda = 1$ gives the technical efficiency under variable return to scale (VRS). Efficiency measurements by the DEA model can be used to determine both pure technical and scale efficiencies (TECRS/TEVRS). The product of these two gives the overall technical efficiency. Efficiency scores in this study are estimated using the computer program, DEAP Ver. 2.1. described in Coelli (1996).

## Results and Discussion

### *Economics of Potato Production*

Economics of potato production in the district has been analysed and the same has been presented in table 1. A close perusal of the table reveals that potato cultivation in a hectare needs about Rs. 53955 investment. Seed alone contributes more than 31 percent of the total cost. The study shows that on an average, farmers were using 26.3 quintal of seed per hectare. They got seed mainly from the neighbouring state West Bengal. Seed quality was not good as West Bengal is not the ideal seed potato producing state. There was no organized and reliable seed supplying agencies in the state. Hence farmers had to depend on the private traders for their seed requirement. Moreover, farmers purchased seed every year, as they could not store their own seed due to lack of cold storage facilities in the district. The next important cost item was labour (23.70%). Mechanization was limited in the study area and all the operations except field preparation were done manually and hence, about one-fourth of the total cost was incurred in paying the labour cost. Further, it was observed that potato production in the study area required about 230 man-days per hectare excluding the labour requirement in marketing, which indicated that potato cultivation is one of the good avenues for rural employment generation. Therefore, labour is a very important input for the potato production and labour availability at the right time is crucial.

There was no proper irrigation facility in the area. Hence, as an alternative, farmers do mulching just after planting using locally available materials and the cost for the same was found to be Rs. 10800 per hectare (20.02%). This cost includes that of mulching material (paddy straw, water hyacinth etc.) and labour cost. Further, farmers did not do any earthing up or weeding operations subsequently. Hence, mulching was one of the significant cost items and its share in the total cost was about 20 percent.

Fertilizers and FYM cost was found to be around 11%. The recommended dose of NPK in the region is 125:120:60 (Singh et al., 2003) but very few farmers applied this dose. In fact farmers were not aware of the recommended dose and the importance of balances dose of fertilizers. Hence, there was a need to intensify proper extension efforts to educate the farmers on the importance of balance dose of fertilizers.

**Table 1: Economics of Potato Production**

| *Items of cost* | *Cost (Rs./ha)* |
|---|---|
| Seed | 16841.15 (31.21) |
| Fertilizer | 5950.52 (11.03) |
| Micronutrient | 573.79 (1.06) |
| Plant protection | 1337.5 (2.48) |
| Labour | 12789.58 (23.70) |
| Hired machinery | 5662.5 (10.49) |
| Mulching | 10800 (20.02) |
| Total cost | 53955.04 (100.00) |
| Yield (Q/ha) | 211.29 |
| Production cost (Rs/Q) | 255.36 |
| Gross return (Rs/ha)* | 84516 |
| Net return (Rs/ha) | 30560.96 |
| B:C | 1.57 |

*Note:* * based on average actual market price (Rs. 400/q) received by them. Figures in the parentheses indicate percentage to the total cost of cultivation.

Majority of the farmers hired tractor for land preparation and bullock labour for harvesting and the hired machinery cost was worked out to be about 10.49%. As far as plant protection is concerned, the farmers were mainly applying mancozeb for controlling the late blight. Majority of the farmers applied mancozeb @ 2 kg/ha giving 3-4 sprays. The per hectare plant protection cost was estimated to be Rs. 1337.50 (2.48%). Thus, it is necessary to demonstrate the usage of latest, more effective systemic fungicides like Metalaxyl in the farmers' field. Cost of production was found to be Rs. 255.36 per quintal with the average yield of around 211 quintal per ha. The production cost could further be reduced if necessary steps are taken to boost the yield. However, potato grower in the area could get a net return of about Rs. 30561 per ha and the benefit cost ratio was estimated to be 1.57, which indicates a gross return of Rs. 1.57 in investing Re. 1. Hence, potato cultivation in the area is remunerative as the local demand is more than the supply.

### *Factors Affecting Potato Yield*

Among the eight factors considered for regression analysis,

the significant ones were labour, plant protection, micronutrient and mulching. Labour cost emerged out to be the most crucial input for the yield (Table 2). One percent increase in labour cost would lead to 1.82 percent increase in yield. In the study area, mechanization was limited and almost all the operations were carried out manually. Hence, labour availability at the right time is of vital importance. But it was found that the labour shortage was a major problem during critical phases of production resulting in low yield of the crop. The results of regression analysis given in Table 2 clearly emphasize the importance of labour in improving the yield. Another important factor was mulching. In the study area, due to lack of irrigation facility, farmers practised mulching for preserving moisture. Potato growers used locally and abundantly available paddy straw or water hyacinth to cover their field just after planting and earthing up as well as weeding operations were not carried out. Hence, in absence of irrigation, mulching was an essential operation in potato production. The fitted model showed that 1% increase in mulching cost would lead to almost 1% increase in yield.

**Table 2: Factors Affecting Potato Yield**

| *Parameters* | *Intercept* | *Labour* | *Mulching* | *Plant protection* | *Micronutrient* |
|---|---|---|---|---|---|
| Coefficient | 0.019** | 1.819*** | 0.908** | 0.07118* | 0.03767* |

*Note:* R2 = 0.60. ***, **, * denote significant at 1, 5 and 10% level of significance.

Plant protection and micronutrient costs were the other factors, which affected yield, but their contribution was found to be relatively low. The return to scale was found to be 2.84, i.e. when all these factors are increased by 1%, the yield will be increased by 2.84 percent. Hence, potato production in the study area was operating under increasing return to scale, indicating the existence of scope to increase the potato production provided the farmers acquire the scientific potato production technologies and government extend the necessary infrastructural supports like irrigation, cold storage facilities etc.

### *Technical Efficiency*

Data envelopment analysis employed has produced the

following results. Overall 63% of the farms are less than 25% technically inefficient (Table 3). Further, majority of the farms are scale efficient and the number of highly inefficient farms are negligible. Overall technical efficiency is 78% whereas scale efficiency and pure technical efficiency are 94% and 83%, respectively. Efficiency can be improved by operating at an optimal scale and by adoption of best practices. Average efficiency scores indicate that pure technical inefficiency (17%) is primarily responsible for the overall technical inefficiency (22%) as compared to scale inefficiency (6%). This implies that in the absence of environmental differences, pure technical inefficiency would reflect departures from best practices.

**Table 3: Efficiency Measures of Potato Growing Farms**

| *Efficiency* | *Less than 25%* | *25 to 50%* | *50 to 75%* | *75% and above* | *Average* | *Minimum* |
|---|---|---|---|---|---|---|
| Pure | 1 (2.78) | 1 (2.78) | 7 (19.44) | 27 (75.00) | 0.83 | 0.219 |
| Overall | 2 (5.56) | 2 (5.56) | 9 (25.00) | 23 (63.89) | 0.78 | 0.084 |
| Scale | 1 (2.78) | 0 | 0 | 35 (97.22) | 0.94 | 0.084 |

### *Constraints in Potato Production*

During the course of investigation it was found that getting quality seed was the major problem. Gogoi (2002) also found the same problem in Jorhat, another district of Assam. Seeds are mainly coming from West Bengal in the name of Bengal Jyoti, the quality of which was not guaranteed. West Bengal is not a conventional seed potato producing state. Good quality seed is also very expensive and the small and marginal farmers cannot afford it. In Assam, bulk of the cultivable land (82%) belongs to the small and marginal farmers and the average size of the operational holding is only 1.37 ha (Bhattacharyya et al. 2001). Hence, government should make available good quality seed to farmers at low price. Lack of irrigation facilities was another major problem. Though the mighty Bramhaputra flooded the area almost every year, there is hardly any proper infrastructure for irrigation. Hence, farmers use the mulching technique, which was expensive, and at the same time yield gets reduced because no weeding operation and earthing up was done. Lack of scientific knowledge regarding potato production was also a reason for the low yield. Farmers were unaware of importance of balanced dose of fertilizers, use of effective fungicides and insecticides, right seed size and seed rate, seed treatment practices etc. Hence, proper extension programmes may be implemented to keep them aware of the latest development of potato production technologies. Inadequate marketing infrastructure at the village level and poor transportation network were major marketing problems in the area (Bhattacharyya, 2001). In addition, lack of cold storage facilities compelled the growers to dispose off the potatoes immediately after harvest. The study found that 97% of the produce was sold just after harvest. Moreover, local storage methods were also not developed. Government or some cooperative societies should take initiatives to establish cold stores so that farmers can take the temporal price advantage.

## Conslusions and Suggestions

The economic analysis of potato production revealed that seed, labour, mulching were the major items of cost of potato cultivation and lack of quality seed, imbalance dose of fertilizers, lack of irrigation facilities and lack of scientific knowledge regarding potato production practices were some of the reasons for poor yield in the study area. The regression analysis showed that for increasing the yield, importance should be given on availability of labour, good mulching, proper plant protection and micronutrient application. Data envelopment analysis revealed that pure technical inefficiency makes greatest contribution to overall inefficiency necessitating emulation of best practices of efficient farms. Hence extension efforts should be made to educate the farmers regarding scientific potato production practices. Further, there is a need of government intervention in providing quality seed and establishing cold storage and irrigation facilities.

## REFERENCES

Banker, R.D., A. Charnes, W.W. Cooper, J. Swarts and D. Thomas (1989), "An Introduction to Data Envelopment Analysis with Some of Its Models and Their Uses", *Research in Governmental and Nonprofit Accounting*, Vol. 5, pp. 25-64.

Bhattacharyya, R.K. (2001), Horticultural Crops—Fruits and Vegetables. In: Agriculture in Assam. Eds: A.C. Thakur (chief editor), Ashok Bhattacharya and D.K. Sharma. Dir. of Extn Education. Assam Agricultural University, Jorhat, Assam.

Bhattacharyya, H.C., K. Borkakoti and R.S. Saikia (2001), *Agriculture in Assam*, Eds: A.C. Thakur (chief editor), Ashok Bhattacharya and DK Sharma. Dir. of Extn Education. Assam Agricultural University, Jorhat, Assam.

Bist, B.S. and H.C. Sharma (1992), "Potato Statistics—India and the world", *Technical Bulletin* No. 40, Central Potato Research Institute, Shimla, pp. 124.

Coelli, T.J. (1996), "A Guide to D.E.A.P. Version 2.1: A Data Envelopment Analysis (computer) Program", C.E.P.A. working paper 96/08, Department of Econometrics, University of New England, Armidale.

D.E.S. (2005), Agricultural Statistics at a Glance, Directorate of Economics and Statistics, Ministry of Agriculture, Government of India.

Gogoi, A. (2002), Production and marketing of potato in Jorhat district of Assam. Unpublished M.Sc. Thesis. Assam Agricultural University, Jorhat, Assam.

Singh, P.H., V.K. Chandla, S.M.P. Khurana, I.D. Garg, K.C. Thakur, Sarjeet Singh and S.S. Lal. (2003), "Potato Cultivation in North Eastern India", *Extension Bulletin*, No. 34 (E). Central Potato Research Institute, Shimla, p. 25.

Vadivelu, S. U. Baruah, B.P. Bhaskar, J. Thampi, D. Sarkar, C.S. Walia, D.C. Nayak and A.K. Maji. (2003), "A Land Use Plan for Jorhat District of Assam State", *Agropedology*, Vol. 13, No. 1, pp. 1-10.

# 19

# FARM SIZE AND TENURIAL STATUS OF FARMER AND PRACTICE OF MODERN FARMING IN ASSAM

*Niranjan Roy*

## Introduction

Although HYV seed–fertiliser technology is said to be scale neutral, there was some apprehension that only the bigger farmers would be able to successfully exploit the prospective economic potential of improved practices. The apprehension was based on the fact that adoption of this technology package require higher amount of investment than the traditional varieties and hence the small and marginal farmers would not be in a position to use this technology package because of their lack of access to financial resources. Early studies on adoption of seed-fertiliser technology indeed found a positive correlation between farm size and adoption of this technology package (Lionberger, 1963; Acharya, 1969; Schulter and Mellor, 1972). However studies related to later years found small farmers to be catching up with the large and medium one in adoption of seed-fertiliser technology (Rao, 1969; Bhalla, 1979). The small farmer with a relatively smaller piece of land will try his best to get the maximum output from his land. The small farmers also enjoyed the advantage of family labour. This lagged response of small farmers' to seed-fertiliser technology has been explained by Castilbo (1975) along the following lines, "while the early adopter large farmers used partial adoption as a means of reducing risks, the small farmers had a `wait and see' attitude with demonstrated results from the large farmers, the small farmers were then willing to go all the way with new seeds". Under the impact of the new technology which is essentially capital based the productivity advantage of small farmers with relative advantage of family labour started moving in favour of the large farms which had more resources and also a more easy access to external finance (Rudra, 1982; Dyer, 1998; Soni, 2007)

So far as adoption of technology and tenurial status of farmers are concerned a study related to Punjab, Tamil Nadu and Maharastra by Mukharjee (1970) found no significant difference in the rate of adoption between the owners and the tenants. Similar findings of uniform adoption by tenants and owner were also confirmed by Muhiah (1971) based on a study for Tamil Nadu and Mandal and Ghosh (1976) based on sample selected from West Bengal, Bihar and Orissa. But these findings regarding adoption and tenurial status of farmers contradict those of Parthasarathy and Prasad (1978) related to farmers in Andhra Pradesh. It was observed that the `association between tenure and adoption was significant' and that `the owners had an edge over the tenants'. In this regard from the personal experience of this researcher from the field survey in Assam, it is probably more judicious to conclude that it is not tenancy as such but the nature of the terms of tenancy which affects the rate of adoption of improved technology. When tenancy takes the form of sharing of output in some pre-agreed proportion (which is 50:50 in most cases of tenancy) without similar sharing of costs between landlords and tenants, the tenant farmers incentives for adoption of improved technology is likely to be adversely affected. This is because the improved technology promises larger output that is to be shared with landowners, while its application involves extra costs which are primarily borne by the tenants. However, in cases where landowners not only share the output but also the cost of cultivation with the tenant farmers, tenancy need not be a discouraging factor for adoption of improved technology. The West Bengal experience seems to lend support to this view. The successful implementation of land reforms protecting the interest of small and marginal operators including tenants in the post 1981 period must have had a significant motivating force in reducing the backlog of seed-fertiliser technology in West Bengal (Khasnabis, 1994; Sawant, 1997).

In light of the conflicting findings regarding adoption and farm size and tenurial status of farmers it would be interesting to investigate the situation in Assam, where the practices of new agricultural strategy yet to gain momentum due to several constraints.

## Agricultural Scenario in Assam

Geographically the state of Assam is comprised of three physical divisions, viz, the Brahmaputra Valley, the Barak Valley and the Hill Range. The Brahmaputra Valley comprising the northern part is largest in terms of both area and population. The Barak Valley in the south is separated from the Brahmaputra Valley by the Hill Range of Karbi-Anglong and North Cachar Hills in the middle. There are six agro-climatic sub-zones in Assam, viz, Upper Brahmaputra Valley (North), Upper Brahmaputra Valley (South), Central Brahmaputra Valley, Lower Brahmaputra Valley (North-West), Lower Brahmaputra Valley (North), Lower Brahmaputra Valley (South), Barak Valley and Hill Zones. Except in Hill Zones (where farmer practices shifting cultivation), the agricultural practices in all other climatic sub-zones are identical since land availability and other physiographic conditions are homogeneous in character.

The state of Assam up to the sixties of the last century was self-sufficient in food supply. But since the seventies she became a net importer from other parts of the country (Baisya, 2003). As per 1995-96 agricultural census the average size of holdings in the state is 1.17 hectares. The distribution of operational holdings in the state is highly skewed with the concentration of more than half of the holdings in the lowest size class of below one hectare. Though the seed – fertiliser technology is said to be technically scale neutral, it is well known that too small a holding size may render adoption of modern agricultural operations economically unviable (Vyas, 1968; Rao, 1989). Accordingly many of the farmers in this smallest size class of holding may be currently handicapped with regards to adoption and effective utilisation of improved agricultural practices. A striking feature of the cropping pattern of the state is the predominance of paddy, which accounts for more than 90 percent of total cropped area. The uses of high yielding varieties of seeds and fertilizers are much less than that of the country as a whole. The irrigated area is only about 13 percent of total cropped area of the state.

## Objectives of the Study

The main objective of the study is to analyse whether the small size of farm and tenurial arrangements in practice have been responsible for slow growth of diffusion of agricultural technology vis-a-vis agricultural productivity in the state of Assam. The study mainly concentrates on rice crop since 90 percent of total cropped area is under rice. The study also excludes shifting cultivated area of the state (hill areas) whose mode of cultivation is different from that of plain areas.

## Hypothesis to be Tested

The institutional factors, such as tenurial arrangements in practice and relatively smaller size of farms are restricting the effective use of the available technology package by the farmers in the state.

## The Empirical Study

### *Sample Design and Methodology*

Since the main aim is to identify whether farm size and tenancy have been responsible for limited adoption and utilisation of the improved technology by the farmers in the state, it was decided to limit the number of observations to be covered and instead to collect data on each observation in some details. As such as a first stage of selection the Barak Valley region of the state is selected for field study location. It was decided to select the sites for sample survey from all the agricultural sub-divisions of that part of the state. Districtwise sub-divisions are: Karimganj district—

(1) Karimganj Sub-division, (2) R.K. Nagar Sub-division; Cachar district—(1) Silchar Sub-division, (2) Sonai Sub-division (3) Lakhipur Sub-division; Hailakandi district—(1) Hailakandi Sub-division. From each of these six sub-divisions one Agricultural Extension Officer (AEO) circle was selected for field study. The sample design used in selecting the sample in each AEO circle was that of two stage random sampling, in which villages constituted the primary sampling units and the farm households as the secondary and ultimate sampling units. In the first stage of selection, from each circle, three (3) villages were selected at random subject off course to the condition that at least in one of the selected villages agricultural infrastructure mainly some amount of irrigation was available for the use of the technology package of HYV seeds and fertiliser. In the second stage, about ten (10) percent of farm households in each selected village were then selected at random. The lists of farming households of the selected villages have been obtained from the Village Level Extension Worker (VLEW) of the concerned village. Two hundred and eighty two (282) farm households selected in this manner from the six selected AEO circles of Barak Valley constituted the whole sample of the field study.

For explaining the variation in the use of HYVs the linear functional form was not found quite suitable. The relevant dependent variable AH being contained in the range 0 to 1, a more appropriate functional form would be one in which the estimated value of the dependent variable will be contained within the range 0 and 1. Accordingly the following logistic functional form was chosen.

$$AH = \frac{1}{1 + e^{-Z}} + U \quad \text{-------- } 1.1$$

Where,

$$Z = a_1+a_2C_2+a_3C_3+a_4C_4+a_5C_5+a_6C_6+a_7IR+a_8EX+a_9OH+a_{10}TL+a_{11}LL \text{ ........} 1.1$$

The meaning of different variables in the model are,

AH => the proportion of area under high yielding varieties of rice in the total rice acreage of the sample farms.

IR=> the ratio of net irrigated area to operational holding of the farm.

EX=> the variable used to capture the effect of close contact with the personnel of government agricultural extension agencies.

OH => size of operational holdings of the farm.

TL=> extent of tenant land to operational holding of the farm.

LL=> percentage of low land to operational holding of the farm.

U => random disturbance term.

$C_2, C_3, C_4, C_5$ and $C_6$ represent dummies which have been used to capture the effect of variations in the locations across the six circles. The dummy $C_i$ (i = 2, 3....6) takes the value 1 for the circle number i and 0 for the other circles. The co-efficient of Ci captures the differential effects due to variations in local conditions between circle i and circle 1.

The ai's are the parameters to be estimated.

It may be noted that as Z goes from $-\alpha$ to $+\alpha$, expected value of AH increases from 0 to 1.

It is to be noted that adoption of HYVs by sample farmers was almost universal in the sense that out of 282 sample farmers 268 (about 95 percent) have adopted high yielding paddy varieties in the six circles. Despite adoption of HYVs by almost all the sample farmers, the spread in the area under these varieties has been found to remain fairly limited, that is only 41.1% of total cultivated area of sample farms.

As in the case of HYVs, the adoption of chemical fertilisers was almost universal among sample farmers. There is, however, great variation in the dose of fertiliser application. The consumption per hectare of N+P+K has been recorded to have varied from 0 to 185 kg. This wide variation in the rate of consumption of fertiliser among farms needs to be explained. Accordingly, a multiple regression analysis of fertiliser consumption per hectare in sample farms has been carried out. Thus the regression equation model used for the purpose of estimation of co-efficients is,

$FC=b_1+b_2C_2+b_3C_3+b_4C_4+b_5C_5+b_6C_6+b_7IR+b_8EX+b_9OH+b_{10}TL+b_{11}LL+b_{12}PF+U$ .........1.2

The variables OH, LL, TL, IR and EX represents the same meaning as in previous section.

FC=>the consumption of chemical fertilser (Nitrogenous,

phospheric and Potassic) in kg. per hectare of the sample farms.

PF => the proportion of area under fertilser intensive crops, like HYV paddy and vegetable to total cropped area of the sample farms.

The bi's are the parameters to be estimated.

The co-efficients of the variables OH, IR, PF and EX are expected to be positive, while the co-efficients of other variables LL and TL are expected to be negative.

### The Size of Operational Holdings of Sample

The distribution of the sample farms according to the size of their operational holdings is shown in table 1. The figures in the table show that 184 out of 282, that is about 65 percent of sample farmers had operational holdings below 2.0 hectares. Only 2 farmers had operational holdings between 5.0 to 8.0 hectares, while 75 farmers, that is about 27 percent were having land holdings between 2.0 to 3.0 hectares and about 7 percent of sample farmers (21 out of 282) fall in the category of land holdings between 3.0 to 5.0 hectares. The combined area of all 282 operational holdings comes to 487.2 hectares, which indicates that the average size of operational holdings of the sample farmers in the six circles was 1.73 hectares.

**Table 1: Distribution of Sample Farms According to Size of Operational Holdings**

| *Size class of operational holdings (in hectares)* | *Number of Sample Farms in* | | | | | | *Total* |
|---|---|---|---|---|---|---|---|
| | *Fakira Bazar* | *R.K. Nagar* | *Narshing-pur* | *Salcha-pra* | *Banska-ndi* | *Panch-gram* | |
| (1) | (2) | (3) | (4) | (5) | (6) | (7) | (8) |
| Below 0.5 | 1 | 4 | 4 | 2 | 4 | 2 | 17(6.0) |
| 0.5 to 1.0 | 5 | 4 | 12 | 4 | 5 | 10 | 40(14.2) |
| 1.0 to 2.0 | 12 | 23 | 26 | 21 | 21 | 24 | 127(45.0) |
| 2.0 to 3.0 | 21 | 11 | 14 | 12 | 13 | 4 | 75(26.6) |
| 3.0 to 5.0 | 8 | 1 | 1 | 4 | 1 | 6 | 21(7.5) |
| 5.0 to 8.0 | 1 | 0 | 0 | 1 | 0 | 0 | 2(0.7) |
| Total | 48(17.0) | 43(15.3) | 57(20.2) | 44(15.6) | 44(15.6) | 46(16.3) | 282(100.0) |

Figure in the parentheses represent percentage of the total sample.

### *Incidence of Tenancy Among Sample Farmers*

The field survey data shows that 135 out of 282, that is about 48 percent of sample farmers, had land on lease as part of their operational holdings. In 62 out of 135 cases of tenancy, the land leased in constituted more than half of the farmers' operational holdings. Indeed, incase of 40 cases of tenant farmers even 75 percent or more of operational holdings were found to have comprised of leased in land. The details of distribution of tenant farmers are shown in Table 2.

The sample data also indicates that tenancy was generally temporary, usually of one year and also completely informal. The mode of payment of rent was not found to be uniform in all cases. In most of the cases the practice was to share the output between the landlord and the tenant farmer, usually at 50:50 ratio. In some cases the cost of cultivation was shared equally by both landlord and the tenant and in some other cases the landlord shares only the cost of inputs, mainly fertilisers, while there are also instances in which a large section of the owner did not share any cost. The sample survey shows that out of 135 tenant farmers, in 48 cases the owner shared cost equally with the tenant and in the remaining 87 cases the owner did not shares any cost.

**Table 2: Distribution of Tenant Farmers in the Sample According to the Percentage of Leased Area in Their Operational Holdings**

| *Percentage of leased area in operational holdings* | *Number of Tenant Farmers in* | | | | | | *Total* |
|---|---|---|---|---|---|---|---|
| | *Fakira Bazar* | *R.K. Nagar* | *Narshing-pur* | *Salcha-pra* | *Banska-ndi* | *Panch-gram* | |
| (1) | (2) | (3) | (4) | (5) | (6) | (7) | (8) |
| Below 25 | 5 | 2 | 0 | 6 | 2 | 3 | 18 |
| 25 to 50 | 2 | 5 | 4 | 3 | 10 | 9 | 33 |
| 50 to 75 | 5 | 6 | 5 | 7 | 4 | 3 | 30 |
| 75 and above | 8 | 8 | 16 | 2 | 11 | 9 | 54 |
| Total | 20(41.7) | 21(48.8) | 25(43.9) | 18(40.9) | 7(61.4) | 24(52.2) | 135(47.9) |

Figure in the parentheses represents percentage of respective sample.

### *Discussion of Results*

Equation 1.1 for explaining variation in the use of HYVs by sample farmers has been estimated by non-linear least square method. The ordinary least square estimates of co-efficients of linear regression of AH on explanatory variables of the equation 1.1 were used as the initial trial estimates for starting the iteration process of the non-linear linear square method. The results that converged after five iterations are presented in table IX.1.

**Table 3: The Results of Non-linear Least Square Estimation for Explaining Variations in the Use of HYVs Across Sample Farms**

| *Variables* | *Estimated co-efficients* | *Standard Error of co-efficients* | *t-value* |
|---|---|---|---|
| Constant | 1.06800 | 0.20317 | 5.2564*** |
| C2 | -0.75598 | 0.19440 | -3.8887*** |
| C3 | -0.53084 | 0.18212 | -2.9148*** |
| C4 | -1.19500 | 0.21370 | -5.5919*** |
| C5 | -0.64876 | 0.20563 | -3.1550*** |
| C6 | -0.62723 | 0.21862 | -2.8691*** |
| IR | 0.32521 | 0.40528 | 0.80243 |
| EX | 0.03512 | 0.03098 | 1.1335 |
| OH | -0.13783 | 0.06189 | -2.2269** |
| TL | -0.28932 | 0.14577 | -1.9848** |
| LL | -0.88919 | 0.21162 | -4.2018*** |
| $R^2$ | 0.26131 | | |
| $F(n_1=10, n_2=271)$ | 9.5865*** | | |

*, **, *** indicates significant at .10, .05 and .01 level respectively.

Though the $R^2$ –value of 0.26131 is not particularly high, it cannot be dismissed as too low in this context of a fairly extensive cross-section study as the present one. Moreover, nine out of the eleven parameters of the function have come statistically significant showing the relevance of the corresponding factors in explaining the variation in the extent of use of HYVs by the farmers. The F-statistics for over-all regression is also highly statistically significant. Thus on the whole, results obtained from the analysis are fairly robust and credible.

Besides the Coefficients of the circle dummies Ci s and the intercept of Z, the Coefficients of the variables OH, TL and LL have come out statistically significant. All these three coefficients are negative. The negative sign of the coefficient of LL is expected as higher proportion of low land leaves a smaller part of operational holdings suitable for the use of HYVs. The negative sign of the coefficient of the variable OH indicates that the smaller farmers actually put larger proportions of paddy acreage under HYVs. In other words smaller farm size is not a handicap as far as the use of HYV is concerned. However, the coefficient of TL being negative implies that tenant farmers tend to use HYVs less extensively. The coefficient of the variables IR and EX, representing respectively the effects of availability of irrigation and access to service are expectedly positive. But neither of these coefficients is significant.

The results of the regression analysis of equation 1.2 as estimated by using the method of Ordinary Least Square are shown in table 4. The model gives a moderate fit to the sample data in terms of the R2 value (R2 = 0.25296). But the F – statistics for overall significance of the fitted regression has been found to be statistically highly significant. Apart from one locational dummies, the co-efficient of four variables have been found to be statistically significant. These are TL representing tenancy factor, PF representing extent of fertiliser intensive crops, IR indicating availability of irrigation and EX representing level of contact with extension agencies. As expected the co-efficients of the latter three variables have come out positive. The positive and significant co-efficient of the variable IR implies that the fertiliser consumption in a farm tend to increase with the increase in irrigation coverage for the farmland. From the significantly positive co-efficient of EX, it can be inferred that farmers' having better contact with extension agencies also tend to use fertiliser at a higher dose. The co-efficient of TL is negative indicating that as in case of extent of use of HYVs, tenant farmers are in a disadvantageous position for applying fertiliser in adequate quantity. Contrary to expectation the co-efficient of the variable OH has come negative but not found statistically significant. Thus there is no evidence of increasing fertiliser consumption per hectare with the increase in the farm size. The remaining variable LL has a negative co-efficient as expected, but the co-efficient is statistically not significant.

**Table 4**

**The Results of the Multiple Regression Analysis of Consumption of Chemical Fertiliser by Sample Farms**

| *Variables* | *Estimated co-efficients* | *Standard Error of co-efficients* | *t-value* |
|---|---|---|---|
| Constant | 27.75843 | 9.60450 | 2.890*** |
| C2 | -18.70817 | 6.75258 | -2.771*** |
| C3 | 9.69494 | 6.33316 | 1.531 |
| C4 | -1.82444 | 7.04266 | -0.259 |
| C5 | -3.98064 | 7.10509 | -0.560 |
| C6 | 14.27853 | 7.53480 | 1.895* |
| IR | 28.47990 | 13.79943 | 2.064** |
| EX | 3.83794 | 1.04593 | 3.669*** |
| OH | -0.36407 | 1.92993 | -0.189 |
| TL | -15.01916 | 4.83939 | -3.104*** |
| LL | -0.01860 | 7.39790 | -0.003 |
| PF | 19.85436 | 9.14840 | 2.170** |
| $R^2$ | 0.25296 | | |
| $F(n_1=11, n_2=270)$ | 8.31167*** | | |

*, **, *** indicates significant at .10, .05 and .01 level respectively.

## Conclusions and Suggestions

Besides farm size and tenancy the study also found out some other factors like lack of irrigation facilities, low lying areas, poor extension service etc as constraints towards increasing agricultural productivity in the state. But these non-institutional factors can also be taken care of by private participation and market forces as it happens in case of irrigation development in the Brahmaputra valley of the state. But institutional factors need governmental intervention in the form of reforms in existing land laws.

Tenancy has been found to be fairly extensive among sample farmers; nearly a half of sample farmers had leased-in land as part of their operational holdings. Total area leased-in constituted about 30 percent of total area under operational holdings of sample farms. The tenancy was generally temporary, usually of one year and also completely informal in the form of share cropping mostly without sharing of cost. Only in 36 percent cases of share cropping the cost was also shared between the tenant farmer and the landlord. Thus tenancy has come out to be major hindrance for adoption and use of improved agricultural practices by farmers. Tenant farmers have been found to be in disadvantage in extending their area under HYVs and application of fertilisers in adequate quantity.

Hence among institutional factors, tenurial arrangements in practice have been found to be restricting the effective use of available technology package by the farmers in the region. However, another factor in this category, namely the relatively small size of farms has not been found to be constraint in adoption and use of improved inputs and practices for better farming.

Since the existing tenurial set up in practice has been identified as a major constraint on agricultural development of the region, tenancy reforms are also calls for to waive the same. The need of tenancy reforms was recognised at the time of independence and has been reiterated in a succession of five-year plans. Four important guidelines laid down in the five year plans for the reform of tenancy were: *(i)* The rent should not exceed the level of one-fifth to one-fourth of gross products, *(ii)* The tenant should be accorded permanent rights in the lands they cultivate, subject to limited right of resumption to be granted to the landowners, *(iii)* In respect of non-resumable lands, the landlord—tenant relationship should be ended by conferring ownership rights on the tenants, and *(iv)* Abolition of tenancy by law in most states, except for certain categories of people. In case of fixation of rent, Punjub, Haryana and Andhra Pradesh achieved the targets preseribed in the five-year plans, while except in West Bengal, sharecroppers, by and large beyond the benefit of tenancy reforms (Throat, 1997).

Incidentally, legislations for tenancy reforms are already in existence for quite sometime in the state. The Assam Adhiar Protection and Regulation Act, 1984, for instance, seem to have enough provisions to protect the interest of the sharecroppers by fixing the maximum rent paid by them. Share croppers are required, according to the provision of this Act, to share their principal crop only, which should provided some incentive to them to raise additional crop too. Ejecting adhiars was also made more difficult (Goswami, 1986).

But despite the provision of the Act, sharecropping is still practiced largely informally in the state of Assam. In this context the Crash Programme in the state of West Bengal popularly known as `Operation Barga' may be cited. The `Barga' programme in West Bengal effectively implemented the provisions of West Bengal Land Reforms Act, 1955, thus safeguarding tenant farmers interest in sharing of crop vis-à-vis cost of cultivation with the landowner. The tenants in West Bengal could not be protected from the illegal extortion and eviction by the landlords because the tenancy arrangement was often oral in nature with no legal validity. In order to protect the tenants, there had been a provision in West Bengal Land Reforms Act. 1955, such that tenants could record their names with the land records department of the government so that they could get legal protection in case of eviction. The land records department did not take the recording business seriously till the Left Front Government in the state took up a `Crash programme' in 1978 on the issue. This is known as "Operation Barga". Under the West Bengal Land Reforms Act 1955, the produce of any land cultivated by the bargadar shall be divided as between the bargadar and the person whose land he cultivates:—*(a)* in the proportion of 50:50, where plough, cattle, manure and seeds necessary for cultivation are supplied by person owning the land, and *(b)* in the proportion of 75:25 in all other cases (Khasnabis, 1994).

A similar `Crash Programme' can also be taken up in the state of Assam so that the provisions of the legislations aimed at protecting the interest of the tenant farmers are enforced in letter in spirit.

## REFERENCES

Acharya, S.S. (1969), "Comparative Efficiency of HYV—A Case Study of Udaypur District", *Economic and Political Weekly*, November 1.

Bhalla, Surjit S. (1979), "Appendix A (on India)", in R.A. Berry and W.R. Cline (Eds.) (1979), *Agrarian Structure and Productivity in Developing Countries*, John Hopkins University Press, London

Castilbo, Galia T. (1975), "Diversity in Unity—the Social Component of Change in Rice Farming in Asian Villages", in *Changes in Rice*

*Farming in Selected Areas of Asia,* International Rice Research Institute, Philippines.

Dyer, G. (1998), "Farm Size and Productivity—A New Look at Old Debate Revisited", *Economic and Political Weekly,* June 27.

Goswami, Atul (1986), "Land Reforms in Assam: A Critical Review", in Atul Goswami (Eds.) (1986), *Land Reforms and Peasant Movement,* Omsons Publications, Guwahati.

Khasnabis, Ratan (1994), "Tenurial Conditions in West Bengal, Continuity and Change", *Economic and Political Weekly,* Vol. XXIX, No. 53, December, 31, pp. A189 – A199.

Lionberger, H.F. (1960), *Adoption of New Ideas and Practices,* Iowa State University Press, Iowa.

Mandal, G.C. and M.C. Ghosh (1976), *Economics of the Green Revolution, a study in East India,* Asia Society, NewYork.

Mukharjee, P.K. (1970), "The HYV Programme—Variables That Matter", *Economic and Political Weekly,* March 28.

Muhiah, C. (1971), "The Green Revolution, Participation by Small Vs. Large Farmer", *Indian Journal of Agricultural Economics,* January – March.

Parthasarathy, G. and D.S. Prasad (1978), "Response to the Impact of New Rice Technology By Farm Size and Teunre in Andhra Pradesh", in Interpretive Analysis of Selected Papers from `Changes in Rice Farming in Selected Areas of Asia', International Rice Research Institute, Philippines.

Rao, A.P. (1969), "Report on HYVP in Shaharanpur District (U.P.), Kharif 1968 with Special Reference to IR-8", A.E.R.C., Delhi University.

Rao, V.K.R.V. (1989), "Inaugural Address of Golden Jubilee Conference of the Indian Society of Agricultural Economics", *Indian Journal of Agricultural Economics,* Vol. 44, No. 40, October–December, pp. 380.

Rudra, A.(1982), Indian Agricultural Economics-Myths and Realities, Allied Publishers, New-Delhi.

Sawant, S.D. (1997), "Performance of Indian Agriculture with Special Reference to Regional Variations", *Indian Journal of Agricultural Economics,* Vol. 52, No. 3, July-September.

Schulter, M. and J. Mellor (1972), "New Seed Varieties and Small Farmers", *Economic and Political Weekly,* Review of Agriculture, March 25.

Soni, R.N. (2007), *Leading Issues in Agricultural Economics*, Vishal Publishing Company, Jalandhar.

Throat, Sukhadeo (1997), "Trends in Land Ownership, Tenancy and Land Reform", in Bhupat M. Desai (Eds.) (1997), *Agricultural Development Paradigm for the Ninth Plan Under New Economic Environment*, Indian Institute of Management, Ahmedabad, p.684–711

Vyas, V.S. (1968), "Economic Efficiency on Small Farms of Central Gujarat", *Seminar Series, No. 7*, Indian Society of Agricultural Economics.

# 20

# AGRICULTURE IN RIVERINE SITUATION: ANALYSIS OF TECHNOLOGY ADOPTION AND CONSTRAINTS OF DEVELOPMENT OF CHAR AREAS OF NORTH BANK PLAINS ZONE IN ASSAM

*R.N. Barman & R. Das*

## Introduction

Increase in farm income is possible either through increase in the size of farm or by intensive cultivation through adoption of modern production technologies. Crop production environments vary over geographic locations due to variability in physiography, soil characteristics, climatic conditions and cropping systems. The crop production environments in the riverine situation are quite different to that of the general upland, medium land, low land and foothill situations. The mighty river Brahmaputra and a number of its tributaries flow through the entire state of Assam. According to estimates of the Central Ground water Board (CGWB) for the Brahmaputra valley; the total ground water recharge is 9148 million cubic meters (MCM). Further, the valley has considerable amount of surface discharge during the monsoon. The rainfall and the river water keep the valley saturated with about 5000 MCM water. In the catchments of Brahmaputra and its tributaries a number of riverine land forms are available. The 'Char' and 'Chapari' lands are such riverine land forms that are periodically formed and eroded due to meandering, braiding and course changing of the rivers. These are flat lands surrounded by river waters on one or more sides and are subjected to frequent flash flooding for varying length of time as well as to varying depth. In other parts of the country the vernacular terms describing this situation is 'diaraland' in U.P. and 'lanka lands' in Andhra Pradesh. The Char areas are mainly characterized by flood prone plain areas with sandy loam and silty loam soil textures which is under laid by thick layers of relatively coarse sands. The ground water is available at shallow depth. Inundation of cultivated fields during monsoon season has been the regular phenomenon. The riverine topography consists of three types of cultivable land situations viz., upland char, medium land char and the low land char. For operational convenience the riverine situation can be divided in to three seasonal rhythms viz., active monsoon or flood season (June to September), post monsoon or post flood season (October to February) and the pre flood season (March to May). The post flood season is utilized for rabi cultivation and the pre flood season is used for cultivation of summer crops including short duration ahu rice. The farmers of the Char areas mainly grow different oilseed crops, pulses and vegetables and among cereal crops ahu rice.

The agricultural technologies are mostly capital intensive as well as location specific in nature. For example, to obtain the maximum yield from the high yielding varieties of a crop, a package of practices in terms of quantity, quality, method and time of application of the crucial inputs such as seed, fertilizers, plant protection chemicals, irrigation etc., along with the farmer's knowledge are the prerequisites. The term 'technology adoption' means selecting and applying a particular technology or a set of technologies in the production process over a period of time. Both socio economic and environmental factors influence the adoption of technology in agricultural production process. Most of the Char areas have been suffering from laggardness as far as adoption and spread of new technology are concerned. Not many references are available regarding adoption of technologies and problems and prospects of Char area farmers. The present study therefore tries to analyze these aspects with the expectation that the findings would help the policy makers to formulate suitable policies for the economic upliftment of the Char farmers.

## The Study Area

North Bank Plains (NBP) zone of Assam is selected as the study area. The riverine situation of this zone has elevation ranges from 34 to 87 meters above mean sea level (MSL). The average annual rainfall of this zone is about 2741 mm. The average annual temperature ranges from 7-9degree centigrade. Soils are formed on the riverine alluviam deposits brought down from the adjacent Assam Himalayas by the river Brahmaputra and its' tributaries. Majority of the Char lands of this zone are scattered in the riverine flood plains of the Brahmaputra and along the courses and terminus of a number of tributaries viz., Dihang, Ghagar, Boginadi, Subansiri, Ranganadi, Jiabharali, Gabhuru, Sonai-Rupai, Pachnoi, Bornoi and Jiadhansiri. The NBP zone of Assam consists of the districts of Darrang, Sonitpur, Lakhimpur and Dhemaji. In total four Char areas/villages representing one from each district were randomly selected for the study. The selected Char areas/ Villages are No1 Arimari (Darrang Dist.), Suaguri (Sonitpur Dist.), Bholukagiri (Lakhimpur Dist.) and Pipulguri (Dhemaji Dist.). In total 200 numbers of farmers (50 from each selected Char village) were randomly selected for detail investigation.

## Analytical Tool

The pattern of resources used has been studied in terms of land, labour (human labour and bullock labour) and working capital used per hectare for different crops. Data collected pertained to the year 2004-05. To study the levels of technology adoption, the average gap index for the selected set of technology components for each of the selected crops was computed.

The gap index (GI) for a particular technology component was computed by using the formula

$$GI = [(R-A)/R] \times 100$$

Where, R = Recommended package

A = Adopted package by the farmer

The average gap was computed as done earlier by Ajore and Singh (1994)

$$\text{Average gap} = \frac{\Sigma \text{ gap index } i}{n}$$

Where, i = gap index in a particular major package

n = total number of major packages

Based on the computation of adoption gap, the farmers were categorized as: *(a)* less progressive having the average adoption gap of more than 30 percent (technology level-1 i.e.T1) ;*(b)* moderately progressive having the average adoption gap of less than 30 percent but more than 20 percent (technology level-2 i.e. T2), and *(c)* progressive having the average adoption gap of less than 20 percent (technology level-3 i.e. T3). As no farmer was found to fulfill the criteria of progressive farmers (T3) hence the study considered only the first two categories while computing the extent of technology adoption gap. The following components of technology have been identified for computation of gap index.

1. Seed bed and field preparation (number of hoeing/ tillage operations)
2. Seed rate (Kg/hec.) and use of proper quality seed
3. Sowing time and method of sowing/planting
4. Manuring and fertilization
5. Number of irrigation applied and percentage of area irrigated
6. Interculture and weed control (number of wee dings)
7. Plant protection chemicals—insecticides/pesticides/ weedicides.

A maximum score of '1' is given for the recommended package for each component. Based on deviation from the recommended package due weight age are given for the adoption of each of the technology component. Variability in the extent of adoption per hectare of the selected crops was also estimated through computation of coefficient of variations. In this study extent of adoption has been defined as the aggregate expenditure on key inputs per hectare (as done in an earlier study by Mukhopadhyay and Pal, 1990).

## Results and Discussion

### *Profile of Riverine/Char Areas*

The socio-economic status of the selected char villages of the four districts is shown in Table1. It is observed that in all the char villages the average area per family is only about 1 hectare. Because of periodic erosion by rivers the farmers in the unstable char areas have to keep on shifting their farming sites by settling on newly developed chars. In these areas the farmers cultivate the lands with periodic patta (ownership). Majority of the riverine areas of Lakhimpur and Dhemaji districts are inhabited by tribal population where as those in Sonitpur and Darrang districts, the farming communities are of mixed ethnic composition.

More than 50 percent of families are below the poverty line. The literacy percent is also very low which is in the range of 15-27 percent only in the villages under study. There is only 1 LP school in the No1 Arimari char under of Darrang district. In the Suaguri char village under Sonitpur district there is 1 LP school and 1 ME school. Similar is the case of Pipulguri char village under Dhemaji district. In the Bholukagiri char village under Lakhimpur district there is 1 LP, 1 ME and 1 High school. In none of the Char villages medical facilities in terms of primary health centers/ dispensaries/medical sub centers exist. The distribution of sample farmers is shown in Table 2. Only 16 to 22 percent of the sampled farmers were found to be progressive with technology adoption gap of less than 25 percent. A conclusion can easily be drawn by looking at the profile of the selected char areas that how under developed the char areas are. The crops mainly grown by the riverine farmers are rice viz., *ahu* rice, *sali* rice, deep water rice, *boro* rice (in less proportion), oilseed crops (rapeseed, mustard, niger, groundnut, sesamum), kharif maize, various pulses (pea, black gram, green gram, lentil etc.), spices (onion, garlic, chillies etc.) and vegetables like ridge gourd, pointed gourd, pumpkin, cucumber, radish, carrot, brinjal, lady's finger and potato. In the relatively upland char areas of Lakhimpur and Dhemaji districts with high tribal concentration farmers adopt mixed farming practices through integration of crops with livestock's and fish farming. People in these areas depend on livestock's like buffaloes, piggeries, poultry farming and natural fisheries (including river fisheries) as means of supplementary income generation.

## Cropping Systems

It is observed that *ahu* rice based cropping systems are more prevalent in the riverine areas. The cropping systems identified in the riverine villages are shown in Table 3. Pre flood ahu rice- post flood sali rice is the widely adopted cropping system followed by the farmers of the surveyed villages (34-38 percent). Among other notable cropping systems; *ahu* rice-fallow- rapeseed/niger (12-20 %) and *ahu* rice- fallow- rabi vegetables (14-22 %) are mostly adopted by the farmers. In Arimari, Bholukagiri and Pipulguri deep-water rice as mono cropping and in Suaguri and Pipulguri *boro* rice as mono cropping are also observed. Deep-water rice + *ahu* rice as mixed cropping are seen in Arimari and Suaguri (4 and 10 %) char villages. Summer vegetables- oilseed crops- *rabi* vegetables is another cropping system observed in all the surveyed villages.

## Levels of Technology Adoption

Farm income can be augmented either through increase in the size of farm or by intensive cultivation through adoption of modern production technologies. After studying the extent of technology adoption gap, the farmers of each char village were categorized as a) less progressive i.e., farmers adopting technology level T1 and b) progressive i.e., farmers adopting technology level T2. The crops selected for studying the component wise extent of technology adoption gap were ahu rice, rapeseed and mustard, summer vegetables (pumpkin, ridge gourd, cucumber, ladies finger etc.) and winter vegetables (radish, brinjal, carrot, chillies etc.). The results are shown in Table 4.

A perusal of Table 4 indicate that in case of ahu rice production the highest technology adoption gap was observed with respect to irrigation for both less progressive (66.47 percent) and moderately progressive farmers (28.15 percent) of the Suaguri char village. The adoption gap was lowest in the Arimari char village in terms of irrigation. Here also the observation was not satisfactory as the gaps were of the percentages of 60.28 and 24.45 for the less progressive and moderately progressive farmers respectively. Among the other components more than 48 percent of adoption gap was observed for less progressive farmers for all the villages in terms of manuring and fertilization in ahu rice. The adoption gaps were in the lower side as far as seedbed and field preparation in ahu rice production was concerned in all the char villages. For oilseed crops and vegetables also similar results were noticed. Here also the highest technology adoption gap was noticed incase of irrigation followed by manuring and fertilization and lowest technology adoption gaps in terms of seedbed and field preparation. These observations were true for both less progressive and moderately progressive farmers of all the villages. Like the other parts of the state, in the riverine areas also more than 70 percent of the annual rainfall is received during the monsoon period. Thus, the late rabi season and early summer season face a comparative drought with varying degrees of moisture stress at critical growth phases of rabi and summer crops. Hence the importance of irrigation during these critical growth phases of crops cannot be ruled out. On a positive note it can be mentioned that in the riverine situation even during dry season (February-March), ground water Table remains shallow and fluctuates from 1 to 4 meters depending upon topography and distance from the river bed and thus can offer a potential water source for irrigating the rabi crops. Further, drainage system is better in most of the riverine situations other than deepwater situation because of light textured sub surface soil condition. The high water infiltration rate in such soil condition reduces the surface runoff and major quantum of monsoon water thus trapped as potential ground water. In both rice and oilseeds, the technology adoption gap regarding plant protection chemicals ranked third in all the villages and for both less progressive and moderately progressive farmers. In case of vegetable crops the gaps were more than 40 percent (46-49 percent) only in terms of irrigation. In all other components the gaps were less than 40 percent in case of less progressive farmers. This indicated that farmers were more preferred towards growing vegetable crops. This was because most of the vegetable crops are of shorter duration and can be easily grown with lesser effort in the soil situation of the Char areas. Further there was lesser amount of risk of it's damage by flood. The general observation was that farmers prefer those crops, which could be grown during flood free period. Govt. policies need to be directed to tap the potential of the Char areas as non-traditional vegetable and oilseed growing areas.

## Problems and Constraints

The selected farmers were personally interviewed to assess the problems they faced with. On the basis of their opinion the problems and constraints were identified, prioritized and listed in Table 5. It is clearly understood from the Table that the riverine farmers were subjected to even the barest minimum need of infrastructures in the form of road and communication network, irrigation facilities, and input distribution network and sale markets. In most of the riverine villages even a primary wholesale market was not available. The farmers had to travel about 10 to 18 Km to purchase their inputs and also to sale their produce and that too at the minimum prices. The road communication with the nearby township is also not good. The farmers had to travel about 75 percent of distance via cutcha roads to reach the near by townships. Forced sale are the common feature due to the absence of proper storage facilities and poor financial position of the farmers (Tables 1 and 5). On account of their poor economic condition about 56-68 percent of the farmers were forced to perform as casual labourers in other areas. Due to apprehension of flood during summer the farmers had to select those crops, which can escape the fury of flood. Winter season crops are generally preferred by the riverine farmers where timely irrigation during the dry spell is required but almost all the areas lack irrigation facilities. Only about 8 to 16 percent of the farmers apply irrigation (as life saving) in their crops that too in their own effort. Flood risk, soil condition and irrigation facilities were ranked 1st, 2nd and third in terms of prioritization of constraints as these three forced the farmers to make adjustments with the cropping pattern. The farmers may be inspired to adopt cooperative farming and cooperative marketing which will solve most of their problems and also minimize risk in farming. The FMC's and NGO's involvement in the developmental activities are only occasional and very minimal and can play very important role on this front. Strengthening of the govt. machinery through establishment of proper infrastructure of the agricultural development offices, storage structures, input distribution centers, schools, primary health centers, village market centers and construction of irrigation canals are very much the need of the hour.

## Conclusion

The study attempted to address one very much neglected area of Assam's agriculture on which not much references are available. The study analyzed the problems and constraints faced by the riverine/char area farmers. Percentages of progressive farmers were negligible and even in the moderately progressive farmers also substantial gaps (more than 20%) in technology adoption were noticed. There is enough scope to tap the potential of the Char areas which can very well be targeted as non traditional oilseed and vegetable growing areas. Government, NGO's, FMC's and financial institutions need to play big role towards the all-round development of the Char areas.

## REFERENCES

Ajore, R. and A.P. Singh, (1994), Causes of wide yield gaps in rice and wheat on sodic soils in progressive and less progressive districts in U.P., *Agricultural Situation in India*, 49(1): 21-23.

Government of Assam (2004), Socio Economic Survey Report of Char Areas of Assam (2003-04), Directorate of Char Areas Development, Assam.

Mukhopadhyaya, A. and S. Pal (1990), Extent of adoption and its variation in Nadia district of West Bengal, *Economic Affairs*, 35 (4): 234-241.

**Table 1: Socio-Economic Status of Riverine (Char) Areas**

| *Name of District* | *Name of Char village* | *Male population (nos.)* | *Female population (nos.)* | *Total population (nos.)* | *Cultivable land (hec.)* | *Uncultivable land (hec.)* | *Total land (hec.)* | *Total nos. of family* | *Family below poverty line* | *Literacy rate (%)* | *Educational institution* | *Medical facility* |
|---|---|---|---|---|---|---|---|---|---|---|---|---|
| Darrang | Arimari(No.1) | 1289 | 1243 | 2532 | 185 | 125 | 310 | 516 | 334 | 26.66 | LP1 | Nil |
| Sonitpur | Suaguri | 1108 | 1073 | 2181 | 246 | 116 | 362 | 368 | 250 | 15.54 | LP1 ME1 | Nil |
| Lakhimpur | Bholukagiri | 1295 | 1015 | 2310 | 233 | 114 | 347 | 373 | 257 | 18.53 | LP1 ME1 HS1 | Nil |
| Dhemaji | Pipulguri | 437 | 389 | 826 | 103 | 51 | 154 | 173 | 123 | 15.74 | LP1ME1 | Nil |

*Source:* Socio Economic Survey Report (2003-04) of Char areas of Assam. Published by the Directorate of Char Areas Development Assam, Guwahati.

**Table 2: Distribution of Sample Farmers in the Riverine (Char) Areas**

| *Name of Char village* | *Dev. Block* | *Less Progressive (Nos.)* | *Moderately Progressive (Nos.)* |
|---|---|---|---|
| No.1 Arimari | Dalgaon Sialmari | 39(78.00) | 11(22.00) |
| Suaguri | Biswanath | 42(84.00) | 8(16.00) |
| Bholukagiri | Bihpuria | 41(82.00) | 9(18.00) |
| Pipulguri | Sissibargaon | 42(84.00) | 8(16.00) |

*NB:* Figures within parentheses indicate percentage to total sample drawn from each village.

**Table 3: Cropping Systems Followed in the Riverine Areas (Percent)**

| *Cropping systems* | *No1 Arimari* | *Suaguri* | *Bholukagiri* | *Pipulguri* |
|---|---|---|---|---|
| Pre flood *ahu* rice-post flood *sali* rice (late transplanted beyond mid Sept.) | 34.00 | 36.00 | 34.00 | 38.00 |
| *Ahu* rice- fallow-rapeseed/niger | 20.00 | 20.00 | 12.00 | 18.00 |
| *Ahu* rice- fallow-rabi vegetables | 18.00 | 18.00 | 22.00 | 14.00 |
| *Ahu* rice- sesamum | 06.00 | 04.00 | -- | -- |
| Deepwater rice (broadcast) as mono cropping | 04.00 | -- | 14.00 | 08.00 |
| Deepwater rice + *ahu* rice as mixed cropping(with 1:1 and 1:2 seed proportion) | 04.00 | 10.00 | -- | -- |
| Boro rice as mono cropping | -- | 06.00 | -- | 08.00 |
| Summer vegetables- e groundnut/niger - rabi vegetables /potato/maiz | 06.00 | 04.00 | 10.00 | 08.00 |
| Kharif maize- rabi vegetables | 04.00 | | -- | 06.00 |
| Fallow - pea/ blackgram/greengram | -- | 02.00 | 02.00 | -- |
| Green manuring crops - pea/potato | 04.00 | -- | 06.00 | -- |

### Table 4: Extent of Adoption Gap in the Production Technologies of Important Crops Grown by the Char Farmers

| *Crops and components of technologies* | *No1 Arimari* | | *Suaguri* | | *Bholukagiri* | | *Pipulguri* | |
|---|---|---|---|---|---|---|---|---|
| | *T1* | *T2* | *T1* | *T2* | *T1* | *T2* | *T1* | *T2* |
| Ahu rice | | | | | | | | |
| Seed bed and field preparation | 31.46 | 20.29 | 33.37 | 22.26 | 32.87 | 21.04 | 33.24 | 21.43 |
| Seed rate and use of quality seed | 35.66 | 22.87 | 36.44 | 23.47 | 36.25 | 22.94 | 36.81 | 23.16 |
| Sowing time and method of sowing | 38.68 | 22.31 | 39.83 | 23.66 | 39.38 | 22.46 | 39.93 | 22.83 |
| Manuring and fertilization | 48.54 | 25.36 | 50.07 | 26.91 | 48.32 | 26.11 | 49.46 | 26.37 |
| Irrigation | 60.28 | 25.45 | 66.47 | 28.15 | 64.56 | 27.75 | 65.11 | 27.92 |
| Interculture and weed control | 34.44 | 21.15 | 35.89 | 21.91 | 35.47 | 21.86 | 36.28 | 22.13 |
| Plant protection chemicals | 36.27 | 24.37 | 38.35 | 24.63 | 36.78 | 24.39 | 37.36 | 24.66 |
| Rapeseed and mustard | | | | | | | | |
| Seed bed and field preparation | 30.46 | 20.28 | 33.19 | 22.17 | 31.77 | 21.49 | 32.42 | 22.12 |
| Seed rate and use of quality seed | 31.26 | 21.17 | 34.36 | 22.42 | 32.63 | 21.66 | 33.18 | 22.84 |
| Sowing time and method of sowing | 38.37 | 21.24 | 31.27 | 22.49 | 39.04 | 22.13 | 30.04 | 23.68 |
| Manuring and fertilization | 53.44 | 26.04 | 54.47 | 27.01 | 54.28 | 26.36 | 55.16 | 26.77 |
| Irrigation | 61.77 | 27.15 | 63.13 | 28.46 | 62.33 | 27.71 | 63.48 | 28.41 |
| Interculture and weed control | 34.39 | 21.07 | 35.24 | 22.18 | 34.87 | 21.56 | 34.96 | 22.25 |
| Plant protection chemicals | 41.57 | 25.19 | 42.65 | 26.31 | 41.84 | 25.23 | 42.21 | 25.17 |
| Vegetables | | | | | | | | |
| Seed bed and field preparation | 26.86 | 20.32 | 28.32 | 21.43 | 27.13 | 19.73 | 28.22 | 21.26 |
| Seed rate and use of quality seed | 38.23 | 20.43 | 29.04 | 22.79 | 38.95 | 20.16 | 29.37 | 22.23 |
| Sowing time and method of sowing | 37.43 | 20.25 | 31.26 | 21.81 | 37.69 | 19.86 | 29.18 | 20.63 |

| | | | | | | | | |
|---|---|---|---|---|---|---|---|---|
| Manuring and fertilization | 37.27 | 24.23 | 29.13 | 23.91 | 38.15 | 22.19 | 29.56 | 22.56 |
| Irrigation | 46.86 | 28.27 | 48.55 | 23.33 | 47.09 | 22.68 | 48.47 | 24.17 |
| Interculture and weed control | 38.58 | 21.35 | 31.43 | 20.23 | 39.27 | 19.24 | 31.27 | 20.74 |
| Plant protection chemicals | 37.34 | 20.38 | 30.66 | 22.61 | 38.19 | 21.66 | 31.82 | 23.52 |

## Table 5: Prioritisation of Constraints of Development of Riverine Areas

| *Particulars* | *No.1 Arimari* | *Suaguri* | *Bholukagiri* | *Pipulguri* |
|---|---|---|---|---|
| 1. Risk of flood damage (%) | 84.00 | 92.00 | 88.00 | 88.00 |
| 2. Soil condition | Sandy/ silty loam | Sandy/ silty loam | Sandy/ silty loam | Sandy/ silty loam |
| 3. Irrigation | | | | |
| a. Govt. constructed canals | Nil | Nil | Nil | Nil |
| B. Personal lift irrign./ shallow tube well (%) | 16.00 | 12.00 | 16.00 | 8.00 |
| c. Pumping (personal) from river (%) | 12.00 | 8.00 | 8.00 | 8.00 |
| 4. Infrastructures | | | | |
| a. Road linkage with townships | | | | |
| i.Pucca (%) | 25.00 | 30.00 | 25.00 | 30.00 |
| ii.Cutcha (%) | 75.00 | 70.00 | 75.00 | 70.00 |
| b.Storage facilities | | | | |
| I. Cold storage | Nil | Nil | Nil | Nil |
| ii.Individual farm level | Negligible | Negligible | Negligible | Negligible |
| c.Agril.dev .office | Not available | Not available | Not avai lable | Not avai lable |
| d.Village wholesale market | 01 | Not available | 01 | Not available |
| e.Input distribution/ sale center | Not available | Not avai lable | Not available | Not available |
| 5.Distance from the nearby township(km) | 11 | 14 | 12 | 14 |
| 6.Distance from the input distribution/sale center(km) | 11 | 14 | 12 | 14 |
| 7.Distance from banks and post offices(km) | 5-11 | 5-14 | 5-12 | 5-14 |
| 8.Involvement of NGO's/ FMC's | Occasional | Negligible | Occasional | Negligible |
| 9. Casual works in other areas (%) | 56.00 | 64.00 | 58.00 | 68.00 |
| 10.Incidences of crop diseases/pests/nematodes | Regular & moderate | Regular& moderate | Regular& moderate | Regular & moderate |

# 21

# FACTORS AND CONSTRAINTS AFFECTING ADOPTION OF NEW AGRICULTURAL TECHNOLOGY IN ASSAM WITH SPECIAL REFERENCE TO BARPETA DISTRICT: AN EMPIRICAL STUDY

*Aftab Uddin Ahmed*

## Introduction

Sustained growth of agriculture in the long run depends on the improvement of farming technology in the country. Introduction of new agricultural technology seems to offer an opportunity to increase output and income substantially. For this reason adoption of technological innovations in backward agriculture has been drawing attention of development economists. But the point that needs special attention is that till now the introduction of new agricultural technology has met with only partial success as measured by the observed rate of adoption. Indian agriculture is still in traditional character. It is technological innovation and adoption that can change traditional agriculture into modern one. The importance of technological change in the context of growth and development of traditional agriculture has been analyzed by T.W. Schultz (1964) in his book 'Transforming Traditional Agriculture'. According to him, the distinguishing character that sets it apart from modern agriculture is the type of inputs and technology in use and not its cultural and institutional attributes. Citing various pieces of empirical evidence, Schultz argues that farmers in traditional agriculture do respond to market signals and that they allocate their resources efficiently in a rational manner.

Various studies [e.g. Dasgupta (1979); Bhati (1976); Palmer (1976); Hansan (1980) Asaduzzaman (1979)] highlight that there is a positive association between the farm size and the adoption of HYV technology. In general, large farmers because of their higher income, economic power, social prestige and links with local political leadership, have more assured supply of modern inputs including credit facilities necessary for fruitfully utilizing the potential of new technology [Rahman (1933)]. Given the capital constraints, the land allocated to modern varieties will be positively associated with farm size [Feder (1980)]. But farm size is a surrogate for a number of facts such as access to credit inputs and information. As such, large farmers enjoy preferential treatment in obtaining inputs. They are generally the large adopters of HYV seed technology and its other components. Studies also indicate that prevalence of right conditions may greatly facilitate the adoption of new productive technology and thereby improve the production and productivity to a large extent [Bezbaruah (1989); Sarap (1980); Saikia (1982); Vaidhyanathan (1994); Narain (1998)].

## Objectives and Methodolgy

The present paper is an empirical study on the determinants and constraints of adoption of new agricultural technology in Assam with special reference to Barpeta district. The specific objectives of the present study are: to identify the factors those influence the adoption of new technology in agriculture in the study blocks and to find out the constraints to adoption of new agricultural technologies by different categories of farmers i.e., large, medium and small farmer, the ultimate unit of sample.

This paper is based on farm level primary data collected personally from 10 Community Development Blocks (CD Block) out of 12 blocks of Barpeta District of Assam. Data were collected through a questionnaire. From each block one revenue village was selected for field study. The selection of

village was done based on two considerations. First, the selected village should be fairly representative of the whole block i.e. it should contain the basic characteristic feature of the block. Secondly, the necessary infrastructure for the use of the new agricultural technology should be available at least in some households of the selected village.

A multistage random sample method was used. In the first stage of selection from each community development block one village was selected at random. In the second stage 11 farm households from each selected village were selected at random for study. Therefore, in all 110 farm households from 10 villages (11 from each village) were selected for study as ultimate units of observation.

Data on the general background and data relating to the adoption of agricultural technology by each farm household in the sample have been collected by interviewing a senior member, usually the head of the household, of the farm family. For carrying out these interviews and for recording the information a schedule of questionnaire finalized after a number of test and checks in the field.

## Results and Discussions

### *(1) Factors Affecting Adoption of New Agricultural Technology*

Our field investigations reveal that the distribution of adopter households is not uniform across the size groups of farm households. It was observed from field survey that there are some factors, which are mainly responsible for the adoption of new technology in agriculture.

These factors are:

1. There is a positive relation between farm size and adoption of agricultural technology. The greater the size of farm the higher is the adoption rate of new agricultural technology.
2. Income of the farmer is the main factor affecting the adoption of new agricultural technology. It is an important decision making factor in relation to adoption or non-adoption of new technology in agriculture. Income of the farmers is positively

correlated to the adoption of new technology for all groups of farmers.

3. It was observed from field investigation that risk bearing ability of the farmers is an important factor for the adoption of new technology in agriculture.
4. Irrigation facility is an important factor for the adoption of new technology in agriculture. Most of the sample farmers particularly non-tribal farmers reported that lack of adequate and timely supply of assured water is the main obstacle in adoption of modern agricultural technology.
5. Field survey showed that use of fertilizer especially chemical fertilizers (NPK) is extremely essential for adoption of HYV seed which constitutes an important part of modern agricultural technology. Therefore, fertilizer is an important factor affecting adoption of modern technology in agriculture.
6. Credit facility, particularly institutional credit facility also affects the adoption pattern of agricultural technology by the farmers. Farmers who got credit facility were found to be more responsive to the adoption of improved agricultural technology.
7. Education level of the respondents was found to be positively related to the adoption of modern technology. It was found that respondents who are office holders or learned persons are more responsive towards the adoption of new technology since they have direct contact with extension workers and development agencies at block offices.
8. Field study made it clear that market price of agricultural produce is an important factor affecting adoption of new technology in agriculture. Higher price encourages farmers to adopt improved technology in agriculture for higher production. Many respondents reported that prices of agricultural produce like vegetables, jute, rice etc. go down to a considerable extent particularly during the harvesting season. Low price of agricultural produce discourages the farmers from adopting costly new technology.

9. The study revealed that the middle age farmers (between 30 and 40) are more responsive to the adoption of improved agricultural practices.
10. Source of acquiring new knowledge or information about agricultural technologies and their cost effectiveness are also important factors affecting technology adoption. Most of the farmers particularly tribal farmers are unaware of many programmes sponsored by the govt. to promote adoption of improved technologies because of source of acquiring new knowledge or information regarding agricultural activities are either insufficient or non-existent.
11. Traditionalism and ethnicity also play an important role in affecting the adoption pattern of agricultural technology by the farmers. Tribal farmers being more tradition ridden and conservative are found to be less responsive towards the adoption of new agricultural technology.

### *(2) Constraints to Adoption of New Technology in Agriculture*

Transfer of technology plays a vital role in the process of agricultural development. Transformation necessitates that farmers should be convinced to accept and work for the change. They need to be prepared mentally and emotionally to accept the new agricultural technology with the continuous effort of the government and other extension agencies. Most of the farmers are well informed about the new development in agriculture and they are ready to adopt the new farming technology but are not in a position to adopt the improved technology at full scale due to certain constraints faced by them in day-to-day life.

In order to find out various constraints of different size group of farmers in the present study, open-ended questions were asked. The constraints mentioned by them (both non-tribal and tribal respondent farmers) were noted which can be classified into the following groups:

*(i)*Economic constraints, *(ii)* input constraints, *(iii)* guidance and training constraints, and *(iv)* general constraints

1. Economic constraints comprises the following items:
a) Lack of own capital
b) Inadequate institutional credit
c) High rate of interest of non-institutional credit
d) Lack of collateral of farmers and complex procedure in obtaining credit.

2. Input constraints comprises the following items:
a) Impurity of seeds
b) High cost of HYV seeds
c) High cost of fertilizers
d) Inadequate irrigation facilities
e) High rate of irrigation charge of private pumping sets
f) Erratic supply of electricity
g) Insufficient availability of effective and pure chemicals, pesticides, insecticides etc.

3. Guidance and training constraints consists of the following items:
a) Lack of information about new technology
b) Lesser contact with extension agencies
c) Lack of knowledge of various development programmes to uplift the poor farmers.

4. General constraints comprise the following issues:
a) Low price of agricultural produce
b) More proneness of HYV seeds to diseases
c) Lack of cold storage
d) Transport problem
e) Problem of regulated market
f) Lack of education
g) Problems of soil testing, water testing, soil salinity etc.
h) Environmental problem.

The important constraints to adoption of technology in agriculture as emerged from our field investigations can be summarized in a tabular form as follows:

## Table 1
## Constraints to Adoption of New Technology in Agriculture in the District of Barpeta, Assam

| *Constraints* | *Large farmers N=80* | | *Medium farmers N=80* | | *Small farmers N=80* | | *Total* |
|---|---|---|---|---|---|---|---|
| | *Frequency* | *Percentage* | *Frequency* | *Percentage* | *Frequency* | *Percentage* | |
| 1. Economic constraints | | | | | | | |
| a) Lack of own capital | 12 | 15.00 | 30 | 37.5 | 61 | 76.25 | 103 |
| b) Lack of Institutional credit | 33 | 41.25 | 42 | 77.5 | 58 | 85.00 | 185 |
| c) High rate of interest of non-institutional credit | 5 | 6.25 | 22 | 27.5 | 34 | 42.5 | 61 |
| d) Lack of collateral of farmers and complex procedure in obtaining credit | 13 | 16.25 | 28 | 35.00 | 32 | 40.00 | 73 |
| 2. Input constraints | | | | | | | |
| a) Impurity of seeds | 17 | 21.25 | 18 | 22.5 | 23 | 28.75 | 58 |
| b) High cost of HYV seeds | 32 | 40.00 | 35 | 43.75 | 50 | 12.5 | 117 |
| c) High cost of fertilizers | 25 | 31.25 | 35 | 43.75 | 65 | 81.25 | 125 |
| d) Inadequate irrigation facility | 7 | 8.75 | 22 | 27.5 | 36 | 45.00 | 65 |
| e) High rates of irrigation charges of private pumping sets | 6 | 7.5 | 20 | 25.00 | 43 | 53.75 | 69 |
| f) Erratic supply of electricity | 3 | 3.75 | Not ment. | - | Not men. | - | 3 |
| g) Inadequate availability of pure chemicals, pesticides, insecticides etc. | 12 | 15.00 | 15 | 18.75 | 21 | 26.25 | 48 |
| 3. Guidance and training constraints | | | | | | | |
| a) Lack of information about new technology | 30 | 37.5 | 36 | 45.00 | 57 | 71.25 | 123 |
| b) Lesser contact with govt. extension agencies | 67 | 83.75 | 27 | 33.75 | 80 | 100.00 | 174 |

| | | | | | | | |
|---|---|---|---|---|---|---|---|
| c) Lack of knowledge of various development programmes | 30 | 37.5 | 50 | 62.5 | 61 | 76.25 | 141 |
| 4. General constraints | | | | | | | |
| a) Low price of agricultural produce | 70 | 87.5 | 61 | 76.25 | 20 | 25.00 | 151 |
| b) More proneness of HYV seeds to disease | 48 | 60.00 | 49 | 61.25 | 21 | 26.25 | 118 |
| c) Lack of cold storage | 70 | 87.5 | 61 | 76.25 | 25 | 31.25 | 156 |
| d) Transport problem | 52 | 65.00 | 50 | 62.25 | 46 | 57.5 | 148 |
| e) Problem of regulated market | 26 | 32.5 | 25 | 31.25 | 15 | 18.75 | 66 |
| f) Lack of education | 41 | 51.25 | 46 | 57.5 | 49 | 61.25 | 136 |
| g) Problem of soil testing, water testing etc. | 60 | 75.00 | 64 | 80.00 | 74 | 92.5 | 200 |
| h) Environment problem | 19 | 23.75 | 20 | 25.00 | 11 | 13.75 | 50 |

*Source:* Field Survey, 2004.

Among the economic constraints, lack of capital counts the most. Since majority of the farmers are poor, they have not internal sources of their own to purchase the entire or any of the components of new agricultural technology. It is mainly responsible for their low income, which does not permit them to adopt more remunerative agricultural technology. As many as 12 large, 30 medium and 60 small farmers reported about lack of own capital for which they could not adopt new agricultural technology at full scale.

Since many farmers do not have their own capital for adoption of new agricultural technology, they have to depend on borrowed capital. Since non-institutional sources of capital are insufficient and bear exorbitant rates of interest, farmers have to depend on institutional credit. But field survey showed that only 65 (81 percent) of large farmers out of 80, 58 (73 percent) medium farmers out of 80 and 42 (53 percent) small farmers out of 80 had access to institutional credit. So, 19 percent large farmers, 27 percent medium farmers and 47 percent small farmers reported credit constraints. The main reason of this inadequacy of bank credit, as reported by sample farmers, is their lack of collateral due to their poor asset base and complex loaning procedure. Most of the farmers, being

poorly educated, do not know the procedure of borrowing from nationalized banks. The main problem, therefore, is to evolve a suitable agricultural credit system so that credit is made available to the farmers on easy terms at reasonable rate of interest and without rigorous requirement of security and without irksome formalities. Non-institutional credit sources charge high interest rate. As many as 6 percent large, 30 percent medium and 43 percent small farmers reported about high rate of interest charged by village money lenders and rich men. So far as input problems are concerned, out of 240 sample farmers, 17(i.e., 7 percent) large farmers, 18(i.e., 8 percent) medium farmers and 23(i.e., 10 percent) small farmers reported that many high yielding variety seeds are not pure and very often taking the advantage of ignorance and illiteracy of farmers, many unscrupulous business men sell impure seeds of high yielding varieties at high prices. As many as 32 (13 percent) large farmers, 35 (15 percent) medium farmers and as high as 50 (21 percent) small farmers reported about high prices of HYV seeds.

During the field survey, it was found that big farmers possessed more pump sets to irrigate their fields. Some of large and even medium farmers hire out their pump sets to irrigate the fields of their neighbouring small and medium farmers, who do not possess pump sets, at a quite high rental. As many as 6(8 percent) large, 20(25 percent) medium and 43(54 percent) small farmers reported about high charge of pump sets. It was also found, during field investigation, which the supply of irrigation water by govt. scheme is not sufficient and convenient for the beneficiary farmers. This is mostly because of erratic supply of electricity. The shortage of power prevented the govt. tube wells from functioning properly. A sizeable number of farmers, i.e., 12 (15 percent) large, 15(19 percent) medium and 21(26 percent) small farmers complained that the chemicals they purchase from market are not pure and effective. So, the chemicals and pesticides, they reported, are not fully effective in protecting their crops from pests and diseases. As regards fertilizer input, which is the most important component of new farming technology, as, many as 25 (31 percent) large, 33(41 percent) medium and 65(81 percent) small farmers reported about high price of fertilizer.

Due to high price of some fertilizers, many sample farmers could not afford to purchase recommended doze of fertilizers.

As regards the guidance and training constraints, many farmers, especially small and illiterate farmers of non-tribal category and most of the farmers of tribal category complained about lack of information relating to new technology. As many as 30 (38 percent) large, 36(45 percent) medium and 57(71 percent) small farmers reported the lack of information regarding new agricultural technology. Lesser contact with extension workers and lack of knowledge of various development programmes, which are meant for the upliftment of the poor farmers, also stand on the way of adoption of new technology in agriculture. It was complained by many farmers (particularly tribal farmers) that block officials do not disseminate the necessary information to them regarding the latest improved agricultural practices for which they can not adopt new practices in due time.

As far as the question of general constraints are concerned, out of 240 sample farmers as many as 70 (29 percent) large, 61 (25 percent) medium and 20 (8 percent) small farmers reported about low price of high yielding variety produce. During field survey, many farmers, specially producing marketable surplus, reported that almost every bumper production is followed by lower price and many time particularly during the harvesting season they have to make distress sale.

About 50 percent of the sample farmers, i.e., 48 (20 percent) large, 49 (20 percent) medium and 21 (9 percent) small farmers expressed that high yielding varieties are more prone to diseases. Similarly 156 sample respondents, i.e., 70 large, 61 medium and 20 small farmers felt the need of cold storage. Field survey showed that more than 80 percent roads are unmetalled and most of them become muddy and unsuitable for transportation of agricultural produce during the entire rainy season. Most of the farmers in the study villages use handcart, van rickshaw, bullock carts etc. as means of transportation of agricultural produce and input. Out of 240 sample farmers, 26(11 percent) large, 25(10 percent) medium and 15(6 percent) small farmers reported about inadequate number of regulated markets to sell their surplus output to and purchase required input there from. Due to lack of adequate number of regulated markets, price of most of the agricultural produce go down during the harvesting season. Due to non-remunerative prices, income of the farmers cannot increase reasonably which obviously discourage them to adopt new agricultural innovations. There are about 75, 80 and 92.5 percent of large, medium and small farmers respectively who face the problems relating to test of soil, water and soil salinity.

Besides the above constraints, there are some other constraints, which also deserve mentioning. Among them are more proneness of HYV seeds to diseases, problem of skilled labour, lack of education, lack of cold storage, transport problem, problem of regulated market etc.

Some farmers reported about environment problem and ecological imbalances caused by excessive use of chemical fertilizers, pesticides etc. Some educated farmers reported that excessive use of pesticides and insecticides for plant protection measures cause environmental degradation. For example, they mentioned that due to the excessive use of these chemical inputs, some pet animals like cows, goats etc. are affected and even sometimes many useful living beings like frogs, fishes, earthworm etc. which are natural protector of crops die due to the poisonous effect of the chemicals used. Excessive and non-judicious use of plant protection chemicals has not only resulted in environmental pollution but also developed resistance to several pests. Residues of DDT, aldrine, lindane etc. have been found in different food samples, which are harmful to health. Moreover, increase in the use of chemical fertilizers and water for increasing agricultural yield brings about a steady depletion of micronutrients from the soil, erosion of topsoil, spread of salinity and waterlogging etc. Imbalances in nutrient status lead to significant deficiency of N.P.K., Zn, S and disturbance of soil texture and its physico-chemical properties.

Environmental degradation like depletion of stratospheric ozone, nitrate toxications etc. are causing health hazards like cancer, methamoglobinemia, respiratory illness, hypertension etc. (Bhattacharya and Bihari, 2003). As many as 19 (8 percent) large, 20 (8 percent) medium and 11 (5 percent) small farmers reported emergence of ecological imbalances.

## Conclusions

On the basis of the results of present investigation it may be concluded that some constraints like lack of capital, low price of agricultural produce, problem of insufficient cold storage, inadequate institutional credit, problem of soil and water testing facility, inadequate irrigation facility, high cost of fertilizers, high rental charges of implements and machines, lack of information and problem of regulated market are the most important and mainly due to these constraints that the process of adoption of new agricultural technology in Barpeta district has been slow and interrupted.

## REFERENCES

Asaduzzaman, M. (1979), "Adoption of HYV Rice in Bangladesh", *Bangladesh Development Studies*, vol. 7, No.3, Sept. pp. 23-52.

Bezbaruah, M.P. (1989), Technological Transformation of Agriculture–A Study in Assam (Ph.D.Thesis).

Bhati, U.N. (1976), Some Social and Economic Aspects of the Introduction of New Varieties of Paddy in Malayasia; Village Case Study, United Nations, Geneva.

Dasgupta, B. (1979), *Agrarian Change and The New Technology in India*, United Nations Research Institute of Social Development, Geneva.

Feder, G. (1980), "Farm Size, Risk Aversion and the Adoption of New Technology under Uncertainty", *Oxford Economic Papers*, Vol. 32, No.2, July pp. 263-283.

Hansan, Harald (1980), *The Adoption of Agricultural Innovation, Urban Growth and Rural Stagnation* (Edited by Rothermand, D. Kroppa and Dienemand, G.P.), pp. 73-75.

Narain, Hari (1998) "Rural Development Through Science and Technology", *Journal of Rural Development*, Vol. 17, No. 2, April-Jan.

Palmer, I. (1976), The Rice in Asia: Conclusions from Four Country Studies: Report, United Nations Research Institute of Social Development, Geneva.

Pandey, K.N. (1989), *Adoption of Agricultural Innovation—A Study of Small and Marginal Farmers of Varanasi (U.P.)*, Northern Book Centre, New Delhi.

Rahman, R.I. (1983), "Adoption of HYV; Role of Availability and the supply-side Problems", *Bangladesh Development and Change,* Vol. 21, No. 2, pp. 281-307.

Saikia, P.D. (1987), "Impact of Irrigation on Small Farmers of N.E. Region".

Sarap, K. (1990), "Factors Affecting small Farmers' Access to Institutional Credit in Rural Orissa, India", *Development and Change,* Vol. 21, No. 2, pp. 281-307.

Schultz, Theodore W., (1964), *Transforming Traditional Agriculture,* York University Press, New Haven.

Vaidyanathan, A. (1994), "Agrarian Relations in the Context of New Agricultural Technology; An Issue Paper", *Indian Journal of Agricultural Economics,* Vol. 49, No. 3, January-March.

# 22

# PROBLEMS AND PROSPECTS OF WATERSHED DEVELOPMENT PROJECTS IN MANIPUR

*S.B. Singh & S.V. Ngachan*

## Introduction

Manipur state in northeastern hill region of India has a geographical area of 22,328 sq km in which 90% are of hills. Soil erosion is one of the outstanding ecological problems in the state and it offers a frightening prospect for the coming decades. In most of the hill areas, the soil is lost at the rates of 146, 170, 30 and 8 tons per hectare per year under first year, second year, abandoned jhum and bamboo forest respectively (Singh and Singh, 1978). The situation may further worsen without proper measures to regenerate soils and to halt erosion. The main cause of erosion are a number of land use practices mainly jhuming that attempt to cope with the rapidly increasing population density, which now stands at about 3.6 lakh hectares (NEC, 2002). More than 80% of the population has no other choice but to depend solely on the land for their livelihood because of lack of employment opportunities in the other sector of the economy. But the area of land available to each farmer household is decreasing due to an increasing subdivision among heirs. Consequently agricultural activities have been extended into areas of unsuitable land and these activities have resulted in deforestation of step mountain lands. The farmers cultivate land in the traditional way under most unsophisticated, low yielding condition. Therefore, land productivity and environmental damage is unavoidable with an increasing population. In view of the above facts, Government of India resolves that ending neglect of vast rainfed and mountainous areas would be a major policy concern in the eight five year plan.

The National Watershed Development Project for rainfed areas was restructured and expanded to the northeastern hill states including the state of Manipur with the objective of creating more employment opportunities and provides additional income in an equitable and sustainable manner. There is hardly any scientific study on the impact of watershed project on employment and income growth and its distribution among stakeholders in Manipur. This research paper is an attempt to study Problems and Prospects of Watershed Development Projects in Manipur.

The study also tries to identify the contributing factors in hindering successful implementation of watershed programmes in the state.

## Methodology

Maklang watershed in Senapati district Manipur being one of the 8 watersheds selected under National Agricultural Technology Project (NATP) for evaluation was purposively selected for the study because of its representation of both the hills and valley areas.The watershed project was implemented by the Department of Horticulture and Soil Conservation, Government of Manipur under the centrally sponsored National Watershed Development Project for Rainfed Areas (NWDPRA). The project was implemented during the year 1992-93 to 1997-98 in an area of 500 ha. All the 5 villages falling under the project was purposively selected due to the significant differences in the socio-cultural status of the farm households. Finally, using proportionate random sampling technique 65 farm households were selected as representative sample. Personal interview method using pre-tested questionnaire was applied for the collection of primary data before and after the project during the year 2000-2001. Due to non-availability of proper record on farm sizes and numbers, selection of farms under different categories was not possible. Secondary data were collected from various reports of the project and published and unpublished record of the Government of Manipur. Simple statistical tools such as averages, percentages and frequency distribution were used to arrive at the results related to the study objectives. Gini Concentration ratio and Lorenz Curve were used to examine the impact of the project on household income distribution. The quantitative measure of Gini Concentration Ratio is given as follows:

$$L = 1 - \sum_{i=1}^{n} pi\ (Ii + I\ i\text{-}1)$$

Where,

L = Gini Concentration Ratio

pi = Proportion of population of ith class

I= Cumulative proportion of total income at ith class

I = 1,2,3 .........n

n= Number of classes in the distribution

To study the determining factors that hinder the successful implementation of the project, eight factors were identified after discussion with various resource persons including watershed farmers through group discussion and personal meetings. Frequency and scoring model were use to quantify the factors. The mathematical form of the scoring model was is:

$$Si = \sum_{j=1}^{m} Nij(m\text{-}j\text{+}1)$$

Where,

Si = Score of the ith factor

i = 1, 2, 3 ....n, and n = total number of factors.

Nij = Number of farmers falling on the jth rank of the ith factor

j = 1, 2, 3 .... m, and m = total number of ranks

## Results and Discussion

### *Size of Holdings*

The examination of size of holdings (Table 1) revealed that the distribution of land holdings in Maklang watershed was highly skewed. Majority of the farmers in the watershed were either landless or having holding less than 1 ha. About 3 per cent of the households occupied 25 per cent of total holdings which indicated unfair distributing of holdings among the households. There was no change in distribution of size of holdings before and after the watershed project. The average size of holding per household remained stagnant at 1.23 ha.

### *Land Use Pattern*

Table 2 showed that the total operational area under Maklang watershed increased significantly. It is due to utilization of more of community lands for fruit and forest tree plantations. The utilization of community lands in Maklang watershed increased by 115 per cent after the project. Creating Irrigation facilities in the watershed through construction of check dams and water harvesting structures was mandatory. However, no or little attention had been given for the creation of irrigation infrastructures that resulted into increased in irrigated area a marginal 1.43 per cent and area under wetland paddy cultivation 2.87 per cent. Finally, the area under jhum cultivation increased a significant 21.34 per cent watershed, which was against the objective of the project. The area under orchard aquaculture, and afforestation increased substantially. Ultimately, the current fallow land declined significantly in these watersheds.

### *Cropping pattern*

Table 3 reveals that paddy is the only crop in irrigated land during Kharif season. However, in Rabi season a negligible amount of land is under cultivation for Rabi crops.

There is moderate increase in area under Rabi crops after the watershed project. The homestead lands are used for growing vegetables and cash crops in both Kharif and Rabi seasons. The area under vegetables and cash crops in homestead land also increased marginally after the watershed project. The overall cropping intensity increases negligibly from 103.64 per cent to 104.29 per cent.

### *Livestock Population and Milk Production*

Table 4 depicts the performance of livestock improvement program in the selected watershed. Livestock being an integral part of hill farming system, livestock improvement program was introduced during the watershed project. Artificial insemination with exotic breeds of cattle, distribution of improved breeds of pig and poultry were major components of the program.The table shows that the project has brought moderate change in livestock composition. There is no change in milk production and insignificant increased in meat production. The reason being negligence on the part of implementing agency in development of livestock as one of the activities of the watershed project

### *Changes in Share of Farm and Off-Farm Income to the Total Income*

To assess the impact of the project on farm income, income from various sources was calculated in terms of percentage change in income before and after the project. A perusal of table 5 indicated that annual average income per household increased to Rs.40226 (15 per cent increase) in Maklang watershed. Farm income from agriculture, fruits and livestocks including fisheries increased about 8, 633 and 23 per cent after the implementation of the watershed project which indicates that income from fruits contributed a major share in increasing income of the watershed households. The relative share of farm income in the total household income also, increased marginally from 77.5 per cent to 78.46 per cent. The off-farm income particularly from labour within the watershed increased by 21 per cent. This shows that watershed project is successful in increasing farm income and employment opportunities . Further it can be concluded that watershed project have little impact in improving income from agricultural crops, livestocks including fisheries and off-farm income particularly from village based small scale industries which need more attention for identifying underlying factors.

### *Pattern of Income Distribution*

It was found that the share of the poorest 69 per cent of the households in Maklang watershed before the project was 56 per cent of the total income, which increased to about 62 per cent after the project. On the other hand, the share of the upper 5 per cent in the total income was 18 percent before the project, which decreased to about 14 per cent after the project. This increase in the share of the poorest group and decline in the share of the richest group after the project indicated an improvement in the farm income distribution after the project. The Lorenz curve for farm household in Maklang watershed before and after the project were drawn and is shown in Fig. I. The Lorenz curve of the households after the project lie above that of before the project and the average farm income after the project was higher than that of before the project.

The GCR was lower (0.36173) after the project as compared to before the project (0.38654), indicating that the income was more evenly distributed after the implementation of the project. It shows that the income disparity among the sample households have become better-off from before, i,e. the gap between the rich and poor has brought down. However, the impact of the project on the income distribution cannot be marked significant as shown by the Gini Concentration ratio and Lorenz curve. Therefore, more attention would be given to small, marginal and landless farmers for upcoming watershed projects in Manipur.

### *Changes in Employment Level of the Sample Households*

Employment level of sample households in mandays for the two study period was analysed and found that the per hectare employment level has been marginally reduced. In other words, the per hectare employment which was 311.5 mandays has been decline to 306.8 mandays after the project i,e. a percentage change of (-) 1.5. The reason behind may be due to the increase in crop area which in turn led to the decrease in per hectare utilisation of labour. On the other hand the per family employment pattern shows that the level of employment has been increasing from 369.53 mandays to

457.94 mandays with a percentage change of 24 only during the two period of study. It is mainly contributed by the horticulture components of the Watershed Programme. In short employment opportunities has been created but it is not drastically changed.

## Problems and Constraints

Among the identified factors (Table 6), lack of empowerment of local people and village level institutions comes out to be the most hindering factor in successful implementation of watershed projects in the state. People's participation in watershed development and management programmes is crucial for their successful and cost-effective implementation. Plans such as afforestation, orchard plantation, bunding, leveling, etc. which can be implemented by the farmers involved acting individually and there are many other items such as check dams, waterways that can be implemented only through collective action of the farmers. However, the watershed committee constituted mostly by village heads lacked equal representation of common people and thus, have little impact to mobilize mass participation. The beneficiaries selected under the project are quite discriminatory mostly favouring class relatives and friends. Thus, two groups (one assailing and one hailing) could be observed. It indicates that people's participation is lacking because of lack of empowerment to local farmers and institution. Local institutions must also be made to get themselves involved at all stages of planning. e.g. women organization serve to empower women socially and economically. An important way in which PIA can facilitate the participation of women in community activities is by providing support system, such as childcare facilities that enable the women to attend meetings, workshops and camps. The next hindering factor is lack of appropriate technology intervention. Requirements of sustainable agriculture and appropriate technology are to be carefully evaluated. There are some recommendations propounded by research and extension agencies, which are too general to be adopted at watershed level. Plantation of orange, for example, was one of the major activities of the horticultural development programme in Maklang watershed. But after fruiting, it was realized as a lemon plant. This warrants for prioritizing areas for watershed development; site-specific research or on-farm research and a multi-disciplinary perception of issues by research and development agencies. Lack of institutional support on the part of poor farmers for treatment of arable and non-arable lands in the watershed and development of suitable crop production technology leads to the little realization of the full potentials of watershed development as they lack the needed institutional support such as credit, extension and technical support. For example, micro watershed projects make possible the introduction of new technologies, the use of which often requires a large initial investment but the lack of credit facilities and the high rate of interest of the informal sector act as a disincentive to medium and long-term investment. Arrangement for such support should be made by way of networking the linkages between farmers and relevant institution like Banks and related departments. Another hindering factor perceived by the farmer is lack of operational partnerships.The main stakeholders in the watershed development process include government agencies, NGOs, the private and banking sectors, political institutions and communities. Government responsibilities for watershed development remained dispersed through their multi-disciplinary teams i.e. forestry, agriculture, horticulture, livestock and soil conservation, etc. The integration of these relevant agencies gained technical efficiency, people's participation and transparency in implementing watershed projects. However, every implementing agency experienced their own mode of management without collaborating relevant departments. This resulted into failure in achieving overall determined objectives of the project. For example, although due attention was given in the development of crops and livestock including fisheries, the economic impact is minimal as indicated above. Thus for the project to be successful, there must be agreement on the respective roles of different partners and a shared vision, where partners agree on and share common goals and objectives. Partnership between the concerned agencies need to be structured with each playing a role according to its comparative advantage. The incentive and panel implications are adequately spelled out and codified. Social Issues includes, existing land tenure system i.e. lack of ownership rights, alarming tensions between communities and deteriorating law and problem in the state are some of the problems which hindering successful implementation of the project. It indicates that watershed programmes are most likely to succeed in villages with homogenous communities and where differences in socio-economic status are small.

The guidelines of watershed schemes indicate the importance of soil and water conservation measures as one of the major activities in the watershed programme but fails to apply in the state. The guidelines do not permit the construction of permanent structures for soil and water conservation measures. But a minimal number of structures could be maintained after the project because the temporary arrangements do not fit into the existing topographical and geographical characteristics of the state. Similar observations were noticed in the case of soil conservation structures. It necessitates a change in the guidelines or arrangement for flexibility suitable to the hill conditions. Certain provisions for post watershed period such as maintenance of structures and plantations, marketing and monitoring and evaluation is essential for success of the project and as well as for up coming watersheds. However, the project was completed abruptly without putting any attention on post watershed period. It resulted into poor performance of structures and plantations and low level of technology adoption. Therefore, certain provision should be made in tune of the fund on management of the post project period. It will help to identify the problems in the management, factors of adoption, success and failures of the various activities and also, built confidence to the fellow farmers. Non-integration of various schemes, either central government or state government sponsored, are running in the states from time to time. However, integration of these schemes have little impact on successful implementation of watershed programmes in the state.

## Conclusion and Policy Implications

The above facts clearly indicate that the watershed project

altered the land use system favorably to horticultural crops mostly fruits with little attention to the development of field crops and livestocks including fisheries. The project could increase the income and employment opportunities of the households in the watershed. However, the most crucial thing is that these increased was not distributed uniformly to all sections of people in the area satisfactorily. It necessitates proper attention to the landless, marginal and small farmers while planning for watershed development projects. Self employment schemes such as village level small scale industries, post harvest technologies and livestock and poultry etc. need to be developed. Existence of traditional land institution characterized by predominance of tenant's population reflected uneven distribution of benefits and thus, restricts participatory criteria of watershed management. People's participation in watershed development and management in general is poor due to lack of empowerment to local people and village institutions. Inappropriate technology intervention is observed in most cases due to non-prioritizing areas for watershed development and lack of on farm research and a multi-disciplinary perception of issues by research and development agencies. Institutional support such as credit, extension and technical support, etc. is missing in most of the cases. In most of the cases operational partnership is missing without collaborating relevant departments. Lack of flexibility in the watershed guidelines, the social issues such as traditional land institutions and, tension between communities and poorest low and order problems and finally little or no attention on post watershed period constraints efficient management of the project. Empowerment of local people and village institutions, site specific research and a multi-disciplinary perception of issues by research and development agencies, institutional support such as extension and credit, enactment of suitable land reforms, joint operational partnership, flexibility in watershed guidance and attention on post watershed period are some of the policy option for successful implementation of watershed schemes. Watershed programmes are also most likely to succeed in villages with homogenous communities and where difference in socio-economic status are small.

## REFERENCES

Singh, A. and Singh, M.D. (1978). Effect of Various Sstages of Shifting Cultivation on Soil Erosion from Steep Hill Slopes, *Indian Forester,* Vol.106, No.20, pp. 115-121.

NEC (2002). *Basic Statistics of North Eastern Region,* NEC Secretariate, Shillong

### Table 1: Distribution of Size of Holdings of Sample Farms in Maklang Watershed

| *Particulars* | *No. of farms* | | *Area owned (ha)* | |
|---|---|---|---|---|
| | *BP* | *AP* | *BP* | *AP* |
| Landless | 11(17) | 11(17) | 0(-) | 0(-) |
| <1 | 20(31) | 20(31) | 10.7(13) | 10.7(13) |
| 1-2 | 20(31) | 20(31) | 26(32) | 26(32) |
| 2-5 | 12(18) | 12(18) | 23.5(29) | 23.5(29) |
| >5 | 2(3) | 2(3) | 20(25) | 20(25) |
| Total | 65 | 65 | 80.2 | 80.2 |
| Average size of holding | - | - | 1.23 | 1.23 |

Figures in parentheses indicate percentage to total.

### Table 2: Land Use Pattern of Sample Farms in Maklang Watershed

| *Particulars* | *BP* | *AP* |
|---|---|---|
| Sample Size | 65 | 65 |
| Total operated area | 153.60 | 252.13 (64.15) |
| Private land | 69.17 | 71.70 (2.87) |
| Community Land | 83.9 | 180.43 (115.05) |
| <u>Arable land</u> | | |
| Wet land cultivation | 69.7 | 71.7 (2.87) |
| Jhum cultivation | 27.32 | 33.15 (21.34) |

| | | |
|---|---|---|
| Orchard | 25.10 | 56.9 (126.7) |
| Afforested area | 7.10 | 71.26 (903.7) |
| Aquaculture | 0.01 | 0.5 (49.00) |
| Sub total | 129.23 | 233.51 (80.69) |
| Current fallow/Natural forest | 24.38 | 18.62 (-23.63) |
| Irrigated area | 70 | 71 (1.43) |

Figures in parentheses indicate percentage change.

## Table 3: Cropping Pattern of Sample Farms in Maklang Watershed

| *Particulars* | *BP* | *AP* |
|---|---|---|
| Irrigated area | | |
| (i) Kharif | | |
| Wetland paddy | 69.7 | 71.70 (2.87) |
| (ii) Rabi | | |
| Mustard | 1.88 | 2.47 (31.38) |
| Potato | 0.19 | 0.30 (57.89) |
| Others | 0.24 | 0.31 (29.17) |
| Gross cropped area | 72.01 | 74.78 |
| Cropping intensity | 103 | 104 |
| Rainfed area | | |
| (i) Kharif | | |
| Wetland paddy | - | - |
| Jhum | 27.32 | 33.15 (21.34) |
| Ginger | 0.46 | 0.65 (41.3) |

| | | |
|---|---|---|
| Chilies | 0.16 | 0.16<br>(-) |
| Beans | 0.66 | 0.70<br>(6.06) |
| (ii)Rabi | | |
| Potato | 0.54 | 0.71<br>(31.48) |
| Mustard | 0.33 | 0.45<br>(36.36) |
| Others | 0.67 | 0.63<br>(-5.97) |
| Gross cropped area | 30.14 | 36.44 |
| Cropping intensity | 104 | 104 |
| Total grossed cropped area | 102.15 | 111.22 |
| Overall cropping intensity | 103.64 | 104.29 |

Figures in parentheses indicate percentage change.

**Table 4: Livestock Population and Milk Production in Maklang Watershed**

| *Particulars* | *BP* | | *AP* | |
|---|---|---|---|---|
| | *Imp* | *Local* | *Imp* | *Local* |
| Cow | 1 | 19 | 2(100) | 29(53) |
| Bullock | 7 | 21 | 16(129) | 24(14) |
| Buffalo | 18 | 27 | 26(44) | 31(15) |
| Pig | 20 | 31 | 27(35) | 62(100) |
| Goat | - | - | - | - |
| Total livestock | 56 | 98 | 71(97) | 146(49) |
| Milk production lt/lactation | | | | |
| i. Cow | 450 | 300 | 450 | 300 |
| ii. Buffalo | - | - | - | - |

Figures in parentheses indicate percentage change

## Table 5: Family Income (Rs./family) Under Maklang Watershed, Manipur

| *Components* | *BP* | *AP* | *% change* |
|---|---|---|---|
| A. Farm Income | | | |
| i. Agriculture | 25,855 | 27,890 | 7.87 |
| ii. fruits | 353 | 2589 | 633.43 |
| iii. Livestock including fisheries | 878 | 1083 | 23.34 |
| Sub total | 27084 | 31562 | 16.53 |
| | (77.50) | (78.46) | |
| B. Off Farm Income | | | |
| i. Service including business | 4655 | 4766 | 2.38 |
| ii. Labour within watershed | 2760 | 3325 | 20.47 |
| iii. Selling of forest products | 389 | 412 | 5.91 |
| iv. Small scale industries | 58 | 61 | 5.17 |
| Sub total | 7862 | 8564 | 8.92 |
| | (22.50) | (21.54) | |
| Total | 34946 | 40226 | 15.11 |

BP=Before project: AP=After Project

Fig. 1. Lorenz curve on distribution of income among farms under Maklang watershed, Manipur

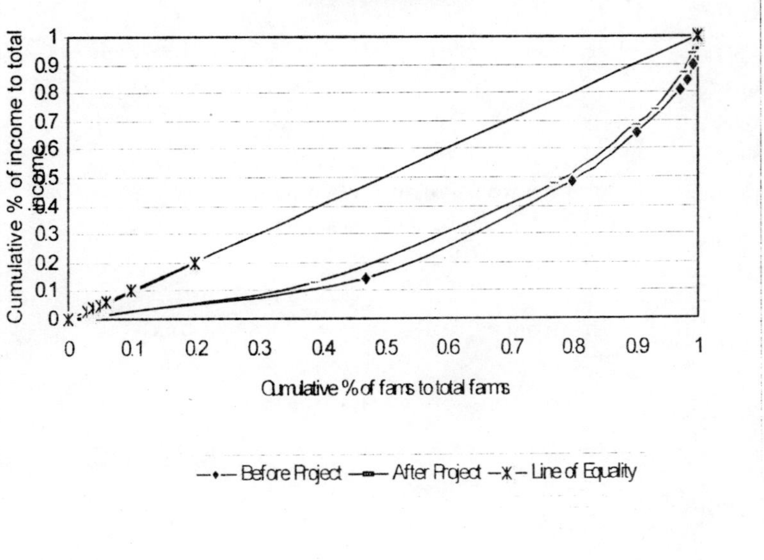

**Table 6: Problems and Constraints in Implementation of Maklang Watershed Development Project, Manipur**

| | *Preference/ % of respondent* | | | | | | | | |
|---|---|---|---|---|---|---|---|---|---|
| | *1st* | *2nd* | *3rd* | *4th* | *5th* | *6th* | *7th* | *8th* | *Score* |
| Problems/Constraints | | | | | | | | | |
| 1. Lack of empowerment of local and village institutions | 100 | - | - | - | - | - | - | - | 520 |
| 2. Lack of appropriate technology intervention | - | 52.31 | 9.23 | 7.69 | 12.31 | 9.23 | 9.23 | - | 386 |
| 3. Lack of institutional support | - | 7.69 | 32.31 | 46.15 | 7.69 | 4.62 | 1.54 | - | 342 |
| 4. Lack of operational partnership | - | 35.38 | 7.69 | 9.23 | 12.31 | 18.46 | 18.46 | - | 313 |
| 5. Non-suitability of watershed guideline | - | - | 16.92 | 15.38 | 46.15 | 12.31 | 9.23 | - | 272 |
| 6. Social problems | - | - | 23.08 | 18.46 | 9.23 | 23.08 | 27.69 | - | 255 |
| 7. No attention on post watershed period | - | 4.62 | 9.23 | 7.69 | 13.85 | 32.31 | 32.31 | - | 223 |
| 8. Non-integration of schemes | - | - | - | - | - | - | - | 100 | 100 |

# 23

# AGRICULTURAL DEVELOPMENT AND STRUCTURAL CHANGES IN THE NSDP OF MEGHALAYA

*S.V. Hariharan & M. Anandan*

## Introduction

This paper analyses mainly the growth of agricultural sector in Meghalaya and compares it with the growth of other sectors based on the available data. Further, the position of Meghalaya in the Indian Economy in terms of Net State Domestic Product and the growth of Net State Domestic Product (in current and constant prices) of Meghalaya are also analysed. In the last part, the performance of the agricultural sector of the Meghalaya state is compared with other North-Eastern States through a few indicators.

Time series Data for the Meghalaya State are not available continuously. Data available in Economic Survey mainly provide information about NSDP and percapita NSDP. With these constraints, this paper attempts to study the development of the agricultural sector and the structure of NSDP of Meghalaya. The important sources of data used in this paper are Economic Survey and the Economy of Meghalaya, Plan Documents available in internet.

## Position of in the Indian Economy

In 1990-91, the value of Net State Domestic Product of the Meghalaya State in 1993-94 prices was Rs. 1014 crores. In this year the value of Net National Product for the Indian Economy in 1993-94 prices was Rs. 6,14,206 crores. The NSDP of Meghalaya forms 1.65 per cent to the NNP of India in 1990-91. The NSDP of Meghalaya has increased continuously and reaches Rs.2403 crores in 2003-04. At the same point of time, NNP for the Indian Economy has risen to Rs. 11,06,786 crores. In 2003-04, NSDP of Meghalaya forms 2.17 per cent to the NNP of India.

Hence, the share of Meghalaya in the national income of India is increasing and it starts exceeding 2 per cent since 2000-01.

In 1993-94, Meghalaya stood in 25th rank in terms of its absolute value of NSDP among the 32 states of the Indian Economy and it has come up to 22nd place in 2003-2004.

In terms of per capita NSDP, Meghalaya was in the 19th place in 1993-94 and it has elevated to 15th place in 2003-04. The actual value of percapita NSDP was less than the percapita NNP of India. In 2003-04, the percapita NSDP of Meghalaya in current prices was Rs.18,135 and for the Indian Economy, the percapita NNP was Rs.21,142. For a long time Meghalaya remains below the national average in terms of percapita NSDP. This explains the need for more concentration by the Union Government for investment and other development programmes in Meghalaya.

### Growth of NSDP in Meghalaya

The Net State Domestic Product of Meghalaya in current prices was Rs.767 crores in 1990-91 and it has increased to Rs.4349 crores in 2003-04. In terms of constant prices (1993-94 prices), the NSDP has increased from Rs.1014 crores to Rs.2403 crores during the same period.

The trend analysis reveals that the NSDP of Meghalaya in current prices increases by Rs.284 crores per year and in constant prices, it increases by Rs.112 crores per year. The compound growth rate of NSDP of Meghalaya in current prices works out to 14.68 per cent and in constant prices, it is 7.28 per cent. The compound growth rate of Net National Product for the Indian Economy as a whole in current prices comes about 12.5 per cent and in constant prices it is 5.06 per cent.

Hence, the growth rate of NSDP of Meghalaya is much higher than the growth rate of NNP of the Indian Economy.

## Growth of Agricultural Sector in Meghalaya

The contributions of agricultural sector in current prices show an increasing trend from Rs.203 crores in 1990-91 to Rs.425 crores in 1996-97 and then to Rs 1026.91 crores in 2003-04. In terms of constant prices, it has increased to Rs.585.34 crores in 2003-04.

The share of agricultural sector is coming down slowly from 26.46 per cent in 1990-91 to 23.61 percent in 2003-04. Hence, the available data point out the fact that agricultural sector is contributing approximately one-fourth of NSDP and the industrial and service sectors contribute the remaining three-fourth of SDP.

Further, the contribution of agricultural sector measured in constant prices is not increasing till 1996-97. However it starts increasing since 2000-01.

## Indicators of Agricultural Development in Meghalaya

To have an assessment about the development of the agricultural sector in Meghalaya, a few indicators, based on the availability of data, are given in Table 5. The average rice yield per hectare has increased from 963 kg to 1829 kg in 2003-04. Average yield of food grains has increased from 1000kg to 1349 kg. However, they remain below the national average.

Further, cultivators as a percentage to main workers have come down from 57.7 per cent to 48.14 per cent in 2001. On the contrary, agricultural workers as a percentage to main workers increase from 10.1 to 17.7 per cent in 2001. Irrigated area has come down from 20.2 to 19.3 per cent. Irrigated area per cultivator has also come down marginally from 0.13 to 0.11 per cent. Compared to all India average, the percentage of irrigated area is much less in Meghalaya.

A careful examination of these data reveal that the agricultural sector of Meghalaya is not developing by an impressive rate, and the state remains below the national average level, even though it has a good potential for the development of agricultural sector.

## Agricultural Development of Meghalaya: A Comparison With Other North-Eastern States

Selected indicators of the agricultural sector for Meghalaya and the Other North-Eastern states, namely, Arunachal Pradesh, Assam, Manipur, Mizoram, Nagaland and Tripura are given in Table 7.

In terms of gross irrigated area, Meghalaya stands in the fifth place with 46,000 hectares. It is below most of the other North-Eastern states. Cultivated area works out to 10.7 per cent to the total area, where as for Assam and Tripura, they are 48.7 and 43.7 per cent respectively. In terms of irrigated area also, it stands in the third place after Manipur and Nagaland. Cultivators as a percentage to total main workers form 52.2 per cent in Meghalaya, and for Arunachal Pradesh, Manipur, Mizoram and Nagaland, it is higher. The percentage of agricultural labourers is higher in Meghalaya (16.3) whereas excluding Tripura, it is much less in other North-Eastern states.

In terms of crops yield, it is above Arunachal Pradesh. However, Meghalaya stands below Assam, Manipur, Mizoram, Nagaland and Tripura. Percapita availability of food grains is the lowest in Meghalaya compared to other North-Eastern states. Percentage of villages electrified is also the lowest in Meghalaya. Only 49 per cent of villages have been electrified in Mehgalaya.

A comparison of Meghalaya with other North-Eastern states also suggests that the agricultural sector of Meghalaya remains to be developed further.

## Structural Changes in the NSDP of Meghalaya

The shares of various sectors in the NSDP of Meghalaya for various years are given in Table 7. The share of agricultural sector is slowly reducing from 26.46 per cent to 23.61 per cent. In most of the cases there are marginal changes. The share of manufacturing sector has come down from 14.05

per cent in 1990-91 to 13.85 per cent in 2003-04. The share service sector is increasing from 51.17 per cent to 54.25 per cent. Only for trade and hotels, and Banking and Insurance services, the share is increasing. In the remaining sectors, the share remains more or less constant or changes marginally. On the whole, the overall structure of NSDP of Meghalaya has not changed very much and agricultural sector seems to be a significant source of NSDP of Meghalaya.

The growth rate of agricultural sector is 6.67 per cent and it is less than the growth rate of NSDP. The growth rates of agricultural sector is lower compared to that of the manufacturing sector (7.96 per cent) and service sector (7.46).

## Conclusion

The agricultural sector of Meghalaya State has a good potential for development. However, the available statistics reveal that the agricultural development of the state is not impressive. Even though the share of NSDP is declining marginally, still agricultural sector remains an important contributor to the NSDP. Hence Central and State Governments need to concentrate more on the development programmes to stimulate the process of development of the agricultural sector in Meghalaya.

**Table 1: Net State Domestic Product of Meghalaya**

*(Rs in Crores)*

| *Year* | *NSDP in Current prices* | *NSDP in 1993-94 prices* | *Share of Meghalaya in India's NNP (in per cent)* |
|---|---|---|---|
| 1990-91 | 767 | 1014 | 1.65 |
| 1991-92 | 884 | 1020 | 1.65 |
| 1992-93 | 979 | 1093 | 1.69 |
| 1993-94 | 1304 | 1304 | 1.90 |
| 1994-95 | 1432 | 1234 | 1.68 |
| 1995-96 | 1729 | 1383 | 1.76 |
| 1996-97 | 1898 | 1396 | 1.64 |
| 1997-98 | 2166 | 1558 | 1.75 |
| 1998-99 | 2579 | 1653 | 1.74 |
| 1999-00 | 2908 | 1841 | 1.83 |

| | | | |
|---|---|---|---|
| 2000-01 | 3338 | 2048 | 2.10 |
| 2001-02 | 3669 | 2184 | 2.15 |
| 2002-03 | 3976 | 2285 | 2.15 |
| 2003-04 | 4349 | 2403 | 2.17 |

*Source:* Economic Survey, 2005-06.

## Table 2: Position of Meghalaya in Terms of SDP and Per Capita SDP in India

| *Year* | *Rank of Meghalaya in NSDP* | *Rank of Meghalaya in PC NSDP* | *PC NNP of India (in Rs)** | *PC NSPP (in Rs)** |
|---|---|---|---|---|
| 1993-94 | 25 | 19 | 7690 | 6893 |
| 1996-97 | 25 | 20 | 11564 | 9243 |
| 2000-01 | 27 | 21 | 16729 | 14632 |
| 2003-04 | 22 | 15 | 21142 | 198135 |

*Source:* Column 2 & 3: computed from the data published in *Economic Survey* Column 4& 5: *Economic Survey* – 2005-06.

* Values in current prices.

## Table 3: Results of Trend Analysis of NNP and NSDP

| *Model* | *Variable* | *a* | *b* | *SE of b* | *R2* | *CGR* |
|---|---|---|---|---|---|---|
| Linear | NNP in cur prices | 239819 | 124777* | 3387 | 0.991 | |
| | NSDP In cur prices | 155 | 284* | 12.17 | 0.978 | |
| Linear | NNP in const.prices | 547687 | 40799* | 1709.01 | 0.978 | |
| | NSDP In const. prices | 760 | 112* | 6.25 | 0.961 | |
| Semi-log | NNP in cur prices | 12.99 | 0.12* | 0.0006 | 0.970 | 12.52 |
| | NSDP In cur prices | 6.56 | 0.14* | 0.004 | 0.988 | 14.68 |
| Semi-log | NNP in const. prices | 13.27 | 0.04* | 0.002 | 0.969 | 5.06 |
| | NSDP In const.prices | 6.81 | 0.07* | 0.003 | 0.982 | 7.28 |

* Significant at one per cent level.

## Table 4: Contribution of Agricultual Sector in NSDP of Meghalaya

| *Year* | *Agricultural Sector* | *Share of Agl. in NSDP (in per cent)* |
|---|---|---|
| 1990-91 | 203.06 | 26.46 |
| 1991-92 | 246.00 | 27.82 |
| 1992-93 | 247.73 | 25.32 |
| 1993-94 | 289.29 | 25.52 |
| 1994-95 | 324.66 | 25.81 |
| 1995-96 | 368.64 | 23.19 |
| 1996-97 | 425.29 | 24.13 |
| 2001-02 | 927.36 | 25.07 |
| 2002-03 | 977.18 | 24.58 |
| 2003-04 | 1026.91 | 23.61 |
| 2004-05 | 1075.51 | 22.62 |

*Source:* www.Economy of Meghalaya.

## Table 5: Few Indicators of Agricultural Development of Meghalaya

| *S.No.* | *Indicator* | *1988-89* | *1993-94* | *1997-98* |
|---|---|---|---|---|
| 1 | Cultivators as % to main workers | 57.7(40.2) | 54.5(37.9) | 52.2(36.9) |
| 2 | Agl. Labourers as % to main workers | 10.1(23.9) | 14.5(26.5) | 16.3(27.0) |
| 3 | Forest Area as a % to total area | 41.93(20.41) | 41.84(20.81) | 41.70(19.27) |
| 4 | % of Irrigated Area to total Cropped Area | 20.2(32.2) | 19.3(36.6) | 19.3(37.5) |
| 5 | % of irrigated area per cultivator | 0.13(0.64) | 0.12(0.62) | 0.11(0.59) |
| 6 | Crop Yield per hectare (in Kgs) | | | |
| | (a) Rice | 963(1690) | 1000(1330) | 1128(1890) |
| | (b) Food grains | 1133(1500) | 1346(1879) | 1349 (1601) |

Figures in parantheses are corresponding values for the Indian Economy.
*Source:* Plan Documents of Meghalaya (www.meghalaya.com)

## Table 6
## Comparison of Meghalaya with Other North-Eastern States

| S.N | Indicators | Meghalaya | AP | Assam | Manipur | Mizo | Nagaland | Tripura |
|---|---|---|---|---|---|---|---|---|
| 1 | Cultivators as a % to total main workers | 52.2 | 53.7 | 50.9 | 60.5 | 55.6 | 72.9 | 34.8 |
| 2 | Agl. Labourers as a % to total main workers | 16.3 | 8.5 | 12.1 | 8.2 | 4.0 | 2.1 | 23.0 |
| 3 | Gross Irrigated Area ('000 hectare) | 46 | 36 | 572 | 75 | 8 | 63 | 60 |
| 4 | Cultivated Area as a % to total area | 10.7 | 3.1 | 48.7 | 8.9 | 4.8 | 43.7 | 43.7 |
| 5 | % of irrigated area as a % to total cropped area | 19.3 | 13.9 | 15.0 | 37.6 | 7.5 | 13.0 | 13.0 |
| 6 | Gross Irrigated Area per cultivator (in hec) | 0.11 | 0.15 | 0.15 | 016 | 0.04 | 0.15 | 0.19 |
| 7 | Rice yield per hectare(in kg) | 1160 | 1150 | 1350 | 2510 | 1540 | 1320 | 2010 |
| 8 | Food grains yield per hectare (in kg) | 1160 | 560 | 1270 | 2510 | 1590 | 1240 | 1940 |
| 9 | Percapita availability of cereals (gm/day) | 198 | 607 | 384 | 452 | 393 | 410 | 406 |
| 10 | Percapita availability of food grains (gm/day) | 202 | 1986 | 408 | 452 | 413 | 433 | 409 |
| 11 | Application of fertilizer per hectare (in kg) | 14.41 | 2.11 | 14.62 | 66.24 | 3.62 | 3.96 | 18.76 |
| 12 | Percentage of villages electrified | 49 | 70 | 99 | 99 | 100 | 99 | 77 |

*Source:* Plan Document of Meghalaya *op.cit.*,

## Table 7: Sructure of NSDP of Meghalaya

*(in percentage)*

| *Year* | *Agricultural Sector* | *Agr&Allied Sector* | *Manufacturing Sector* | *Service Sector* | *NSDP* |
|---|---|---|---|---|---|
| 1990-91 | 26.46 | 34.78 | 14.05 | 51.17 | 100 |
| 1991-92 | 27.82 | 35.68 | 13.08 | 51.35 | 100 |
| 1992-93 | 25.32 | 31.87 | 16.61 | 51.51 | 100 |
| 1993-94 | 25.52 | 31.25 | 16.08 | 52.51 | 100 |
| 1994-95 | 25.81 | 32.98 | 13.85 | 53.16 | 100 |
| 1995-96 | 23.19 | 30.78 | 14.40 | 54.81 | 100 |
| 1996-97 | 24.13 | 31.76 | 15.96 | 52.68 | 100 |
| 2001-02 | 25.07 | 33.62 | 12.66 | 53.72 | 100 |
| 2002-03 | 24.58 | 32.46 | 13.85 | 53.69 | 100 |
| 2003-04 | 23.61 | 31.90 | 13.85 | 54.25 | 100 |
| 2004-05 | 22.62 | 32.30 | 13.02 | 54.68 | 100 |

## Table 8: Structure of NSDP of Meghalaya

| *S.N.* | *Sector* | *1990-91* | *1991-92* | *1992-93* | *1993-94* | *1994-95* | *1995-96* | *1996-97* |
|---|---|---|---|---|---|---|---|---|
| 1 | Agriculture | 26.46 | 27.82 | 25.32 | 25.52 | 25.81 | 23.19 | 24.13 |
| 2 | Forestry & Logging | 1.55 | 1.27 | 1.03 | 1.11 | 1.51 | 1.40 | 1.20 |
| 3 | Fishing | 0.50 | 1.07 | 1.14 | 0.79 | 0.96 | 0.84 | 0.82 |
| 4 | Mining&Quarying | 6.26 | 5.41 | 4.39 | 3.82 | 4.72 | 5.36 | 5.21 |
| 5 | Manufacturing | 3.11 | 3.12 | 4.34 | 3.94 | 3.81 | 3.38 | 3.30 |
| 6 | Construction | 12.32 | 12.00 | 12.80 | 12.31 | 11.07 | 11.46 | 12.85 |
| 7 | Trans & Commu | 7.47 | 7.27 | 6.96 | 6.95 | 7.00 | 6.57 | 6.73 |
| 8 | Trade & Hotels | 11.49 | 11.51 | 10.71 | 12.21 | 13.53 | 16.06 | 15.62 |
| 9 | Banking & Insuran | 2.37 | 2.48 | 2.55 | 2.46 | 2.62 | 2.59 | 2.67 |
| 10 | Real Estate | 5.46 | 5.79 | 5.64 | 5.24 | 5.35 | 4.70 | 4.63 |
| 11 | Pub.Admn | 13.86 | 14.08 | 15.68 | 14.08 | 14.21 | 13.35 | 12.98 |
| 12 | Other sectors | 10.53 | 10.21 | 9.97 | 10.83 | 10.46 | 11.54 | 10.00 |

*Source:* www.meghalaya/statistics.com

# 24

# AGRICULTURE IN THE ARUNACHAL ECONOMY: FROM REPRODUCTION TO GROWTH

*N.C. Roy*

## Introduction

The sequence of agrarian changes in Arunachal Pradesh constitutes an important part of the overall transformation experienced by its economy over the last six decades. The economic change in the state in terms of technological induction, occupational diversification, institutional refiguration, shift of growth trajectory, etc. is so large that it can be called, in the language of Polanyi 'great transformation' (Polanyi 1944). Before independence there was no Arunachal economy; there were a good number of tribal economies based mainly on swidden cultivation locally known as jhuming. Insularity characterized these barter-based economies. The jhum cultivators were also producers of different handicrafts – clothes, baskets, small tools, etc. - order to ensure autarky at the household level. There was no occupational differentiation – the peasantry subsumed all occupational categories within itself. The areas forming Arunachal Pradesh were administered indirectly by the British through offices located in what is now Assam. So there was neither a modern tertiary nor a modern secondary sector, a situation which kept the state's economy as a semi-disjoint set of pure peasant economies.

After independence the Government of India introduced direct administration in Arunachal Pradesh (then known as the North East Frontier Tracts) and launched simultaneously the development programme in order to create administrative and other infrastructures such as roads, transport and communication, hospitals, educational institutions, etc. The Government also mapped out an elaborate programme for extension of permanent cultivation. The Government expenditure brought cash transactions in the tribal economies leading to their progressive monetization. This resulted in two types of integration: one is horizontal—all tribal (peasant) economies were integrated into the Arunachal economy and the other is vertical—the Arunachal economy became integrated with the national economy. Two new sectors—modern secondary and tertiary sectors—appeared which brought in the forces of dynamism. With the appearance of market, the process of what Kautsy called 'agriculturalization of peasant' was set into motion (Kautsy 1966). The motion was and is extremely slow but the process has since been going on with increasing numbers of jhum cultivators making a switch over to permanent cultivation. Some jhum cultivators have made a diversification into horticulture and a few have switched over to plantation of tea, coffee, etc. Expansion of permanent cultivation, development of horticulture, etc. have shifted the growth trajectory of agriculture. This paper highlights the factors determining the agricultural growth in the state. For convenience the paper is divided into three sections. The first section deals with the relative position of agriculture in the state economy. Section II describe the techno-institutional changes underlying the agricultural growth and constructs a conceptual framework which is tested in the third section. Concluding remarks also appear in the third section.

## Agriculture's Contribution to the State's Domestic Production

The community-guided peasant economies of pre-independence era have now merged into the market-driven Arunachal economy, which has witnessed progressive diversification resulting in the gradual decline of the relative importance of agriculture both as the contributor of income and employment.

The limitation of data truncates our period of analysis to 1970-2005, and the non-availability of the data on Gross State Domestic Product at constant prices for this period makes us use the data on Net State Domestic Product (NSDP) at constant prices of 1993-94. Table 1 shows the changing profile of agriculture. In 1970-71 agriculture made a modest contribution of 38.33 percent to the NSDP, which is indeed substantially lower than what it did two decades earlier. (A simple extrapolation puts agriculture's contribution to NSDP at about 85 per cent in 1947-48). Relative contribution of agriculture to NSDP declined to 24.37 per cent by 2004-05. The pattern of fall of relative contribution of the primary sector is almost the same as that of agriculture. In 1970-71 primary sector contributed to 59.19 per cent of NSDP; its contribution by 2004-05 was reduced to a modest 31.09 per cent.. Manufacturing was insignificant before 1970s, its weight gained in the 1970s and 1980s but in the 1990s it paled almost into insignificance. The fall in relative importance of agriculture is due not so much to the expansion of secondary activities but to the growth of the tertiary sector especially public administration and other services. (It may be noted that Arunachal Pradesh has a large-sized Government relative to its economy; in some years the Government expenditure exceeds the State's Domestic Product).

**Table 1**

**Contributions of Agriculture and Other Sectors to NSDP in Arunachal Pradesh: A Comparison**

*(At constant prices of 1993-94)*

| *Year* | *Agriculture* | *Primary Sector* | *Manufacturing* | *Secondary Sector* | *Tertiary Sector* | *NSDP* |
|---|---|---|---|---|---|---|
| 1970-71 | 38.33 | 59.19 | 0.85 | 20.33 | 20.48 | 100 |
| 1980-81 | 36.91 | 47.28 | 6.51 | 22.21 | 30.51 | 100 |
| 1990-91 | 35.09 | 46.19 | 6.04 | 21.56 | 32.25 | 100 |
| 2000-01 | 31.10 | 37.37 | 2.64 | 18.34 | 44.29 | 100 |
| 2004-05 | 24.37 | 31.09 | 2.27 | 25.42 | 43.49 | 100 |

*Source:* Different population censuses of Arunachal Pradesh.

A more concise as well as accurate picture is provided by trend equations estimated on the basis of data of all the years from 1970-71 to 2004-05. Given below are these equations:

$$A = 41.54 - 0.36\,t \qquad R^2 = 0.61$$
$$P = 61.32 - 0.75\,t \qquad R^2 = 0.80$$
$$M = 3.02 + 0.04\,t \qquad R^2 = 0.04$$
$$S = 19.33 + 0.09\,t \qquad R^2 = 0.12$$
$$G = 8.75 + 0.13\,t \qquad R^2 = 0.22$$
$$T = 9.34 + 0.66\,t \qquad R^2 = 0.84$$

(Here A is agriculture's, P is primary sector's, M is manufacturing sector's, S is secondary sector's, G is public administration's and T is tertiary sector's percentage contributions to the NSDP; t is time measured in years with origin 1969-70). During 1970-2005 agricultural contribution to the NSDP declined by 0.36 percent points per annum and primary sector's by 0.75 per cent points per annum. Relative decline in agricultural contribution was not offset much, as stated before, by the expansion of either manufacturing or secondary sector but by the growth of the tertiary sector.

### Agriculture's Contribution to the Growth of NSDP

Though agriculture's relative contribution to the NSDP has markedly fallen over the years, its contribution to the growth of NSDP is significant. During 1970-2005, agriculture's average contribution to the NSDP is 34.99 per cent. Agriculture grew during this period at the rate of 5.95 percent per annum and its contribution to the NSDP's growth was 29.67 percent. Compared with agriculture, manufacturing sector's contribution is very small—only 5.03 percent (table 2).

**Table 2: Agriculture's Contribution to the Growth of NSDP: An Inter-Sectoral Comparison During 1970-2005**

| *Sectors* | *Average contribution to NSDP (in %)* | *Growth rate (in % per annum)* | *Contribution to the growth of NSDP (in %)* |
|---|---|---|---|
| Agriculture | 34.99 | 5.95 | 29.67 |
| Primary Sector | 47.81 | 5.40 | 36.79 |
| Manufacturing | 3.74 | 9.44 | 5.03 |
| Secondary Sector | 21.01 | 7.49 | 22.42 |

| | | | |
|---|---|---|---|
| Public Administration | 11.06 | 8.05 | 12.69 |
| Tertiary Sector | 31.18 | 9.18 | 40.79 |
| NSDP | 100.00 | 7.06 | 100.00 |

*Note:* Sectoral contribution to the growth of NSDP (in percentage) is defined as (Ci * Gi /G) * 100, where Ci is ith sector's average contribution to the NSDP, Gi is ith sector's growth rate and G is the growth rate of NSDP. Growth rate is the estimated value of b (multiplied by 100) in the equation, log y = a + bt, where y is NSDP or a sector's absolute contribution to it and t is time in years. The number of observations used for estimation of growth or average is 35 covering the period 1970-05.

## *Employment in Agriculture*

Prior to 1947, the main occupation of all workers was agriculture. Over the years the relative employment capacity in agriculture has declined as the non-agricultural sectors have expanded. However, the absolute size of employment in agriculture has increased over the years. Not that the size of agricultural employment has only expanded, but the growth rate of this size has also increased. Thus the agricultural employment grew at the yearly rate of 0.68 per cent during 1971-81. This rate rose to 1.08 per cent in the 1980s and to 1.64 per cent in the 1990s (table 3). However non-agricultural employment expanded more rapidly than agricultural and this is reflected in the fall of the proportion of agricultural employment. In 1971 as high as 80.30 percent of all workers were engaged in agriculture but in 2001 the percentage declined to 61.74.

**Table 3: Employment in Agriculture: Its Relative Position and Growth in Arunachal Pradesh**

| *Year* | *% of workers employed in* | | *Yearly Growth of employment (in %)* | | |
|---|---|---|---|---|---|
| | *Agriculture* | *Non-agriculture* | *Agriculture* | *Non-agriculture* | *Total* |
| 1971 | 80.30 | 19.70 | --- | --- | --- |
| 1981 | 73.75 | 26.25 | 0.68 | 5.50 | 1.63 |
| 1991 | 65.49 | 34.51 | 1.08 | 6.40 | 2.47 |
| 2001 | 61.74 | 38.26 | 1.64 | 3.69 | 2.35 |

*Data source:* Different population censuses of Arunachal Pradesh.

*Note:* Data relate to census years. The growth rate is inter-censal average yearly. Dash means not available.

## Simple Reproduction

The swidden cultivation with extremely low capital intensity was so low productive that it could barely meet the subsistence needs of the people. Agriculture was thus a simple reproduction: neither capital nor labour force experienced any significant growth prior to 1947[1]. A sustained growth in population and hence labour force could have led to an expansion of permanent cultivation or what Boserup called intensification of agriculture (Boserup 1965). But it did not happen. Without an endogenously generated growth impulse, agriculture required an exogenous thrust to move on, and this was provided by the structural changes of economy emanating from the establishment of direct administration and implementation of development programmes by the Government of India. The beginnings demand some highlight.

## Initial Growth Impulse

The establishment of direct administration and creation of different infrastructural facilities such as roads, transport and communications, hospitals, etc. required land. But land was community-owned and no market for it existed. The individual jhum cultivator enjoyed only usufructuary rights, the rights which did not vest him with any with power to alienate his land to the members of other community or to the Government. This restriction created a big hurdle for land acquisition. The Government, however, exercised an extreme caution: it did not use any coercive power. Rather it took on the path of persuasion through negotiation, a method which turned the acquisition of land into a game of bargaining between the community leaders and the Government. The bargaining became a positive-sum game yielding benefits to both the parties—the Government got the land and gradually spread its wings peacefully throughout the State and the communities received rent and other benefits such as health services, education, roads, etc. that for the first time appeared on the soil of Arunachal Pradesh. The upward shift of the demand for land created its rent, which, prior to 1947, was unknown to the people of the State, which is richly endowed with land relative to the size of the population.

Accrual of land rent was accompanied with the availability of wage labour from other states of the country and the emergence of market, which set the Arunachal economy on the path to monetization. The receivers of rent who were largely community leaders and people with some entrepreneurial orientation employed newly available wage labour from other states. These wage labourers are considered unskilled in the rest of the country but they became 'experts' of the agricultural practice, traditional elsewhere but new in Arunachal Pradesh. Their services were highly demanded, and though temporary was their presence because of Inner-Line Regulation[2], they diffused the technique of wet-rice cultivation throughout the foothills and river valleys of the State which is 95 per cent hilly. The introduction of the new agricultural practice, permanent cultivation, and the monetization of the economy changed the institutional configuration; the community ownership of land gave way to individual ownership. This institutional shift of focus from the community to the individual is a movement from status to contract (Maine 1906). Institutional refiguration and the appearance of market, albeit highly restrictive in its operation in the initial stage, linked agriculture both forwardly and backwardly with the secondary and tertiary sectors of the economy. The linkage was, however, limited to the areas practising permanent cultivation; most of the jhum areas because of their remoteness remained isolated as before.

Jhuming has still retained its importance though it has lost its preponderance. Even today more than half of the cultivators practise jhuming and 47.3 percent of the net sown area, is under jhum cultivation. The process of diversification of sources of livelihood has tended to change the institutional configuration of jhuming. The individualization of community land was first restricted to permanent cultivation; gradually with the spread of market the jhum land has also come under individual ownership. The property rights are now well established in cultivable land but the Government has not yet formalized it by conducting a cadastral survey. So there is no record of ownership and the absence of such records has stood in the way of advancing institutional credit to the farmers. Land is now property but its mortgage value to the financial institutions is highly restricted.

In spite of restricted credit flow, Arunachal agriculture has experienced substantial growth. The sources of the growth are many: expansion of irrigation facilities use of fertilizer, new varieties of seeds, introduction of new crops, etc. An outcome of agrarian changes, which is rooted in permanent cultivation and property rights in cultivable land, is the large-scale inequalities in the distribution of land holdings, a phenomenon that is unprecedented in the state. In what follows an attempt is made to formalize the agrarian changes. The formalisation and part of estimated results closely follow Roy 2005 (a) and 2005 (b).

## A Framekwork for Growth Analysis

The framework is based more on context and history than on a prior reasoning. The establishment of direct administration entailed the inflow of central funds and migration of workers. Not only labour, but also other inputs were brought from the rest of the country. The process of causation ran thus: the inflow of central funds raised the wage rate in Arunachal Pradesh relative to the rest of the country, a situation which made job attractive in the State causing in-migration of workers. The workers from outside were essential because in Arunachal Pradesh there was no wage labour (3). The relation can be stated as

$$Wi / W = f_1 (G, Li) \quad \ldots\ldots (1)$$

In eq (1), the wage rate is Wi in Arunachal Pradesh and W in the rest of the country; G is the quantity of central funds flowing into the State and Li is the local supply of labour. A value of Wi/W greater than unity is an inducement for migration in Arunachal Pradesh. The first partial derivative of Wi/W with respect to G is positive, but with respect to Li is negative. That is, a rise in G, given Li, raises the ratio. The relationship between migration and Wi/W is straightforward:

$$Lm = f_2 (Wi/W) \quad \ldots\ldots (2)$$

In eq (2), Lm is the number of workers having migrated

from the rest of the country, and $f'_2 > 0$. The migration of labour and learning through experience were instrumental in expansion of permanent cultivation, which created the growth-potentialities of Arunachal agriculture. This relation runs as:

$P_L = f_3$ (H, Lm)............................................... (3)

The permanent cultivation belongs to the same process which involved the accumulation of capital and the transition of the Arunachal agriculture from the circle of reproduction to the path of growth. The accumulation began, as stated before, with the transformation of a part of land rent received from the Government into capital. The Government since the 1950s has been spending liberally on the improvement of agriculture especially extending permanent cultivation. Next to transport and communication agriculture received the highest priority (Government of NEFA, 1964). Subsidy was provided to the farmers for reclamation of land from bushes and jungles, and terracing of jhum land for permanent cultivation. The Government distributed gratis small tools and other inputs such as improved seeds, etc. especially in the 1950s and 1960s. Irrigation was also subsidized. All these made the Government an inseparable part of the accumulation of capital in agriculture. So capital in agriculture (K) can be visualized as a function of the cumulative private and Government investments.

That is

$K = f_4$ (Ip, Ig)............................................... (4)

In eq (4), Ip is cumulative private and Ig is cumulative Government investment in agriculture. The empirical estimation of K in the next section requires some highly simplifying assumptions. Total labour force is defined as

L = a lm + b Li ............................................... (5)

Eq (5) is a definition. L is total workers in agriculture, and a and b are fractions. Lastly, aggregate production function in agriculture is specified as

$Q = f_5$ (K, L, N, T) ............................................... (6)

In eq (6) Q is output, K and L are as explained before, N is the land and T is a measure of technology. There are five endogenous variables, Wi/W, Lm, $P_L$, K and Ø to be determined by five equations.

The number of all workers is determined exogenously, but a part of it, Lm, is determined endogenously. Similarly total land under cultivation (net sown area) is exogenous, though permanently cultivated land ($P_L$) is endogenous. T is purely an exogenous variable. Estimated results are discussed in the next section.

## Growth-Promoting Factors

At the operational level, the propulsive force behind the growth of agriculture—or of any sector or of the entire economy—is the technological progress, a prime-moving factor with various forms and manifestations. In Arunachal Pradesh, its initial form was permanent cultivation, the agricultural practice that is accommodative of different types of innovations. Jhum cultivation, because of its location on the hillside, is closed to most of the techniques such as irrigation, mechanisation, etc. (It may be noted that even today bullock-driven plough is not used in jhuming). So the expansion of permanent cultivation became the main vehicle of application of new inputs, which raised the agricultural productivity.

**Table 4: Technology of Agricultural Production: Levels of Some Indicators: 2004-05**

| *Indicators* | *Value* |
|---|---:|
| Percentage of net sown area irrigated | 22.2 |
| Chemical fertilizer used per hectare (in kg) | 3.1 |
| Percentage of net sown area under jhum | 46.3 |
| Percentage of net sown area under permanent cultivation | 53.7 |
| Percentage of farms using power tillers/tractors * | 2 |
| Percentage of area under rice using HYV seeds | 31.65 |
| Output (in quintals) of paddy per hectare | 11.10 |

* The figure on the use of power tiller/tractor is only approximate.

*Data Sources:* Annual Operating Plan, 2006-07, Directorate of Agriculture, Government of Arunachal Pradesh, Naharlagun and Statistical Abstract of Arunachal Pradesh, 2005, Government of Arunachal Pradesh.

In spite of a sustained growth, the overall level of agricultural productivity remains still low as shown in table 4. Irrigation has covered only 22.2 percent of the net sown area. In Arunachal Pradesh only about 10 percent of the area is double-cropped which corresponds to an intensity of cropping of 1.1, chemical fertilizer used per hectare is 3.1 kg which is one of the lowest in the country. Nagaland is the lowest user with only 1.6 kg per hectare and expectedly Punjab is at the top applying 192.5 kg per hectare. The average use in the country was 96.6 kg per hectare during 2004-05 (Government of India 2006, p. 168).

Permanent cultivation has been an ardent adopter of new varieties of crops; 58.94 per cent of net sown area under wet-rice cultivation uses high yielding varieties (HYV) of rice. This means 31.65 percent of all net sown area (including *jhum* land) being under HYV rice. In spite of the HYV rice being under increasing use, the yield rate is very low especially because of the falling productivity of *jhuming*. Table 4 shows that yield per hectare was just 11.10 quintals during 2004-05. This is almost one-third of the national average—29.15 quintals per hectare during 2002. (Indian yield per hectare happens to be one of the lowest in the world. Average yield of rice in the world is 39.16 quintals per hectare; USA has the highest yield of 73.72 quintals per hectare (Government of India 2006, p. 156).

## Agricultural Growth

The level of different indicators is indeed very low, but the picture of agriculture growth is rosy (table 5). This is because of a very low base from which the process of modernization started. Output (part of NSDP from agriculture) grew at the yearly rate of 6.40 percent during the 1970s. Productivity accelerated during the 1980s, boosting the growth rate to 9.18 percent.

Growth rate dropped in the 1990s to a mere 2.48 percent and became negative during 2000-05. Productivity growth of both labour and land shows a broad similarity with the overall growth of agriculture. The most impressive growth is shown by labour productivity, which ultimately came from the expansion of permanent cultivation. Separate data on jhum and permanent cultivation being not available, it is not possible to compare the productivity of these two types of agricultural practices. However, all available evidence suggests a stagnating, or even a declining condition of jhum productivity (Roy and Kuri 2001).

**Table 5**

**Growth of Productivity in Arunachal Agriculture**

| *Period* | *Output* | *Labour* | *Land* | | | *Productivity of* | |
|---|---|---|---|---|---|---|---|
| | | | *Jhum* | *Permanent* | *Total* | *Labour* | *Land* |
| 1970-80 | 6.40 | 0.66 | -2.91 | 6.12 | 0.04 | 5.74 | 6.36 |
| 1980-90 | 9.18 | 1.02 | 1.09 | 6.08 | 3.57 | 8.16 | 5.61 |
| 1990-2000 | 2.48 | 1.52 | 1.06 | -1.29 | -0.17 | 0.96 | 2.65 |
| 2000-05 | -0.72 | 1.52 | -1.33 | 0.88 | -0.17 | -2.24 | -0.55 |
| 1970-2005 | 5.95 | 1.13 | 0.14 | 3.31 | 1.43 | 4.82 | 4.52 |

*Data Sources:* Estimated on the basis of data from (1) Estimates of State Domestic Product, Directorate of Economics and Statistics, Government of Arunachal Pradesh, and (2) Different Population and agricultural Censuses of Arunachal Pradesh.

*Note:* The cadastral survey having yet not been conducted, the data on land especially on jhum suffer from a wide margin of error. The data on labour and partly on land for 2000-05 are estimated on the basis of growth rate of the 1990s. The growth rate of Y is the estimated value of b (multiplied by 100) in the equation log y = a + bt, where t is time in years.

## Empirical Estimation

The behaviour of agricultural wage rate in Arunachal Pradesh relative to that in the rest of the country subsequent to the inflow of central funds is shown in the estimated equation:

$$100 * (Wi / W) = 1117.84 - 85.18 \log Li + 5.10 \log G \quad \text{..... (7)}$$

$$(7.4) \qquad (5.9) \qquad (2.7)$$

$$R = 0.91 \qquad N = 33$$

Parenthesised figures are t - values. The period covered is 1971-2003 yielding 33 observations symbolized by N. $R^2$ is high and both the arguments, Li (local labour supply) and G (Central funds) are significant. The fit is too good to be believed. In fact it is auto-correlation (value of D.W. is 0.85) which has bolstered the fitness. That G has a highly regular time-path is shown by

$$Gt = 1190.08 + 1.01\ G\ t\text{-}1 \quad \text{.......................... (8)}$$

$$(8.0) \qquad (25.1)$$

$$R^2 = 0.95 \qquad N = 32$$

Extraction of auto correlation by Cochranc-Orcutt method yields the result:

$$100 * (Wi/W) = 891.09 - 63.94 \log Li + 1.60 \log G \quad .... (7a)$$
$$(5.4) \qquad (4.4) \qquad (0.8)$$
$$R^2 = 0.49 \qquad N = 33$$

There is now a drastic reduction of the explanatory power, but more noticeable is the falling of the coefficient of log G into insignificance. Li retains its significance.

The rise of the relative wage, Wi/W, entailing the migration of workers is shown below:

$$\Delta \log Lm = 0.344 + 0.0032 (Wi/W * 100) \quad .............. (9)$$
$$(6.4) \quad (7.4)$$
$$R^2 = 0.64 \qquad N = 32$$

The period covered is, as before, 1971-2003. The significance of eq (9) testifies the assertion that the migration of labour, the diffuser of new agricultural technology, was entailed by a high wage rate in Arunachal Pradesh compared with the hinterland supplying migrants. Migration led to the extension of permanent cultivation:

$$\log P_L = 7.72 + 0.28 \log H + 0.77 \Delta \log Lm \quad ........... (10)$$
$$(5.6) \quad (2.6) \qquad (0.13)$$
$$R^2 = 0.35 \qquad N = 32$$

In eq (10) $P_L$ is permanently cultivated land and H is cumulated agricultural output used as learning a la Verdoorn (Verdoora 1956). Eq (10) has a low explanatory power; moreover migration does not appear as a significant factor. Rather experience through learning-by-doing carries a strong influence upon the new agricultural practice, permanent cultivation.

The next task is to find a time-series of agricultural capital. This is almost intractable. However, without an estimation of capital (K), no idea can be formed about the nature of Arunachal's agricultural production function. So some methods, albeit crude, devised. One is the use of fertilizer as the representation of K, but its representative ness is questionable in view of its smallness. Unlike agriculture in the rest of the country where private investment is much more than public investment, Arunachal's agriculture received more public than private investment especially in the initial decades. This fact is taken care of by an estimation of public funds flowing into agriculture through transformation into private investment. An important medium is construction part of NSDP. It is public-funded but most of income from construction activities flows into the private hands, a part of which is invested in agriculture.

Over the years the relative share of construction-generated income flowing into agriculture has declined as the alternative sources of investment have appeared. The growth rate of non-construction sectors is used to deflate the construction (C) to generate a part of the K series. Thus K is estimated as

$$K = \tfrac{1}{2} f + \tfrac{1}{2} \exp(-0.5\ t).\ C$$

Here f is aggregate chemical fertilizers used in Arunachal agriculture, t is time in years. K is thus an index having no unit of measure (the unit of f is tonne and that of C is Rupee at constant prices of 1993-94). The estimated production function is as follows:

$$\log Q = -35.925 + 1.556 \log N + 2.066 \log L + 0.089 \log K + 0.015\,T \ldots (11)$$
$$(12.2) \quad (5.8) \quad (7.3) \quad (0.9) \quad (4.4)$$
$$R^2 = 0.98 \quad N = 35$$

Here Q is agricultural NSDP, N is land (both jhum and permanent), L is labour represented by all agricultural workers, K is as defined before and T, proxied by the percentage of land under permanent cultivation, is the measure of technology in agriculture. Naturally there is some multi-collinearity between N and T. A remedy lies in dropping T:

$$\log Q = -44.19 + 2.40 \log N + 1.99 \log L + 0.08 \log K \ldots (12)$$
$$(15.4) \quad (10.3) \quad (5.6) \quad (0.5)$$
$$R^2 = 0.97 \quad N = 35$$

The remedy does not improve the situation much. K still suffers from insignificance. The coefficients of N and L are too big to be much meaningful. Moreover, the constant has a high negative value showing an agriculture which is burdened with all inefficiencies, though the estimated returns to scale in both eqs (11) and (12) are well above unity.

Fertilizers as a substitute for K seems apparently to improve the result, though hardly with any meaningful consequences. Fertilizer as capital gives the following result:

$$\log Q = -20.17 + 1.56 \log N + 0.56 \log L + 0.70 \log f \ldots (13)$$
$$\quad (-5.5) \quad (8.6) \quad (2.0) \quad (7.3)$$
$$R^2 = 0.99 \qquad N = 35$$

All variables are significant at 1 percent level except L whose significance level is 5.3 percent. Instead of Q, if Q/L or Q/N is taken, the result does not improve. The estimated results do not shed very clear light but indicate some points, which are worth consideration. One is the role of technical change as provided by permanent cultivation, which has opened the door for adoption of chemical fertilizers, HYV etc. The growth of labour productivity is the outcome of use of new inputs. Growth rate has turned negative in recent years. Regaining of momentum of growth requires strengthening of new institutions by conducting the much-wanted cadastral survey and ensuring a larger flow of bank credit to agriculture in the state. Diversification of the jhum cultivator's income through orchardization can reduce the incidence of atrophied jhuming.

## Notes

1. It is very difficult, in the absence of data, to make any reliable estimate of the size or growth of the population in Arunachal Pradesh prior to 1961, the year which witnessed the first population census being conducted in the State. (In the rest of the country the first population census took place in 1872). However, anecdotal evidence suggests almost a stationary population prior to 1947. A sustained growth of population began in the state in 1947. So the year 1947 can be called the demographic great divide

for Arunachal Pradesh (in the country 1921 is this great divide).

2. Enacted in 1873 by the British Indian Government, Inner Line Regulation (ILR), which is still in force, restricts the entry of Indian citizens to Arunachal Pradesh. Migrants from other States are allowed to stay during their service but they are not allowed to own any land, or even fixed assets. However, they can operate land as tenants. Most of the tenants, mainly share croppers in the State are migrants from the rest of the country.
3. In the traditional agriculture of Arunachal Pradesh there was mobilitization of community labour based on mutualism or reciprocity. This mutualism was characterized by equality. It was not a hierarchical interdependence found in serfdom. Community based labour mobilization was a form of risk-coping mechanism, it also underlay redistribution of community's labour resources (Roy and Kuri 2001).

## REFERENCES

Baserup, E. (1965): *The Conditions of Agricultural Growth*, London: Allen and Unwin.

Government of North East Frontier Agency (NEFA) (1964): *Statistical Handbook of NEFA*, Shillong.

Government of Arunachal Pradesh (2006): *Statistical Abstract of Arunachal Pradesh, 2005*, Directorate of Economics and Statistics, Itanagar.

Government of India (2006), *Economic Survey 2005-06*, New Delhi.

Kautsky, Karl (1966): *The Agrarian Question*, Manchester: Manchester University Press. Quoted in Hamza Alavi: Peasantry and Capitalism: A Marxist Discourse, in: Teodor Shanin (ed) (1988) *Peasants and Peasant Societies*, London: Penguin (2nd edition), pp. 183 –196.

Maine, Henry (1906): *Ancient Law, its connection with early history of society and its relation to modern ideas*, London: John Murray. (With introduction and notes by Frederick Pollock). First published in 1861.

Polanyi, Karl (1944): *The Great Transformation*, Boston: Becon Press.

Roy, N.C. (2005 (a)): Sources of Growth in The Arunachal Economy, *Rajiv Gandhi Research Journal*, Vol. 8, No. 1, pp. 1-13.

Roy, N.C. (2005 (b)): *A Study of Inflow of Funds from the Centre and Economic Development in Arunachal Pradesh*, The Report submitted to the ICSSR, New Delhi.

Roy, N.C. and P.K. Kuri (2001): *Land Reform in Arunachal Pradesh*, New Delhi: Classical Publishing Company.

Verdoorn, P. J. (1956): Complementarity and Long-range Projections, *Econometrica*, 24; pp. 429-450.

# 25

# HORTICULTURE IN ARUNACHAL PRADESH: HOW TO INITIATE AGRO-PROCESSING IN THE STATE?

*S.S. Sangwan*

## Agro-climatic Conditions

The state of Arunachal Pradesh can aptly be called a depository of bio-diversity in India. It is the largest State of North Eastern Region (NER) with geographical area of 83743 sq km. Its area is entirely hilly and gradually rises towards the Himalayas up to 7,300 meters at some places. The state is spread over an area of and endowed with dense topical forests; high snow capped peaks, torrential rivers, difficult hilly terrain as well as some beautiful plateaus. The major rivers in the State are Kameng, Subansiri, Siang, Dibang, Lohit, Diyum and Dihing that merge with the great Brahmputra in Assam. The state is divided in 16 districts that are named with a prefix or suffix of the above rivers. The state can be recognized into five sub-regions on the basis of altitude, rainfall pattern, temperature, etc. (Table 1).

**Table 1: Sub-climatic Regions in Arunachal Pradesh**

| *S.No.* | *Sub-Region* | *Areas Covered* |
|---|---|---|
| 1. | Humid Hypothermic (Foot Hills) | 10, 9 and 10 % areas of west, east Kameng and Subansiri (lower) districts |
| 2 | Humid pre Hypothermic (foothills) | 100% area of Tirap & Changlang districts and 10 to 15% of Siang and Dibang Valleys |
| 3. | Thermic-pre-humid (Mid hills and valleys) | About 80% area of the Kameng and 85% area of the Subansiri valleys. |

| | | |
|---|---|---|
| 4 | Thermic Humid (Mid hills and valley) | About 80% area of Siang and 50% area of Dibang and Lohit valleys. |
| 5 | Alpine Zone (High hills) | 100% area of Tawang, 8 to 12% of Kameng, Subansiri and Siang valleys and 30% of the Dibang and the Lohit valleys. |

*Source:* Ninth Five-Year Plan (1997-2002), Govt. of Arunachal Pradesh, p. 23.

The normal annual rainfall in the State varies from 1000 mm to as high as 5194 mm with average of around 4000 mm in the Siang, Dibang and Lohit valleys. These areas are suitable for tea cultivation as rainwater flows down in the Brahmaputra due to higher sea level. Though, about 2/3rd rainfall is received during the months of May to September but sufficient rainfall also takes place between January to April that enables cultivation of Rabi crops and sustenance of horticultural and plantation crops under rain fed conditions. Minimum rainfall is received during the months of October to December. Seasonal pattern of rainfall is almost similar all over the state. The soil types range from loam to sandy loam mixed with pebbles and stones in the foothill zones. The lower reaches of valleys and plains are clayey aluminous and rich in organic contents. Owing to these agro climatic conditions only 40 per cent of the sown area (NSA) is under food crops and the remaining NSA is occupied by horticultural and net plantation crops as discussed ahead.

## Objective of the Study

The state has achieved rapid increase in area under fruits during 1980s and 1990s when special schemes for orange and pineapple were implemented in Siang valley. But can you imagine three pieces of Orange for one rupees and pineapple at throw away prices in November 2002 in Upper Subansiri district of the state. Kivi fruits were being sold @ Rs 50 per kg in 2004 at Itanagar. This scenario looms large on the sustainability of the approach of development through horticulture. Distance from the main consumption centers is the main problem. The state government have emphasized for agro-processing in its industrial policy and the annual & five year main feature of the latest Industrial Policy was that outside entrepreneurs may be plans. The allowed to hold land for Project on lease for a period of 30 years, which can be further, renewed upto for 30 years. This lease may be used as security for bank loan. However, the achievement in industrialization especially fruit processing is almost nil so far.

This paper seeks to suggest a practicable strategy, ceteris paribus, for agro-processing to sustain the present tempo of horticulture growth. To start with, the paper discusses the existing status of horticulture and its potential in the state and the strategy for agro-processing is specified thereafter.

## Data Base

The status and other information are based upon secondary data from the Statistical Abstracts and Five-Year Plan Documents of Arunachal Pradesh. In spite of some discrepancies in the data of over the years, the publication of the Abstracts and plan documents provide updated data in this widely spread out state especially since 2000. Besides the data, the author's direct experience of travelling all over the State during 2002-2004 and interaction with the concerned officials and farmers is the basis for the interpretation and suggestions made in the paper.

## Status of Horticultural Production

The long-term approach of horticultural development in the state has been replacing of Jhum cultivation by suitable fruit and plantation crops for generating higher income to the individuals and communities. Pineapple, orange and guava have been planted since 1980s and tea plantation was also started in some places in Siang and Dibang valleys. The apple, kiwi fruits have emerged as important crops in last 10 years. The setting up of Medicinal Plant Board and opening of the office of Spices Board of India has popularized the medicinal and spices crops especially during the Tenth five-year plan. Aromatic crops like Citronella (Cymbopogom winterianus), Geranium (Pelargonium graveolens), patchouli (Pogostemon cablin benth) and mentha (Menthe arvensis) are coming up due to their comparative advantage in marketing from this distant location. Among the spices, large cardamom and black pepper are important crops grown in the state. There is an

Orchid Research Centre at Bhalukpong (Tipi) in west Kameng district and a few commercial units have already come up in the Lower Subansiri district. The total area under fruits during 2003-04 has been reported as 56132 ha with total production of 103594 MT and 14564 ha under spices with 41137 MT of production. Area and production of important horticultural crops are given in Table-2.

**Table 2: Area and Production of Main Fruits/Spices Crops**

| *Fruits crops* | *Area (ha)* | | *Production (MT)* | | *Main Production* |
|---|---|---|---|---|---|
| | *2000-01* | *2003-04* | *2000-01* | *2003-04* | *Areas/Districts* |
| Apple | 6733 | 7852 | 8513 | 9288 | Tawang and West Kameng districts |
| Walnut | 2285 | 3054 | 51 | 56 | - Do - |
| Citrus | 19147 | 21864 | 24000 | 26716 | Siang, Dibang, and Subansiri Valleys |
| Pineapple | 7329 | 7849 | 31980 | 35598 | - Do - |
| Banana | 3538 | 4858 | 13200 | 14526 | Siang, Dibang, Subansiri, Lohit & Tirap Valleys |
| Kiwi | 25 | 173 | 7 | 60 | West kameng, Lower Subansiri & dibang valleys |
| Other fruits | 5096 | 15340 | 5771 | 15016 | - Do - |
| Sub-Total | 44153 | 56132 | 93084 | 103594 | |
| Big Cardamom | 2142 | 3842 | 507 | 560 | Siang, Dibang, Tirap area |
| Ginger | 6449 | 7174 | 29059 | 35947 | - Do - |
| Black Pepper | 635 | 1320 | 117 | 130 | Kameng, Subansiri and Siang valleys |
| Turmeric | 1219 | 571 | 1872 | 2200 | East Siang, Lohit and Changlang |
| Chillies | 1317 | 1657 | 2097 | 2430 | Lohit Changlang and Siang valley |
| Sub-Total | 11762 | 14564 | 33652 | 41137 | |
| Total | 55915 | 65985 | 126736 | 144731 | |

*Source:* Statistical Abstracts of Arunachal Pradesh for various years.

Besides the fruit and spices crops area under vegetables has also increased from 12811 ha in 2000-01 to 17123 ha in 2003-04 ha and kharif potato is also grown in Tawang region.

It is to be noted that the above progress has been achieved without the substantial support of bank credit. Technicalities and procedures involved in the process of identification of *(a)* temperate, *(b)* sub-temperate, *(c)* subtropical fruits zones, and the kind of fruit gardens to be established, and management practices to be followed in various locations are sorted out by the personnel of agriculture and horticulture departments. In fact, the absence of land records is a major constraint in extending bank credit with land as security. Besides the fruit and spice crops, area under tea plantation is also increasing in Siang and Dibang Valleys. The Tenth five-year plan gave an allocation of Rs 90 crore for all horticulture crops and Rs 80 crore for plantation crops. A perusal of the table brings out that the area and production of fruits increased by 5.49 % and 3.76 % per annum during 2000-01 to 2003-04 as compared to 7.94 % and 7.41 % for spices. Some area has also been covered under medicinal and aromatic crops. Small extraction plants for aromatic oil have also been established at Naharlagun in Itanagar and Dhirang in West Kameng districts. Further potential for horticulture development in the state is discussed ahead.

## Ongoing Schemes and Programmes

The Horticultural Department is maintaining own Farm of Nurseries & arranging saplings from other states. Training farmers for new activities like Mushroom farming. Kiwi Cultivation, kitchen gardening and floriculture. Holding marketing Fairs & Exhibitions. A annual budget of about 10 crore is given for horticulture in this small state.

Among the important centrally sponsored programmes are Integrated Programme for development of Spices, Commercial Floriculture, Mushroom cultivation, Integrated Development of Fruits, Development of Horticulture through Plasticulture Intervention and Coconut Development Programme etc. In post-harvest management in the state, communication is being the major bottle beck for most of the district headquarters during rainy season and the markets located through the plains of Assam for further flow to other parts of the region.

## Potential for Horticulture Crops

Horticulture Department of Arunachal Pradesh has estimated overall scope of 4.31-lakh hectare in the temperature, sub-temperate and sub-tropical fruit zones of the State (Table-3).

**Table 3: Potential for Horticulture and Vegetable in Arunachal Pradesh**

| Sl. No. | Horticultural Zone/ Altitude | Suitable Crops | Potential Area in ha |
|---|---|---|---|
| 1 | Foothills and valleys (170 to 915 m) | Citrus, Guava, Banana, Mango, Litchi, Pineapple, Plum, pear and seasonal vegetables | 145,000 |
| 2 | Mid hills (915 to1803 m) | Apples, Plum, Apricot, Pear, Almond, Pomegranate, Olive, grapes and seasonal vegetables for seeds & trunk gardening | 140,000 |
| 3 | High hills (Above 1830 m) | Apples, Cherry, Walnut, Pomegranate and off-season vegetables and temperate vegetable seeds | 115,000 |
| 4 | Rain Shadow Area with < 40″ rainfall & wide range of chilling requirement and temperature | Apples, Pear, Plum, Peach, Apricot, Almond, Walnut and Seasonal vegetables | 31,300 |

*Source:* Tenth Five-Year Plan of Arunachal Pradesh.

Though, the state may be self-sufficient in overall vegetable production but the state capital is still purchasing from markets in Assam. The New agricultural Policy of the state has projected area to be brought under fruits and spices as 83402 ha by 2010. The salient features of strategy being adopted by horticulture department of the state are: *a)* ensure quality planting material of Horticulture crops, *b)* reduction in losses by establishment of market system for both outputs and inputs, *c)* Collection and dissemination of market intelligence relating to horticulture crops shall have added emphasis, *d)* development of quality plants and seeds as well as quality testing laboratory, *e)* to develop small scale processes and technology, to map up small surpluses of Horticulture produce in remote, less communicated area under low volume high cost consideration, and *f)* to ensure credit support and risk management for production and processing.

## The Challenges

As against this scope, the challenges are construction of markets, cold storage and setting up of processing units from almost non-existent base notwithstanding the various policy announcements so far and provision of bank credit for all viable horticultural activities from almost negligible present base upto the genuine requirement. The announcement of New Agricultural and the Industrial policies have not made any headway so far in the agro-processing units. It indicates some other latent reasons, which do not allow the normal laws of profitability to attract entrepreneurs. Perhaps, in decision making for capital investment, the benefits from the project shall come into consideration after all other conditions like law and order situation, quality of governess, right to ownership of project site/property, the physical infrastructure etc. are comparable with other states of the country.

## Prevailing Marketing Infrastructure

The State has been historically a subsistence economy. It is still deficit in foodgrain production. There are small periodical hats at some places where the local people purchase for their own requirements. There are a few functional marketing societies in eastern districts but there is no regulated market throughout the state. At present the State does not allow free entry of traders/others by imposing the restriction of the inner lines permit for non-resident of Arunachal. If one goes to various parts of the State by road, a large number of shops are seen just outside the entry points of the State at Bhalukpong, Banderdeva, Likhabali, Ruksin, Namsai, Bordumsa, Deomali, etc. All these shops/dealers are in territory of Assam State whereas the customers are from Auranchal Pradesh. In this way, State is losing sizable revenue in terms of sales tax and other taxes while the Arunachalies have to cross Assam boarder for sale of their goods and purchase of their requirements.

Entry barriers of the inner line permit and unlawful coarsen by some local gangsters are cited among the reasons for non-development of marketing infrastructure in the state. In the absence of local market, it is difficult to procure a truck load of a product at one place that makes almost impossible for outside traders to participate. The small growers are worst sufferer as they are unable to bargain with individual buyers due to lack of awareness and less quantity. The traders/agents have also to engage local toughs who exploit the farmers in connivance with the traders. The author could get the feedback of the prevailing coarsen during 2004 when ginger price has gone up in Delhi and Mumbai. The State government has to realize that it is no more a subsistence economy especially in terms of fruit and spices production and therefore, regulated markets are sine qua non to sustain the growth rate in horticulture. The concept of regulated market may be more relevant in the initial stage instead of private market due to low level of awareness. The regulated market will ensure truck load of products at one place which will invite more outside traders and reduce their profit margin. These markets inside the state will also mobilize additional revenue for Government and generate employment.

## Status of Agro-processing

The need for fruit processing was felt by the State Government about twenty years back when a big processing unit was set up in upper Siang district, an area known for pineapple and oranges. As usual, the unit was closed due to its losses under government management. After 20 years, the same unit with its rusted machinery and dilapidated civil structure was leased out to a private entrepreneur. The unit has become operational. There is hardly any other big processing unit throughout the State despite a huge increase in area under various fruit crops in the last 15 years or so. Apples, pineapples, citrus and ginger have adequate level of production to start a few processing units. A few tea-processing units can also be feasible with present level of production. Patchouli and citronella processing units can also be run in two three places. The success of the old unit and significant increase in acreage under fruit crops do indicate the scope for more processing units especially in East Siang district near Pasighat, in West Kameng district near Bhalukpong or Dhirang and near Namsai in Lohit district, etc. The regulated markets, cold storage and the processing units need to be developed at these places as an integrated project. However, the strategy to take up the project has to be somewhat unusual due to circumstances discussed ahead.

### What is Required for Investment in Agro-processing?

An industrial investor is one who is going to invest a substantial amount of his funds unlike traders who start with almost nil capital. Therefore, an outside investor would not take personal risk like a trader whose lively hood was at stake before he came to Arunachal or any other part of NER. Secondly, a big investor expands his industry empire, considering not only profits but also safety and social recognition too as it is being accorded by the states all over the country. Thirdly, a big investment is not possible without the support of bank credit for capital cost as well as working capital, which in turn need security. Right to ownership of private property is not allowed in the state to the non-Arunachalies and a prudent investor would not like giving security of outside assets along with the margin he is investing in any project. Therefore, a pragmatic policy has take care of all these requirements to attract investors from outside the state. All these requirements of investors are difficult to meet in the short period, though there is way out to mitigate some of these. The state government has to step in the shoes of an investor for time being to realize the true potential of horticulture and other resources. In the Meiji Period in Japan from 1868 to 1912, a number of large capital projects were funded and grounded by the Government and then the units were sold or leased out to the private entrepreneurs. Almost the same lesson can be drawn from a few units, which were leased out by Government of Arunachal Pradesh in recent years. This strategy for operation is being further explained in the next paragraphs.

### Built and Lease Out Policy

The state can take up setting up of some units of markets, cold storage and processing in the strategically identified areas on its own. The starting point of this approach is that the state government will take an irrevocable decision to lease out these units to experienced private entrepreneurs from within and outside the state in transparent manner on competitive basis. The leasing out process is to be inbuilt in conception of the project proposal to avoid delay and loss to the government. The immediate question is the availability of resources with the State government. In this regard, loan from banks, North Eastern Financial Development Corporation, Rural Infrastructure Development Fund and Fund of National Bank For Agriculture and Rural Development can be taken by the government at cheaper rate of interest than individuals. Besides, subsidies are available for all these investments under the ongoing Central Government schemes. The lease period can be as per the industrial policy with annual payment of lease amount. If a well thought out detailed plan is workout by professionals than most of the repayment may come from the lease amount itself.

To illustrate, let us assume a market yard with 30 shops cum residence, a small cold storage of 3000 tons and a godown of 5000 MT at some strategic place with expected construction period of 2 years. In this project, cost of Cold storage will be about Rs150 lakh and after subsidy Rs. 100 lakh is to be repaid. The repayment period will be 7 years including 2 years of grace period. The annual rent can be about Rs10 to 15 lakh and hence the state has to bear a portion of repayment burden for initial seven years or so and thereafter, all rent will be profit to the government. In the shops of market yard the percentage of repayment burden may be even less due to more competition among the lessees. Moreover, the generation of higher state domestic product in this process will give additional value added tax and other taxes, stamp duty, etc., to the state government.

This strategy will be successful due to the following reasons. Firstly, it will mitigate the risk of capital investment and loan requirement of an individual investor. Secondly, the problem of bank loan for capital investment will be taken care by the state government at lower rate of interest. Thirdly, the lessee will be able to take working capital loan on the basis of lease document of the state government. These investments will generate employment and the income multiplier effect of such projects involving construction works will be very high on rest of the state economy. However, planning and implementation of this strategy requires strong political will and a competent team of a few technocrats and financial experts to take up from project formulation, obtain loan from banks and financial institutions, secure release of Central Government subsidy under the concerned purpose, commissioning the project and leasing it out immediately on completion. A few projects can be taken to start with purely on the basis of availability of raw materials and better linkage of the project site to the producers.

*Note:* The author is thankful to the Chief General Manager, NABARD, Maharashtra Regional Office, Pune, who reviewed the draft of this paper and gave valuable suggestions in finalizing the same. Comments from Dr D.C. Sharma, GM were also useful in improving the draft. However, the interpretations and explanations in the paper are the personal views of the author and not of his employer.

## REFERENCES

Government of Arunachal Pradesh, *Statistical Abstracts of Arunachal Pradesh for 2004 and earlier Years,* Directorate of Economics and Statistics, Itanagar.

-------- (2001), *New Agricultural Policy,* Department of Agriculture, Itanagar.

--------- (2001), New Industrial Policy 2001, Department of Industries, Itanagar

----------(2002), Drafts of Tenth (2002-2007) and Ninth (1997-2002) Five Year Plans, Department of Planning, Itanagar.

Meiji Period (1868-1912): http://www. Japan Guide.com

Osik, N. N. (1996), *A Brief History of Arunachal Pradesh,* Omsons Publications, New Delhi.

Sangwan, S.S. (2002), Arunachal Pradesh the Land of Rising Sun, the State Focus, *Technical Digest,* National Bank for Agriculture and Rural Development, July-September, pp. 31-34.

---------------(2003), "Arunachal Pradesh—Prospects for Development", *Rural Banker,* State Bank Institute of Rural Development, Hyderabad, Issue No. 35, April-June, pp. 20-24.

# 26

# CURRENT STATUS AND ROLE OF AGRICULTURE IN THE ECONOMIC DEVELOPMENT OF SIKKIM

***Subhabrata Chakrabarti***

## Introduction

Development experience of any country, state or region cannot deny a given set of situations, determined by nature, its history, culture, etc. Human activities along with these given conditions try to achieve economic growth under three basic sectoral performances, viz, Primary sector, Secondary sector and Tertiary sector. Sikkim with extreme natural beauty, a rich wealth of alpine meadows, rhododendrons, orchids, butterflies, birds and cultural diversities, is not an exception in this sense.

The objective of the paper is, analysing structural changes in the economy of Sikkim keeping in mind the fundamental idea on structural change. Structural change is basically characterized by the changes in the values of different parameters. These changes in the parameters must be guided by the "development with human face". The entire research is mainly dependent on the population values of different target variables, based on Sikkimese economy - not on the sample values. Percentage contribution of three broad sectors (primary, secondary and tertiary) and sub-sectors in State Domestic Product, proportions of working population in different sectors and sub-sectors, per capita GSDP etc. will be treated as variable. In the last section of this paper three statistical tests have been presented. Firstly, individual growth trend equations for agriculture and manufacturing sector are formulated. Secondly, to examine the strengths of the correlation-ships among the sub-sectors Pearson Correlation matrix is constructed. Thirdly, Granger Causality test is done among the primary, secondary and tertiary sectors. This test simply reveals whether any sector has taken any role to flourish other sector or not.

## Sectoral Dynamics and Agricultural Status

The transition from underdeveloped to developed status of an economy can be explained in terms of a series of stages, which all countries must follow. As for American historian W.W. Rostow: " It is possible to identify all societies, in their economic dimensions as living within one of five categories: the traditional society, the preconditions for take- off into self sustaining growth, the drive to maturity, and the age of high mass consumption......... These stages are not merely descriptive." In the early stages of development the inspiration for industrialization must come from the agricultural sector which creates the demand for industrial good. The path of economic development is characterized by a change in the composition of aggregate output, with a decline in the share of agriculture and the rise of industry to begin with, and then a boost in the share of services at the cost of industry, further on. Traditional societies are not much influenced by saving-investment based monetary economics. On the way of development, investment was essential in two ways. Firstly it helps to stop the drainage of money through saving and to bring back this money into the common pool of national income. Secondly investment is net addition to the physical stock of capital. Thus it brought growth of secondary and tertiary sector through a new series of production, income and employment opportunities. Obviously the desire for more and the new inventions are the guiding forces behind this growth.

**Table 1: Sectorwise Distribution of Gross Domestic Production of Sikkim at Current Price**

*(Rs. in lakh)*

| *Sector* | *1980-81* | *83-84* | *86-87* | *88-89* | *93-94* | *96-97* | *99-00* |
|---|---|---|---|---|---|---|---|
| Primary | 2546 (48%) | 3863 (48%) | 6694 (46%) | 8220 (44%) | 13237 (35%) | 19897 (36%) | 19382 (24%) |
| Secondary | 1077 (20%) | 1546 (19%) | 3003 (20%) | 3432 (18%) | 7171 (18%) | 10481 (19%) | 16747 (20%) |
| Tertiary | 1583 (32%) | 2604 (33%) | 4823 (34%) | 6813 (38%) | 17335 (47%) | 24884 (55%) | 45566 (56%) |
| Total GSDP | 5206 | 8013 | 14520 | 18465 | 37743 | 55262 | 81695 |

*Source:* a) State Domestic Product of Sikkim 2001; Planning & Development Department, Govt. of Sikkim

b) State Domestic Product of Sikkim 1992; Planning & Development Department, Govt. of Sikkim.

[Figures within the 1st brackets show percentage of total GSDP of corresponding year]

Table 1 (Annexure) shows that the share of secondary sector in the GSDP is almost stagnant between 18%-20% during 1980-2000. During 1990's there is rising share of tertiary sector at the cost of declining share of primary sector. One point is to be noticed that the trend of share of secondary sector in the total GSDP is neither decreasing nor increasing. It seems that the economic activities of secondary sector have no relation with the economic activities of primary sector and tertiary sector. So the share of primary sector has been falling without influencing secondary sector. Again the tertiary sector has taken an important role from the very beginning and it's rising trend is independent of the secondary sector.

Development experiences of Sikkim did not follow the common rules of stages of development in two ways. Firstly even if the share of primary sector is falling over the years, secondary sector is not expanded significantly. Secondly the growth of service sector is not brought by falling share of secondary sector. Share of tertiary sector has been rising at the cost of decline in the share of agriculture sector, not of industry. The components of primary sector are given, during 1980-2000 in Table 2.

## Table 2: The Contribution of Primary Sector and Its Components

*(Rs. in lakh)*

| *Sector* | *1980-81* | *83-84* | *86-87* | *88-89* | *93-94* | *96-97* | *99-00* |
|---|---|---|---|---|---|---|---|
| Agriculture & Allied | 2501 (47%) | 3752 (47%) | 6544 45%) | 8097 (43%) | 12310 (33%) | 18727 (34%) | 17961 (22%) |
| Forestry & Logging | 31 (0.61%) | 91 (1%) | 117 (0.81%) | 45 (0.25%) | 785 (3%) | 1021 (2%) | 1304 (2%) |
| Fishing | 1 (0.02%) | 8 (0.11%) | 14 (0.10%) | 15 (0.08%) | 49 (0.13%) | 75 (0.14%) | 58 (0.07%) |
| Minning & Quarrying | 13 (0.23%) | 12 (0.12%) | 19 (0.12%) | 63 (0.34%) | 94 (0.25%) | 74 (0.13%) | 59 (0.08%) |
| TOTAL | 2546 | 3863 | 6694 | 8220 | 13237 | 19897 | 19382 |
| Total GSDP | 5206 | 8013 | 14520 | 18465 | 37743 | 55262 | 81695 |

*Source:* Bureau of Economics & Statistics; Planning and Development Department; Govt. of Sikkim.

[Figures within the brackets show percentage of total GSDP].

From the Table we observe that agriculture and allied component is the major component of primary sector. During 1980-2000 the share of agriculture and allied activities in GSDP has been falling from 47% to 22%. The dynamics of the shares can not be the only clue to the causes, underlying economic progress. It must be accompanied with quantum rises in the aggregate value of GDP for economic development to take a firm hold. According to the report of world development indicators 2004 the share of agriculture sector of Japan, UK, and USA are only 1%, 1% and 2% respectively. Keeping in mind the issue of self-sufficiency in food supply, development can take place, even if there is low and declining share of primary sector. Sikkim with 32.86% decadal population growth (6$^{th}$ in India) has not a strong hold at this point. Per hectare productivity of paddy, wheat, ragi in Sikkim are lower than national average. To satisfy the total demand of some basic items, the state has to depend on the food grains allocated by the Union Government. Allocations of rice and sugar have been of the following order in the last few years and mentioned in Table 3.

**Table 3: Allocation of Foodgrains by Union Government (Tonnes)**

| | *1993-94* | *1996-97* | *1999-2000* | *2001-2002* |
|---|---|---|---|---|
| Rice | 4500 | 4800 | 15660 | 10906 |
| Sugar | 174 | 176 | 2199 | 2858 |

*Source:* The Glory of New Sikkim, published from Information and Public Relations Department; Government of Sikkim.

Moreover in Sikkim vegetables are taken mostly as soup in rural areas as rural people cannot afford to take dal as their regular diet. Hence the per capita requirement of vegetables in the state is very high. Again a considerable quantities of fresh vegetables are supplied to army canteens and neighbouring states. The estimated annual shortfall in the state is about 30,000 MT. However Sikkim became a part of India in May 1975 at the end of Fourth Five Year plan. It is commonly said that agricultural development in Sikkim accelerated after the merger. The production of pulses, oilseeds, soybean, dry ginger, and cardamom are higher than national average. The production levels of some of the main items are given Table 4.

**Table 4: Production of Different Crops ('000 Tonnes)**

| | *1975-76* | *1995-96* |
|---|---|---|
| Rice | 10.00 | 21.88 |
| Wheat | 0.15 | 15.3 |
| Maize | 16.5 | 56.6 |
| Total Oilseed | 0.70 | 7.63 |
| Total Fruits | 4.7 | 12.00 |
| Vegetables | 5.00 | 28.00 |
| Potato | 8.00 | 24.00 |
| Ginger | 2.00 | 29.00 |
| Large Cardamom | 2.30 | 3.60 |

*Source:* Sikkim—Perspective for Planning and Development, (Article, Economic Structure of Sikkim, H.R. Pradhan).

During 1975-95 production of maize, total oilseed, fruits, potato, ginger and large cardamom are increasing by 243%, 990%,155%, 200%, 1350%, 57% respectively. Depending on domestic demand Sikkim has also become a large exporter of some crops like cardamom. Actually in the post merger period the strategy was to provide a package of services to improve peasant economy. This included land reforms, agriculture credit, marketing provision of inputs like seeds, fertilizers, minor irrigation and encouragement to horticulture and cash crops. All these are related to the basic sources of agriculture backwardness in Sikkim. However budget allocation on agriculture in Sikkim is not satisfactory enough to solve these problems (Table 5).

**Table 5: Budget Expenditure on Agriculture (Rs. '000) 1980-00**

| *Year* | *State Budget* | *Total Budget* | *%* |
|---|---|---|---|
| 1980-81 | 5437 | 574593 | 0.95 |
| 1985-86 | 32347 | 828055 | 3.91 |
| 1990-91 | 43710 | 1666615 | 2.62 |
| 1995-96 | 48997 | 6287808 | 0.77 |
| 1999-2000 | 81075 | 6645067 | 1.22 |

*Source:* Department of Agriculture, Government of Sikkim, Gangtok.

Next, we cannot deny the influence of nature in positive or negative sense on the agriculture system of Sikkim. The ecological conditions from subtropical to alpine range determine the cropping pattern and its intensity in different zones of Sikkim. Most of the cultivated areas lie below 1800 meters. But various kinds of fruits, vegetables and commercial crops can be cultivated at different altitude (Table 6). Secondly, Sikkim, being a hilly state, soil erosion, deforestation, landslides and other natural calamities create problems for cultivation. Thirdly irrigation in the hills is difficult. The springs and small rivers become over-flooded in the summer but get dry during *rabi* season. Again the capillary movement of soil moisture is horizontal in the hills unlike vertical in the planes. Once the soil gets dry there are no means of irrigation unless there is a rain. Fourthly mechanization in the hills is almost impossible due to sloppy hill terrain.

**Table 6: Crop Pattern at Different Altitudes**

| *Altitude* | *Crops* |
|---|---|
| 300-1500m [Tropical and Subtropical] | Banana, Mango, Guava, Sapota Mosambi, Mandrain,Orange, Avocado, Strawberries |
| 1500-2700m (Temperate) | Peach, Plum, Kiwi Fruit,Passion Fruit |
| 2700-4000m [Sub-Alpine and Alpine] | Apple, Walnut, Almond etc. |

*Source:* Sikkim Study Series, Vol. 1, published from Department of Information and Public Relation, Government of Sikkim, Gangtok.

Primary sector has been unable to create sufficient demand for industrial product (luxury, semi-luxury, production input etc.). As a result primary sector did not take any role in favour of secondary sector. This is why rural Sikkim is the main origin of primary sector workers. On the other hand over the past three decades the share of rural population in Sikkim is 85%-90%. So a large number of total population are dependent on primary sector of which the contribution is gradually falling. In fact rate of transfer of workers from primary to other sector is less than rate of falling of primary sector's share. We can easily conclude that primary sector is the main source of the below poverty line group of people with low per capita income and consumption level. In the absence of proper banking system there is no possibility of induced consumption through loan or any other ways. Exploitations of private money lenders in many ways have reduced the economic power of farmers. In one system "Chakurey and Phakurey" farmers have to provide all types of services to the lenders. Thus primary sector is vain to create a market demand for industrial product and to provide an impetus for secondary sector. MarkeTable surplus of some crops could not bring sufficient cash money in the hands of farmers due to absence of proper marketing system. A study of Bhatnagar (1990) estimated market surplus of some major crops. The estimates are Ginger (72.2%), Cardamom (98.5%), Chili (71.6%), Potato (52.6%), Cabbage (76.1%), Tomato (67.6%), Orange (90.8%), Banana (72.2%), Mango (67.8%) and Pulses (58.8%). Actually money economy and the expansion of peasant export product can be interlinked in the two ways: "a) bringing an increasing number of peasant family into export production b) there may be specialization in export sector through devotion of whole resources and thus setting up a cash demand for goods and services produced locally or in outside" (Myint, 1973).

### Regional Disparity through Employment Pattern

Trend of sectoral distribution may explain the extent of rural urban inequality. Before that we have to notice that percentage of rural population in Sikkim always remained much higher than national level (Table 7). This is why altogether urbanization in Sikkim is based on eight towns only (viz.Mangan, Gangtok, Singtam, Rangpo, Namchi, Jorthang, Nayabazar, Gyalshing).

**Table 7: Urban Population as Percentage of Total Population**

| | *1971* | *1981* | *1991* | *2001* |
|---|---|---|---|---|
| Sikkim | 9.37 | 16.15 | 9.10 | 11.10 |
| India | 19.91 | 23.34 | 25.73 | 27.80 |

*Source:* Census Report 1971, 1981, 1991, 2001

Relatively there is no significant change of urban or rural population. This means if the inequality between rural and urban area is a given situation then the economic benefit of growth has to be brought into the rural area to a higher extent than urban area to reduce the rural-urban gap. Without significant change in rural-urban population, a structural shift of labour force within the sectors is an interesting demographic issue. It suggests that a part of the secondary sector and tertiary sector both may be situated in rural area. Moreover

Sikkim is a small state, so living in rural area but working in urban area in secondary or tertiary sector is not impossible.

Table 8 shows that most of the main workers are coming from rural area. Rural main workers are basically engaged in agriculture based field of which the share of GSDP is falling over the years. We have already noted that rural population in Sikkim always remained higher than national average and this scenario is almost unchanged during 1971-2001. A very negligible percentage of urban workers are engaged in agriculture based field and others are working in combined field of secondary and tertiary sector with rising share of total GSDP. This clearly explains the logic behind the existence of rural-urban inequality.

**Table 8: Rural and Urban Workers Engaged in Agriculture Field**

| *Year/Workers* | *1971* | *1991* | *2001* |
|---|---|---|---|
| Main Workers | 111607 | 164392 | 212478 |
| Rural Workers (Agriculture Labourers+ Cultivators) | 94291 (84%) | 107795 (65%) | 148314 (70%) |
| Urban Workers (Agriculture Labourers + cultivators) | 296 (.27%) | 134 (.09%) | 47 (.02%) |

*Source:* Census Report, 1971, 1991, and 2001.

[Figures within the brackets show percentage of Main Workers.]

**Table 9: Comparative Growth Rates of Population, Agriculture Based Workers**

| | *Growth Rate of Rural Population* | *Growth Rate of Rural Workers (Agriculture+ Cultivation)* | *Growth Rate of Urban Population* | *Growth Rate of Urban Workers (Agriculture+ Cultivation)* |
|---|---|---|---|---|
| 1971-1991 | 94% | 14% | 88% | -55% |
| 1991-2001 | 30% | 38% | 62% | -65% |

*Source:* Calculated from Census Report 1971, 1991, 2001.

From the Table 9 we conclude that growth rate of rural workers (agril. labourers + cultivators) is rising but the growth rate of urban workers of same category is falling. Corresponding to growth rates of rural population 94% (1971-91) and 30% (1991-2001), growth rates of agriculture based rural workers are 14% and 38% respectively. This means some of the workers who migrated from agriculture to other field, came back again to their origin and restarted a rural agri-based life. It is also possible that new generation is less interested to switch over from primary to secondary or tertiary sector. The logics are supported also by the steady decline of proportional urban population after 1981 (Table given). Along with large and rising share of service sector these phenomena helped to increase rural-urban inequality. This inequlity is not brought at the cost of growth exactly. Actually an economy and it's agents are not sufficient brave to play the necessary expansionary roles of growth.

Let us examine the relative status of four districts (North, East, West, South) in the light of sectoral distribution and its employment trend. In the Table 10 growth rate of corresponding variable over the period 1971-2001 is given within the bracket. Among the four districts growth rates of workers and population both are highest in the North district. In terms of work participation rate (2001) also North district ranked highest (.58). But this does not prove North district as most economically advanced region in Sikkim.

**Table 10: District-wise Employment Pattern**

| *Year* | *Total No. of Workers* | *No. of Agriculture labour & cultivators* | *Population* | *% of Agriculture labour & cultivators to total workers* |
|---|---|---|---|---|
| 1971 | 7755 (N) | 7251(N) | 13014 (N) | 94% |
| | 41890(E) | 29257(E) | 85621 (E) | 70% |
| | 31642(W) | 29938(W) | 58785(W) | 95% |
| | 30320 (S) | 28141 (S) | 52423(S) | 92% |
| 2001 | 23634(N) | 11271(N) | 41023(N) | 48% |
| | (205%) | (55%) | (215%) | |

| | | | |
|---|---|---|---|
| 116802 (E) (179%) | 46044 (E) (57%) | 244790(E) (186%) | 39% |
| 53131 (W) (41%) | 39844 (W) (33%) | 123174(W) (109%) | 74% |
| 69753 (S) (130%) | 51202 (S) (82%) | 131506 (S) (151%) | 73% |

Note: N- North District; S- South District; E- East District; W- West District

As per census report 1971 percentage of agriculture labour and cultivators to the total main workers was lowest in East district (70%). Census report 2001also reveals the same status of East district with lowest percentage of agriculture labour and cultivators to the total workers (39%). From the Table it is clear that over the period 1971-2001 a much higher proportion of workers are engaged in East district than any other district in the non-agricultural field of which the share in total GSDP is rising. On the other hand East district with highest employment elasticity (.96) has provided 179% employment growth against 186% population growth rate. So, greater participation in non-agriculture field has helped East district to hold a better economic position compared to other districts.

The post merger reform in agriculture field has not taken place in a homogeneous manner among four districts. For example as per report of Sikkim Statistical Profile 2002, advance in agriculture sector for East is 182 lakh rupees (as on 31st March, 2001), where as this amount is 19 lakh (North), 68 lakh (South), and 38 lakh (West). If we consider simply the negative relation between firm size and productivity, then also East district holds a better position than other districts. Average area per holding in East is 1.7 hectare, where as this figure is 3.01 hectare (North), 2.2 hectare (South) 2.2 hectare (West). With a lowest percentage of workers engaged in agricultural sector over a long period, East district did not show any declining trend in total agricultural output or productivity. Recent records suggest that East district has highest level of total output and productivity of some major crops among the four districts (Table 11). During eighties in many parts of India the observed rise in real wage was basically explained by the shift of labour from agriculture to the more remunerative non-agricultural sector accompanied with rising marginal productivity of labour. A significant performance with insignificant labour employment in agriculture sector in East district proves that the technological change after the merger with India (1975) in Sikkim influenced the agricultural labour of East district to a higher extent than the rest. This again may be the explaination of the inequality in the relative status of agri-based workers of different districts. Actually labour transfer from primary sector to any other sector had least negative influence in East district among all districts, although post merger technical progress could not overcome the constraints in agricultural field completely. From Table 12 we notice the productivity of some major crops has followed a decreasing trend over a considerable period of time in Sikkim.

**Table 11: Total Production and Productivity of Some Major Crops (1999-2000)**

| | *Area (Hectare)* | *Production (Tonnes)* | *Productivity* |
|---|---|---|---|
| Rice | 1.40(N), 6.72(E), 5.38(W), 2.40(S) | 2.05(N), 9.95(E) 8.0(W), 3.43(S) | 1.46(N), 1.48(E), 1.48(W) 1.42(S) |
| Wheat | 1.04(N), 2.55(E), 2.55(W), 1.96(S) | 1.64(N), 4.31(E) 4.30(W), 2.59(S) | 1.57(N), 1.69(E), 1.68(W) 1.68(S) |
| Maize | 3.09(N), 9.70(E), 13.30(W), 13.3(S) | 4.16(N), 13.4(E) 18.2(W), 17.4(S) | 1.34(N), 1.38(E), 1.36(W) 1.30(S) |
| Barley | 0.15(N), 0.34(E), 0.45(W), 0.20(S) | 0.20(N), 0.50(E) 0.67(W), 0.30(S) | 1.33(N), 1.47(E), 1.48(W) 1.5(S) |
| Pulses | 0.08(N), 1.74(E), 2.54(W), 2.35(S) | 0.06(N), 1.55(E) 2.28(W), 2.05(S) | 0.75(N), 0.89(E), 0.89(W) 0.87(S) |

N- North District; S- South District; E- East District; W- West District

*Sources:* Department of Agriculture, Govt. of Sikkim; Gangtok.

**Table 12: Trend of Total Production & Productivity of Some Common Crops (Rice, Wheat, Maize, Total Cereals, Pulses, Oil Seeds, Fruits) in Total State**

| *Year* | *Production ('000 tonnes)* | *Productivity (kg/ha.)* | *Growth Rate of Production* | *Growth Rate of Productivity* |
|---|---|---|---|---|
| 1975-76 | 33.75 | 667.00 | - | - |
| 1985-86 | 124.50 | 1223.00 | 268.9% | 83.3% |
| 1995-96 | 138.64 | 2067.00 | 11.3% | 69% |
| 2000-01 | 138.32 | 1460.00 | -0.23% | -29.3% |

*Source:* Sikkim, A statistical Profile 2002.

## Statistical Tests

Contributions of different sectors over the period 1980-2000 are given below. The main purpose is to apply different statistical tools in the light of following data trend.

| *Year* | *Agriculture* | *Total Primary* | *Manufacturing* | *Total Secondary* | *Total Tertiary* |
|---|---|---|---|---|---|
| 1980-81 | 48.03% | 48.91% | 6.07% | 20.68% | 30.40% |
| 81-82 | 49.92% | 50.81% | 7.12% | 19.91% | 29.29% |
| 82-83 | 50.77% | 51.90% | 6.57% | 19.67% | 28.43% |
| 83-84 | 48.67% | 50.15% | 7.18% | 20.91% | 28.93% |
| 84-85 | 49.36% | 50.28% | 6.78% | 19.39% | 30.32% |
| 85-86 | 44.20% | 45.10% | 6.41% | 21.32% | 33.58% |
| 86-87 | 38.39% | 39.18% | 5.91% | 27.17% | 33.65% |
| 87-88 | 47.43% | 48.20% | 6.35% | 21.84% | 29.96% |
| 88-89 | 44.43% | 45.37% | 7.01% | 20.74% | 33.89% |
| 89-90 | 44.01% | 44.44% | 7.57% | 22.84% | 32.72% |
| 90-91 | 41.67% | 42.09% | 5.11% | 18.51% | 39.40% |
| 91-92 | 39.58% | 40.10% | 5.05% | 18.22% | 41.68% |
| 92-93 | 35.17% | 35.71% | 5.38% | 19.82% | 44.47% |
| 93-94 | 32.61% | 35.07% | 5.38% | 19.00% | 45.93% |
| 94-95 | 31.49% | 34.02% | 5.52% | 18.60% | 47.37% |
| 95-96 | 30.72% | 33.17% | 5.06% | 21.19% | 45.63% |
| 96-97 | 30.57% | 32.80% | 4.65% | 20.85% | 46.35% |
| 97-98 | 28.04% | 29.98% | 4.09% | 23.62% | 46.40% |
| 98-99 | 26.76% | 28.73% | 3.85% | 20.99% | 50.28% |
| 99-00 | 25.58% | 27.52% | 3.64% | 22.73% | 49.74% |

From the above information we get the trend equations of manufacturing and agriculture sector as follows-

Agriculture= 39.38- .69T

SE=.611 SE=.053 R-Square= .904

t=64.486 t= -13.028

Manufacturing= 5.735 -0.0811T

SE=.146 SE=.013 R-Square=.694

t= 39.172 t=- 6.387

The 't' values in each trend equation is high as well as significant. The time coefficients for both manufacturing and agriculture are negative. So, one can not claim that fall in agricultural share has necessarily resulted in the rise of the share of manufacturing sector. Depending on the values of the sectoral contributions of different sectors the correlation matrix was formed which gave the following results.

**Pearson Correlation Between Different Sectors in Sikkim**

| | *Agriculture* | *Primary* | *Manufa.* | *Constr.* | *Secondary* | *Tertiary* |
|---|---|---|---|---|---|---|
| Agriculture | 1.000 | | | | | |
| Primary | 0.998* | 1.000 | | | | |
| Manufacturing | 0.876* | 0.882* | 1.000 | | | |
| Construction | -0.146 | -0.183 | -0.208 | 1.000 | | |
| Secondary | -0.175 | -0.188 | -0.054 | 0.859* | 1.000 | |
| Tertiary | -0.968* | -0.966* | -0.881* | -0.039 | -0.071 | 1.000 |

* Correlation is significant at the 0.01 level (2-tailed).

Table given above shows the correlation coefficients of the different sectors. Agriculture sector shows significant positive correlation with manufacturing, which goes against usual theories and negative correlation with tertiary sector. Agriculture sector is negatively correlated with construction and secondary sector, but the relationship is not significant. Primary sector is not significantly related with secondary sector. Manufacturing sector is negatively correlated with tertiary sector. Construction sector is not significantly correlated with tertiary sector.

Since we have a continuous series of data of over twenty years, effort is provided to have Granger Test to examine whether there is any cause and effect relationship between primary and secondary sector. A unidirectional relationship is observed from the available data. Results show secondary sector is causing primary sector. This means primary sector has not provided an impetus for the growth of secondary sector. Same fact has been discussed before. The outcomes of the Granger Test are given below (Table a, b, c and d)

**Table a: Unit Root Tests Without Trend**

| *Variable* | *Level* | | *First Difference* | |
|---|---|---|---|---|
| | *DF* | *ADF (1)* | *DF* | *ADF(1)* |
| Primary | -0.177 | 0.217 | -5.363 | -3.998 |
| Secondary | -2.952 | -2.271 | -5.637 | -4.848 |
| Tertiary | -0.662 | -0.399 | -5.342 | -2.854 |

*Note:* Critical values at 5% level of significance of DF and ADF tests are: -3.040 and -3.052.

**Table b: Unit Root Test With Time Trend**

| *Variable* | *Level* | | *First Difference* | |
|---|---|---|---|---|
| | *DF* | *ADF (1)* | *DF* | *ADF(1)* |
| Primary | -3.143 | -2.716 | -5.256 | -4.010 |
| Secondary | -2.859 | -2.192 | -5.444 | -4.692 |
| Tertiary | -5.342 | -2.854 | -5.140 | -2.695 |

*Note:* Critical values at 5% level of significance of DF and ADF tests are: -3.692 and -3.711

**Table c: Co-integration Tests**

| *Regression Equation:* | *DF* | *ADF* |
|---|---|---|
| (a) Regress P on S | -0.088 | -0.231 |
| (b) Regress S on P | -2.972 | -2.274 |
| (c) Regress S on T | -3.056 | -2.368 |
| (d) Regress T on S | -0.612 | -0.585 |

*Notes:* Critical values at 5% level of significance of the DF and ADF cointegration test is: -3.696

P and S denotes primary and secondary sector respectively.

**Table d: Granger Causality Test Between Primary and Secondary Sector**

| *Regression* | *FPE*† | *F-Value** |
|---|---|---|
| (a) Primary as a dependent variable | | |
| i. Regress ΔP on ΔP (m=1) | 0.00597 | 2.547* |
| ii. Regress ΔP on ΔP (m=1) and ΔS(n=2) | 0.00528 | |
| (b) Secondary as a dependent variable | | |
| i. Regress ΔS on ΔS (m=2) | 0.01449 | 1.937 |
| ii. Regress ΔS on ΔS (m=2) and ΔP(n=1) | 0.01591 | |

† FPE represents Akaike's final prediction error.
*Significant at 5% level.

## Conclusion

In this paper structural changes have been identified subject to the trend of State Domestic Product along with sectoral dynamics and employment pattern. The employment pattern in the agriculture sector has not maintained a balance with the share of this sector in total SDP. Actually, the fall in the proportions of the agricultural workers is highly insignificant compared to the fall of the agricultural share in total SDP. This fact has also become a source of economic concentration or regional disparity within the state. Structural retrogression is very much clear in the state economy. Besides, agricultural sector has not been proved helpful for the growth of non-agricultural sector. Some natural barriers in the growth of agricultural sector have been mentioned before. Now, the question may arise about the relevance of fundamental theories on structural change in the light of development experiences of this state. It is the time to answer the question – with a low profile of capital formation and manufacturing sector can a structural change as predicted by empirical and theoretical models, take place in the true sense? Or with all ecological factors and diversities of climate, altitudes, soil and topology of Sikkimese economy, we may have to find out a different dimension of structural change?

## REFERENCES


Dasgupta, Manas (1992); *Sikkim, Problem and Prospects of Development*, Indus Publishing Company; New Delhi.

Government of India; Census Report 1971, 1981, 1991, 2001

Government of Sikkim (1992); State Domestic Product of Sikkim 1992; Planning and Development.

Government of Sikkim (2001); State Domestic Product of Sikkim, 2001; Planning and Development Department.

Government of Sikkim (2002); Sikkim: Manpower review, Published from Planning & Development Department.

Government of Sikkim (2004); Sikkim Study Series, Vol. 1, Published from Department of Information and Public Relations, Kwality Publisher, Kazi Road, Gangtok, 2004.

Government of Sikkim, Budget Analysis, 1979-95.

Government of Sikkim, Budget Analysis, 1995-2002.

Government of Sikkim, *Sikkim: A Statistical Profile 2002*; Directorate of Economics, Statistics, Monitoring & Evaluation.

Joshi, P.C. and C.H. Hanumanth Rao (1980), *Reflections on Economic Development and Social Change*, Allied Publishers, New Delhi.

Lahiri, Ashok K., Saumen Chattopadhayay, Anuradha Vasin(2001); Sikkim ; the people's vision; Indus Publishing Company ; New Delhi.

Lama, M.P. (2001), *Sikkim, Human Development Report 2001*, Social Science Press, Delhi.

Lama, M.P. (Ed) (2003), *The Glory of New Sikkim*, Published from Information and Public Realtions Department; Government of Sikkim; Newstech Publishing; New Delhi.

Lama, M.P. (Ed) (2000), *Articles from Sikkim: Society, Polity, Economy Environment*, Indus Publications, New Delhi.

Meier, G.M. (1964), *Leading issues in Development Economics*, Oxford University Press; NewYork.

Myint, H. (1973), *The Economics of Developing Countries*, Hutchinson Publication; London.

Oberoi, A.S. & H.K. Manmohan Singh (1983), *Causes and Consequences of Internal Migration*, Oxford University Press, Delhi.

Rao, C.H. Hanumanth(1994), *Agricultural Growth, Rural Poverty and Environmental Degradation*, Oxford University Press, New Delhi.

Risley, H.H. (1894), *The Gazetteer of Sikkim*, Edited in the Bengal Government.

Tanzi, Vito & Ke-Young Chu (eds) (1998), *Income distribution and High-Quality Growth*, MIT Press, Cambridge.